Postphenomenology and Technologies within Educational Settings

Postphenomenology and the Philosophy of Technology

***Editor-in-Chief*: Robert Rosenberger**

***Executive Editors*: Peter-Paul Verbeek and Don Ihde**

As technologies continue to advance, they correspondingly continue to make fundamental changes to our lives. Technological changes have effects on everything—from our understandings of ethics, politics, and communication, to gender, science, and selfhood. Philosophical reflection on technology can help draw out and analyze the nature of these changes, and help us to understand both the broad patterns of technological effects and the concrete details. The purpose of this series is to provide a publication outlet for the field of philosophy of technology in general, and the school of thought called "postphenomenology" in particular. The field of philosophy of technology applies insights from the history of philosophy to current issues in technology and reflects on how technological developments change our understanding of philosophical issues. Postphenomenology is the name of an emerging research perspective used by a growing international and interdisciplinary group of scholars. This perspective utilizes insights from the philosophical tradition of phenomenology to analyze human relationships with technologies, and it also integrates philosophical commitments of the American pragmatist tradition of thought.

Recent Titles in This Series

Postphenomenology and Technologies within Educational Settings, edited by Markus Bohlmann and Patrizia Breil

Modernity and Technology: A Philosophical Investigation of Martin Heidegger and Bruno Latour by Søren Riis

Moral Hermeneutics and Technology: Making Moral Sense through Human-Technology-World Relations by Olya Kudina

Postphenomenology and Imaging: How to Read Technology, edited by Samantha J. Fried and Robert Rosenberger

Postphenomenology and Architecture: Human Technology Relations in the Built Environment, edited by Lars Botin and Inger Berling Hyams

Technology and Anarchy: A Reading of Our Era, by Simona Chiodo

How Scientific Instruments Speak: Postphenomenology and Technological Mediations in Neuroscientific Practice by Bas de Boer

Feedback Loops: Pragmatism about Science & Technology, edited by Andrew Wells Garnar and Ashley Shew

Sustainability in the Anthropocene: Philosophical Essays on Renewable Technologies, edited by Róisín Lally

Postphenomenology and Technologies within Educational Settings

Edited by

Markus Bohlmann and Patrizia Breil

LEXINGTON BOOKS

Lanham • Boulder • New York • London

Published by Lexington Books
An imprint of The Rowman & Littlefield Publishing Group, Inc.
4501 Forbes Boulevard, Suite 200, Lanham, Maryland 20706
www.rowman.com

86-90 Paul Street, London EC2A 4NE

British Library Cataloguing in Publication Information Available

Library of Congress Cataloging-in-Publication Data

Names: Bohlmann, Markus, editor. | Breil, Patrizia, editor.
Title: Postphenomenology and technologies within educational settings / edited by Markus Bohlmann and Patrizia Breil.
Description: Lanham : Lexington Books, [2025] | Series: Postphenomenology and the philosophy of technology | Includes bibliographical references and index.
Identifiers: LCCN 2024044891 (print) | LCCN 2024044892 (ebook) | ISBN 9781666939149 (cloth) | ISBN 9781666939156 (epub)
Subjects: LCSH: Educational technology--Philosophy. | Phenomenology.
Classification: LCCLB1028.3 .P674 2025 (print) | LCC LB1028.3 (ebook) |DDC371.3301--dc23/eng/20241119
LC record available at https://lccn.loc.gov/2024044891
LC ebook record available at https://lccn.loc.gov/202404

♾™ The paper used in this publication meets the minimum requirements of American National Standard for Information Sciences—Permanence of Paper for Printed Library Materials, ANSI/NISO Z39.48-1992.

Contents

Tables and Figures

CHAPTER 1

CHAPTER 2

CHAPTER 5

CHAPTER 8

CHAPTER 13

CHAPTER 14

Introduction

Markus Bohlmann and Patrizia Breil

Digital technologies have become smarter, more social, and more regulatory; we rely more and more on their rationality even in highly protected areas. During the COVID-19 pandemic, digital technologies have been deployed in many educational systems around the world. Learning platforms and conferencing software have transformed teaching and learning to an unprecedented degree in areas where technology was previously thought to be barely possible at all. The achievements of artificial intelligence enabled by large language models challenge established ways of working in education. Writing an essay is quite a different cultural technique now. Educators and students around the world use tablet computers, digital projectors, and, in some cases, even virtual reality devices. This volume is a step toward thinking more about how digital technologies are changing education at the level of human–technology relations. How do digital technologies mediate education without giving it some prefixed purpose? How do new technologies sediment while young people develop? What stabilities of technology manifest when there is still an openness in education? What is communication among those present, what is attention, what is performance, and what is the subject in an educational system, where people and technologies can no longer be easily separated? In this volume, conceptual and empirical means are elaborated by which postphenomenology can be used to answer these questions.

ON THE PROBLEM OF POSTPHENOMENOLOGY AND TECHNOLOGIES WITHIN EDUCATIONAL SETTINGS

Postphenomenology is a rapidly growing school of philosophy of technology that stems from the work of Don Ihde (Ihde 1990, 2009). It combines science and technology studies, including the important basic idea of the social construction of technology, with the phenomenological roots of the philosophy of technology, especially in Martin Heidegger and Maurice Merleau-Ponty. There is a pragmatic touch to this philosophy of technology, and human–technology relations are understood in concrete practices (Thompson 2020). Postphenomenology stands after the empirical turn in the philosophy of technology, that is, it is no longer concerned with determining an ever-identical essence of technology, but with describing concrete technologies empirically (Verbeek 2022). Thus, postphenomenology can also be seen as a philosophical methodology for analyzing concrete technologies (Rosenberger and Verbeek 2015). More recently, postphenomenology has developed the four basic human–technology relations going back to Ihde into an empirical repertoire, that is, embodiment, hermeneutic, alterity, and background relations. The concepts of mediation, multistability, sedimentation, and field composition have undergone substantial change; field composition is a rather new addition by Robert Rosenberger (Rosenberger 2021, 2017; Rosenberger and Verbeek 2015; Bohlmann 2022, chapter 10; see this volume and Rosenbergers forthcoming book *Distracted: A Philosophy of Cars and Phones* for the application of these concepts). Recently, postphenomenology has worked on a theoretical background to further define the political implications of technologies. Peter-Paul Verbeek draws the dimensions of his political analysis from Michel Foucault's critique of power (Verbeek 2020). In this project, the concept of subjectivation, which is also widely used in the philosophy of education, is made fruitful for technology studies (Bergen and Verbeek 2021).

At this point, there are three approaches that can make an edited volume on postphenomenology and technologies in educational settings interesting. We also think that these approaches can be interesting beyond the internal boundaries of postphenomenology and even beyond the philosophy of technology. Scholars in educational sciences, philosophy of education, phenomenology in relation to questions of education, and ultimately even every teacher and every student can draw from them some potential for reflection in relation to technologies. These three approaches form the three sections of this volume.

Empirical Evidence. In the first section, our contributors analyze technologies in educational settings through the means of concrete empirical studies. They carried out in-depth interviews, detailed studies of human–technology relations and participant observations. The concepts of postphenomenology are used here to design empirical studies and to analyze empirical material and data. This section is very concrete about the way human–technology relations in the educational systems worldwide are changing at present.

Conceptual Convergence. There is an urgent need to discuss the concepts of postphenomenology with existing concepts in the field of education on a theoretical level. Critical theory, constructivism, phenomenology, cultural theory, and poststructuralism are influential theories in education. Our contributors raise the question of theory convergence with postphenomenology. This section is particularly interesting for an interdisciplinary exchange.

Subject Subversion. The concept of the subject is central to educational theory. This is particularly true of the German discourse on *Bildung*. Whereas the subject of education tends to be passive and shaped according to a socially prescribed ideal, the subject of *Bildung* is empowered to think for itself, to reconstruct, deconstruct, and construct its own relation to society and its lifeworld in general. In short, the traditional goal of *Bildung* is the autonomy of its subject (Koller 2001, 35). In phenomenology, on the other hand, the subject is in question; "the idea of *experience* [in phenomenology] is not that of either common-sense understandings or of the standard misinterpretation of phenomenology as subjective" (Ihde 1990, 21f). Don Ihde articulates technological experiences as human–technology–world relations; here the subject is relativized accordingly. The subject was also already questioned by Merleau-Ponty in his 1945 *Philosophy of Perception*, when he says "my existence as subjectivity is identical with my existence as a body and with the existence of the world" (Merleau-Ponty 2012, 431). But when we talk about technologies in educational settings, the subject seems to be subversive. Our authors discuss in the third section how this central concept of educational theory can relate to postphenomenology. How are subjectivation, practice, mediation, and relation to be understood in educational processes that presuppose a subject?

POSTPHENOMENOLOGY AND TECHNOLOGIES WITHIN EDUCATIONAL SETTINGS SO FAR

This is the first edited volume in the *Postphenomenology and the Philosophy of Technology* series at Lexington Books devoted solely to the topic of technologies in educational settings. Nonetheless, there is certain preliminary work on which this volume builds.

Don Ihde's first involvement with technologies in the educational system dates back to 1975. Even then, Ihde's list of technologies in school was long, ranging from the "tape recorder" to the "public address system," that brings the principal´s voice to all, to the "teaching machines" of the cybernetic age, early computer-others that teach mathematics or logic (Ihde 1975, 189 and 196). "Man-machine relations in our lifeworld are the familiar omnipresence of technological culture," Ihde concludes (Ihde 1975, 189). However, classic examples of technologies in postphenomenology then tend to be navigation, weapons, architecture, or imaging technologies (Ihde 1990, 2009; Rosenberger 2017; Verbeek 2005). This series at Lexington Books has published monographs by Stacey O. Irwin and Galit Wellner dealing with digital media technologies, such as Earbuds and Photoshop, and the most common communication technology, the smartphone, as well as an anthology on digital media (Irwin 2016; Wellner 2015; Van den Enden, Wellner, and Irwin 2017). These technologies can be used in educational settings, but when that happens, they are decisively transformed, modified for that specific context, or even redesigned for it. Galit Wellner aptly described this in the case of the Zoom conferencing software (Wellner 2021). In educational settings, human–technology relations are very different, so a general analysis of technology is not sufficient here. Today, technologies are also being redesigned or even specifically developed for the educational sector.

Some pioneering work on learning has been done by the Danish anthropologist Cathrin Hasse. In a more fundamental understanding, Hasse has incorporated postphenomenologically understood situatedness into an anthropologically grounded theory of learning (Hasse 2015). In a more recent study, together with Oliver A. Tafdrup and Bjarke L. Andersen, she has examined situations of technological breakdown in a Danish primary school and in Australian higher education. Hasse et al. highlight potent and impotent explanatory patterns that teachers use to either explain technological breakdown to themselves, or mystify it (Tafdrup, Andersen, and Hasse 2019). In our volume, by contrast, the researchers focus on the workings of technology, not the certainly still frequent instances of technological breakdown in use in schools and colleges.

The German sociologist Tobias Röhl has set a materialistic focus on the ethnographic consideration of educational processes (Röhl 2012). In this way, technologies also come into view. Röhl's focus, however, is on "quasi-transparent media," such as the blackboard, and "epistemic objects," such as a prism in physics class (Röhl 2018, 38). Röhl understands educational processes in micro-sociological terms. This must always be kept in mind in the contributions to our volume as well; only in rare cases are technologies used in educational settings by one person alone; often there are teachers and learners.

Danish psychologist Jesper Aagaard has introduced an important opposite pair into the discussion of educative technologies from a postphenomenological perspective: attention and distraction (Aagaard 2015b). Aagaard's studies describe how students feel " pulled in" by distracting content, such as Facebook on their laptops (Aagaard 2015a, 2018). Attention is also one of the most important theorems in the discussion of technologies in educational theory today, often referring to Bernard Stiegler (Lewin 2016; Fitzpatrick 2020).

There is a blueprint for using postphenomenology to analyze educational processes by Catherine Adams and Joni Turville in Postphenomenological Methodologies edited by Jesper Aagaard, et al. in this series (Adams and

Turville 2018). A good literature review can also be found here. Adams and Turville open up a range of empirical methods that roughly corresponds to our authors' methods: detailed analysis of human–technology relations, in-depth interview, and participant observation.

In our volume, we want to take on existing research as well as go beyond it. We include new original research and new twists of the already established research approaches. Our contributors analyze the new digital technologies in the educational system and develop new theories from them. An important focus of this volume is the exchange of educational theory.

SUMMARIES OF THE CONTRIBUTIONS

This volume brings together contributions from 22 authors from 11 countries around the world. According to a classical definition by Siegfried Bernfeld, the function of the educational system is to "conserve" a society (Bernfeld 1928, 114). Thus, our contributions also make clear how culturally specific technologies in educational settings can be. This volume has some focus on Germany since philosophical questions arise here to a special degree due to a lively tradition of educational philosophy since the so-called *Geisteswissenschaftliche Pädagogik*. However, some of the most technologically exciting contributions come from Asia. Here are the summaries of the articles in our three sections.

Empirical Evidence

In their chapter "'What To Do With Your Life and How To Spend Your Time?' Life Concepts Mediated through Popular (Educational) Videos," André Epp and Andreas Stock provide an in-depth analysis of an educational video from the YouTube channel *Kurzgesagt – In a Nutshell.* Starting from the postphenomenological premise that technologies mediate our relation to the world and that both the human experiencer and the technological medium are transformed within this relationality, Epp and Stock ask how the perception of self and world is framed by the chosen video.

The chapter "A Note on the Materiality of Educational Frog Dissection," by Robert Rosenberger, takes a postphenomenological look at the materiality of frog corpse dissection. Where arguments against frog dissection tend to be based on ethical or environmental concerns, Rosenberger adds to these arguments the educational downsides of this real-world material practice. This helps to articulate what Rosenberger argues is the main challenge to uprooting such practices: overturning their status as an unquestioned educational ideal.

Kristy Forrest's chapter "Affordances as Solicitations: The Role of Technical Mediation in Academic Care" discusses new mandates for teachers to consider technological implications in their practice of academic care. With reference to the increasing calls for digital well-being, she adopts a postphenomenological perspective to analyze the complex process of teacher risk assessment and deterministic judgment. Drawing on the concepts of multistability and mediation, Forrest identifies implications for teaching practices.

In his chapter "Mediated (Mis)Conduct: Turnitin as an Audience for Academic Work," Eliott Rooke explores the challenges for academic work in the face of similarity detection software such as Turnitin. He argues for the reductive nature of similarity detection software Turnitin, which often flags legitimate work while failing to detect other forms of misconduct. From a postphenomenological perspective, Rooke shows that the non-neutrality of the software actively constitutes the students' understanding of work practices.

The chapter "Equipping Tablets: In-Depth Interviews with Early Adopters on the Sedimentation of Human-Technology Relations in Schools," by Markus Bohlmann and Martin Wilmer, examines the results of a study on human–technology relations with tablets. In a series of in-depth interviews with philosophy students, they explore how the students perceive the technology and what sedimentation processes are evident to them. Bohlmann and Wilmer use a passage from Heidegger's *Being and Time* as an experimental stimulus to get students thinking about human-technology relations.

Lesley Gourlay's chapter "Fugitive Pathways: Sensors, Lines and Knots in the Academic Library" assesses the material, ephemeral, co-present, and embodied aspects of living in a postdigital world that is otherwise associated with the image of a pervasive, ubiquitous, and occluded entity. At the core of her contribution is the premise that an algorithmic gaze is at work in higher education as a constitutive element of a broader culture of surveillance that can be addressed through the use of postphenomenological methodology.

In her chapter "Leveling Up: A Postphenomenological Perspective on Gamification," Stacey O. Irwin explores the use of gamified technology in educational contexts from a postphenomenological perspective. Drawing on Don Ihde's classification of human-technology relations, Irwin identifies different technological stabilities and their respective influence on teachers' and students' understandings of learning as playful practice.

Annie Kurz presents in her chapter "Skype, Zoom, and the Zoombies: Reflections on Artistic Play, Malfunction, and the Traits of the Trade-Offs" key findings from the SkyleLab project, which explored the specificities of designers' and artists' practices of teaching and learning through videoconferencing tools. SkypeLab worked with artistic free play and experimentation. Kurz offers a postphenomenological reflection on both "playing with technology" and online learning scenarios, with the aim of describing the trade-offs and potential gains of videoconferencing for teaching and learning with and through such applications.

Finally, Sou Hee Yang's chapter "Let's Study Together: A Postphenomenological Investigation of 'Study with Me' Content in South Korea" examines various facets of the use of "study with me" content on YouTube from cultural, social, and educational perspectives. Yang explores how the "study with me" content has changed the way Koreans study and engage in social activities, and how these changes, in turn, have affected their use of YouTube as a motivational educational tool.

Conceptual Convergence

In her chapter "Ambiguous Relations. A Postphenomenological Reflection on Technological Multistability in Education" Patrizia Breil explores the ways in which relations to technologies are always necessarily characterized by underlying experiences of ambiguity. Using the postphenomenological notion of multistability, she reveals the ontological roots of the experienced ambiguity and offers a perspective on the pedagogical potential of experiences of ambiguity as a positive lack of understanding that calls for an autonomous approach to technology that is always open to change.

In his chapter, "Freedom through Restriction? Merleau-Ponty, Dewey, and Foucault on Habit-Formation," Jesper Aagaard argues for the need to discipline attention. From a postphenomenological perspective on the relationship between digital devices, attention, and distraction in the classroom, he presents habit-formation as a way to meet the challenges of necessarily technologically mediated attention. Contrary to those who might oppose the idea of disciplining students, Aagaard`s habit-formation in terms of tech habits that enable attention brings disciplining back into educational research as a matter of urgency.

In his chapter "A Critical Examination of Teaching and Learning in Times of Algorithmic Reasoning," Dan Mamlok examines how critical theories of technology and postphenomenological approaches can support pedagogical goals despite their different views of the interplay of technology, society, and the political. In addition to existing attempts to formulate ontological and ethical approaches to theorizing an ultra-technologically mediated world, Mamlok argues for an epistemological and sociopolitical approach.

The chapter "A Postphenomenological-Constructionist Assessment of AI in Education: Toward a Multi-Dimensional Democratic Education," by Galit Wellner and Ilya Levin proposes a novel method for systematic assessment based on the synergies between postphenomenology and constructionism, which allows for a multifaceted understanding of the impact of technologies on students and schools. The assessment method follows the four basic relations identified by Don Ihde and evaluates each of them according to the constructionist principles formulated by Seymore Papert.

Tomoki Sakata's chapter "Diverging Ethical Concepts: Postphenomenological Analysis through the Lens of Japanese Culture" deals with differences in the experience of technological mediation due to cultural background. As an example of such differences, he refers to the results of a small study in which Japanese students were asked about their learning experiences during the COVID-19 pandemic. Comparing their responses with those of German students, Sakata finds that, unlike German students, digital tools are not rooted in the traditional worldview of Japanese young adults.

The chapter "Taking Care for the School in Times of Zoom-Education: A Stieglerian Account," by Joris Vlieghe, addresses modes of attention in the use of conferencing technologies. Drawing on Bernard Stiegler's account of the role of technology in processes of subjectivation, he emphasizes that seeing is deeply shaped by technology. Starting from the premise that learning is an embodied practice of paying attention to and being in contact with the world, Vlieghe advises against seeing digital conferencing tools as a substitute for real, embodied situations of learning in the world.

Subject Subversion

In her chapter, "After Phenomenology? Digital Subjects and the Relevance of Experience," Anke Redecker shows that research on online learning in postdigital scenarios tends to focus on the multiple entanglements of technical and human aspects in order to emphasize the relevance of praxeology. Here, the dominance of practices seems to replace the experiencing and reflecting subject. The latter no longer exemplifies the strong modern subject of knowledge and power, but the vulnerable subject dominated and controlled by algorithms. Instead of claiming the end of phenomenology, Redecker argues for a phenomenological perspective on a subject that builds its own resistance and creativity on the awareness of contingency and vulnerability, as expressed in concepts of uncertainty, intersubjecitivty, and alienness.

Håkon Jakobsen Aaltvedt's chapter "Asking Educational Questions about Technology" combines Gert Biesta's educational philosophy with a postphenomenological perspective on technology to provide a counterexample to the prominent discourse on learning efficiency, in which technologies are reduced to mere instruments. Biesta articulates three educational purposes: qualification, socialization, and subjectification. Using the postphenomenological notions of multistability and mediation, Aaltvedt shows that technologies can take on other than instrumental stabilities that are more relevant to understand in light of the educational domains of socialization and subjectification.

The chapter "Speaking of Education's Technologically Mediated Character(s)," by Jan Peter Bergen, questions the function of educational practice as going beyond the mere task of imparting concrete knowledge and practical as well as intellectual skills. As a primary source of socialization, educational practice also shapes individuals as (moral) subjects, as discussed in terms of so-called "character education." Postphenomenology makes clear that technologies participate in the formation of subjectivity. Rather than focusing on technologically mediated processes in momentary situations, Bergen explores what technologies do to individuals over time. In his contribution, he bridges the gap between the existing postphenomenological vocabulary and character education by introducing insights and concepts from virtue theory.

Finally, in Anne Pesch's chapter, "Rethinking Ihde's Relations Pedagogically: On Learning and Becoming Who We Are in Relations to Things," she combines postphenomenological interrelational ontology with pedagogical reflections on relationality. By contrasting Don Ihde's human-technology relations with Martinus Langeveld's "modes of making sense" (Sinngebungsmodi) and "domains of the intentional object" (Gegenstandsbereiche), she highlights possible benefits for a theoretical approach to learning in technology relations. With a focus on the notion of multistability, Pesch outlines what it means to learn in mediation, alterity, and background relations.

CONCLUDING REMARKS

There is a well-known scene by the German phenomenologist and educationalist Klaus Prange. A man is standing at a traffic light, next to him is a child. No matter what the man does, he cannot *not* educate the child at this very moment (Prange 2003, 26). But the same is probably true for the traffic light. Does it switch to yellow first? Does it blink? Does it count down? Just as this traffic light, technologies in the educational system always educate. With an educational relationship always comes responsibility. But how do technologies rise to this responsibility? With this volume, we wanted to encourage a rethinking of our relations with technologies in educational settings. This rethinking involved a re-evaluation of existing and past human–technology practices, as well as a conceptualization of possible and responsible human–technology relations in the future of education, starting today.

We thank all of our contributors for their inquisitive enthusiasm in exploring the latent structures of everyday human–technology relations and for testing their postphenomenological toolkit on issues of pedagogical relevance or, conversely, for reassessing their pedagogical research from a postphenomenological perspective. We would also like to thank Robert Rosenberger for the great opportunity to publish in his series *Postphenomenology and the Philosophy of Technology* and Jana Hodges-Kluck and Deanna Biondi at Lexington Books for their kindness and support throughout this process.

We sincerely hope that this book will contribute to the further development of the dialogue between postphenomenology and educational theory. By bringing together elementary conceptual investigations, and by providing a nuanced picture of what a theoretical approach like postphenomenology might add to the reflection on specific technology use in pedagogical situations, we hope to lay a foundation for further reflection and action.

REFERENCES

Aagaard, Jesper. 2015a. "Drawn to Distraction: A Qualitative Study of off-Task Use of Educational Technology." *Computers & Education* 87: 90–97. https://doi.org/https://doi.org/10.1016/j.compedu.2015.03.010.

———. 2015b. "Media Multitasking, Attention, and Distraction: A Critical Discussion." *Phenomenology and the Cognitive Sciences* 14 (4): 885–96. https://doi.org/10.1007/s11097-014-9375-x.

———. 2018. "Entering the Portal: Media Technologies and Experiential Transportation." In *Postphenomenological Methodologies: New Ways in Mediating Techno-Human Relationships*, edited by Jesper Aagaard, Jan Kyrre Berg Friis, Jessica Sorenson, Oliver Tafdrup, and Cathrine Hasse, 45–62. Lanham, Md.: Rowman & Littlefield.

Adams, Catherine, and Joni Turville. 2018. "Doing Postphenomenology in Education." In *Postphenomenological Methodologies: New Ways in Mediating Techno-Human Relationships*, edited by Jesper Aagaard, Jan Kyrre Berg Friis, Jessica Sorenson, Oliver Tafdrup, and Cathrine Hasse, 3–26. Lanham, Md.: Rowman & Littlefield.

Bergen, Jan Peter, and Peter-Paul Verbeek. 2021. "To-Do Is to Be: Foucault, Levinas, and Technologically Mediated Subjectivation." *Philosophy & Technology* 34 (2): 325–48. https://doi.org/10.1007/s13347-019-00390-7.

Bernfeld, Siegfried. 1928. *Sisyphos oder die Grenzen der Erziehung*. 2nd Ed. Leipzig: Internationaler Psychoanalytischer Verlag.

Bohlmann, Markus. 2022. *Bildung—Philosophie—Digitalisierung. Eine Curriculumtheorie. Reihe Digitalitätsforschung Band 3*. Edited by Jörg Noller, Malte Rehbein, and Sybille Krämer. Berlin, Heidelberg: J.B. Metzler. https://doi.org/10.1007/978-3-662-65792-8.

Enden, Yoni Van den, Galit Wellner, and Stacey O'Neal Irwin. 2017. *Postphenomenology and Media. Essays on Human–Media–World Relations*. Lanham, Boulder, New York, London: Lexington Books.

Fitzpatrick, Noel. 2020. "Questions Concerning Attention and Stiegler's Therapeutics." *Educational Philosophy and Theory* 52 (4): 348–60. https://doi.org/10.1080/00131857.2019.1625330.

Hasse, Cathrine. 2015. *An Anthropology of Learning. On Nested Frictions in Cultural Ecologies*. London: Springer.

Ihde, Don. 1975. "A Phenomenology of Man-Machine Relations." In *Work, Technology, and Education. Dissenting Essays in the Intellectual Foundations of American Education*, edited by Walter Feinberg and Henry Rosemont, 186–203. Urbana: University of Illinois Press.

Irwin, Stacey O. 2016. *Digital Media. Human-Technology Connection*. Lanham, Boulder, New York, London: Lexington Books.

———. 1990. *Technology and the Lifeworld. From Garden to Earth*. Bloomington, Ind.: Indiana University Press.

———. 2009. *Postphenomenology and Technoscience. The Peking University Lectures*. Albany: State University of New York Press.

Koller, Christoph. 2001. "Bildung und die Dezentrierung des Subjekts." In *Dekonstruktive Pädagogik. Erziehungswissenschaftliche Debatten unter poststrukturalistischen Perspektiven*, edited by Bettina Fritzsche, Jutta Hartmann, Andrea Schmidt, and Anja Tervooren, 35–48. Opladen: Leske und Budrich.

Lewin, David. 2016. "The Pharmakon of Educational Technology: The Disruptive Power of Attention in Education." *Studies in Philosophy and Education* 35 (3): 251–65. https://doi.org/10.1007/s11217-016-9518-3.

Merleau-Ponty, Maurice. 2012. *Phenomenology of Perception*. Florence(UK): Taylor & Francis.

Prange, Klaus. 2003. "Die Form erzieht." In *Form der Bildung—Bildung der Form*, edited by Heinz-Elmar Tenorth, 23–34. Weinheim: Beltz.

Röhl, Tobias. 2012. "Disassembling the Classroom—an Ethnographic Approach to the Materiality of Education." *Ethnography and Education* 7 (1): 109–26. https://doi.org/10.1080/17457823.2012.661591.

———. 2018. "Inviting and Interacting: Postphenomenology and the Microsociology of Education." In *Postphenomenological Methodologies: New Ways in Mediating Techno-Human Relationships*, edited by Jesper Aagaard, Jan Kyrre Berg Friis, Jessica Sorenson, Oliver Tafdrup, and Cathrine Hasse, 27–44. Lanham, Md.: Rowman & Littlefield.

Rosenberger, Robert. 2017. *Callous Objects: Designs against the Homeless*. Minneapolis, MN: University of Minnesota Press.

———. 2021. "Technological Multistability and the Trouble with the Things Themselves." Edited by Shannon Vallor. *The Oxford Handbook of Philosophy of Technology*. Oxford University Press. https://doi.org/10.1093/oxfordhb/9780190851187.013.42.

Rosenberger, Robert, and Peter-Paul Verbeek. 2015. "A Field Guide to Postphenomenology." In *Postphenomenological Investigations: Essays on Human–Technology Relations*, edited by Robert Rosenberger and Peter-Paul Verbeek, 9–42. Lanham, Md.: Lexington Books.

Tafdrup, Oliver Alexander, Bjarke Lindsø Andersen, and Cathrine Hasse. 2019. "Learning to Interpret Technological Breakdowns: A Path to Technological Literacy." In *Human-Centered Digitalization and Services*, edited by Marja Toivonen and Eveliina Saari, 139–52. Singapore: Springer Singapore. https://doi.org/10.1007/978-981-13-7725-9_8.

Thompson, Paul B. 2020. "Ihde's Pragmatism." In *Reimagining Philosophy and Technology, Reinventing Ihde*, edited by Glen Miller and Ashley Shew, 43–61. Cham: Springer International Publishing. https://doi.org/10.1007/978-3-030-35967-6_4.

Verbeek, Peter-Paul. 2005. *What Things Do. Philosophical Reflections on Technology, Agency, and Design*. University Park: Pennsylvania State University Press.

———. 2020. "Politicizing Postphenomenology." In *Reimagining Philosophy and Technology, Reinventing Ihde*, edited by Glen Miller and Ashley Shew, 141–55. Cham: Springer International Publishing. https://doi.org/10.1007/978-3-030-35967-6_9.

———. 2022. "The Empirical Turn." In *The Oxford Handbook of Philosophy of Technology*, edited by Shannon Vallor, 35–54. Oxford: Oxford University Press. 10.1093/oxfordhb/9780190851187.013.4.

Wellner, Galit. 2015. *A Postphenomenological Inquiry of Cell Phones: Genealogies, Meanings, and Becoming*. Lanham, Boulder, New York, London: Lexington Books.

———. 2021. "The Zoom-Bie Student and the Lecturer: Reflections on Teaching and Learning with Zoom." *Techné: Research in Philosophy and Technology* 25 (1): 1–25.

Part I

EMPIRICAL EVIDENCE

Chapter 1

"What To Do With Your Life and How To Spend Your Time?"

Life Concepts Mediated through Popular (Educational) Videos

André Epp and Andreas Stock

Social transformation and change processes, which are negotiated under terms such as risk society, reflexive modernity, individualization, and globalization (cf. Albrow 1996; Beck 1986; Giddens 1990), are not only reflected in the differentiation and pluralization of life situations, living conditions, and life courses of individuals[1] but also have an impact on the diversification and change of educational spaces and places. Nevertheless, formal educational locations, such as schools and universities, are (still) attributed an unbroken significance with regard to the unfolding of educational processes. For this reason, there is also talk of an undisputed leading authority (cf. Rauschenbach 2003, 53), although it is repeatedly pointed out that almost all topics essential for later life are usually not learned in school (cf. Richter 1999) or that 70% of all educational processes are carried out outside formal educational institutions—at least for adults (cf. Dohmen 2001).

Despite the social relevance of formal places of education and learning, in recent decades there has been an increasing interest in informal and nonformal places and spaces of education and learning. These places and spaces of education often compete with each other and are interwoven (cf. Seipold 2014). Further, it is often emphasized that places and spaces of education and learning are becoming more and more blurred (cf. Epp 2019, 107) or even hidden (cf. Bach 2019).

With regard to the acquisition of knowledge, skills, abilities, and key qualifications, it is emphasized that all places and spaces of everyday life make contributions to educational processes. However, these are not always targeted, regulated, and visible (e.g., in the form of grades) (cf. Rauschenbach 2003, 53). With regard to these developments, there is also talk of a temporal and spatial expansion of the pedagogical (cf. Krüger 2019) or of a comprehensive pedagogization of (everyday) life (cf. Kade, and Seitter 2007). Against this background, a variety of places and spaces have been empirically scrutinized in recent decades and scanned for their educational potential. For example, for youth work (e.g., Epp 2011, 2012; Wendt 2010), civic engagement (e.g., Düx et al. 2009; Epp and Barth 2021; Hinrichsen 2020), music scenes (e.g., Epp 2021a), (protest) movements (e.g., Deutschmann 2021), dance schools (e.g., Beyeler and Patrizi 2012), theatres (e.g., Bolwin 2012), art associations (e.g., Baumann 2012), and sociocultural centers (Bode, Hesse, and Nagel 2012), studies documented that educational processes take place in such exemplary fields.

The differentiation of formal, nonformal, and informal educational venues has also been further differentiated in the recent past by the megatrend of digitization. Digital learning and educational spaces are not only found in schools and universities (e.g. Epp 2021b; Höfer-Lück et al. 2021) but also in nonformal and informal contexts. For example, in recent years, animated (educational) videos in particular, which present scientific and socio-political topics, have enjoyed increasing popularity: such as the animated video series *Kurzgesagt—In a Nutshell*. It is one of the top three most-subscribed YouTube channels in Germany and also attracts a great deal of international attention with its English-language videos (cf. Socialblade 2022).

Since the digitalization of learning and educational spaces and places is omnipresent (in everyday life) and considered to be extremely important for educational processes (cf. OECD 2021), as well as nonformal learning—both are often assumed to counteract educational inequity (cf. Täubig 2018, 8)—we take, due to its international popularity, a look at the aforementioned YouTube channel from a postphenomenological perspective. With this approach, we want to reveal what is hidden and being transported through it. In this context, we will focus on a video that deals with the lives of people (Kurzgesagt 2021). First, we introduce the postphenemenological instruments, we want to apply (2), illustrate the focused video, and take a postphenemenological look at it (3). We conclude the paper with a critical conclusion and outlook (4).

TOWARD A POSTPHENOMENOLOGY OF MOVING IMAGES

The research perspective of postphenomenology, a philosophical approach developed by the American philosopher Don Ihde, focuses on human–world relations, which are mediated through technologies. As a transformation of classical phenomenological branches of Husserl, Heidegger, and Merleau-Ponty, postphenomenology blends different phenomenological angles—cognitive, praxical, existential, and linguistic-hermeneutic—with American pragmatic philosophy to allow for a deeper description of human–technology relations (cf. Ihde 2009; Ihde 2016; Rosenberger 2020; Rosenberger and Verbeek 2015; Van Den Eede, Irwin, and Wellner 2017). Rather than seeing technologies as separated functional or instrumental objects or alienating entities, postphenomenology enables us to analyze those technologies and media as transformative mediators of human–world relations that affect the way we approach certain (social) phenomena and challenges (cf. Hauser et al. 2018, 460; Ihde 1990, 2009, 22). This mediation has *revealing* and *concealing* functions and is therefore *not neutral*. In order to take a differentiated look at the transformative power of videos as a medium, we will take up aspects of Merleau-Ponty's phenomenology of (visual) expression to provide a theoretical framework for analyzing individual videos under specific aspects.[2]

Phenomenology of (Visual) Expression

With Merleau-Ponty, we can say that images are not just pictures that present the world to us, but more profoundly that we see "according to, or with" these images (Merleau-Ponty 1964a, 164), which is equally true for videos as moving images.[3] This raises the question of what influence this seeing *with* images has on the way we relate to ourselves and the world, and which aspects, in particular, are obscured (*concealed*) and which are emphasized (*revealed*) by specific (moving) images. We can see a strong parallel between technologies, as conceived by postphenomenology, and videos as a medium—especially those videos that convey concepts of life—considering Don Ihde's following statement: "Technologies transform our experience of the world and our perceptions and interpretations of our world, and we in turn become transformed in this process" (2009, 44; for a distinction between technology and media see Wiltse 2017). We will first approach Merleau-Ponty's interwoven phenomenology of language, in which meaning constantly emerges anew.

Since the sign has meaning only in so far as it is profiled against other signs, its meaning is entirely involved in language. Speech always comes into play against a background of speech; it is always only a fold in the immense fabric of language. [. . .] Like a charade, language is understood only through the interaction of signs, each of which, taken separately, is equivocal or banal, and makes sense only by being combined with others (Merleau-Ponty 1964b, 42).

This "immense fabric," which Merleau-Ponty describes, in which the words and forms we are perceiving unfold their meaning only through their interaction, is not limited to speech or particular arts, but also applies to movies (cf. Sobchack 1992) and videos, as we will argue. If we agree with Merleau-Ponty that "a movie is not thought; it is perceived" (2019, 110), then this is precisely where the philosophical interest lies, argues Oliver Fahle: "it is not the reduction of the world to meaning, but the reattachment of meaning to its expression, from which it cannot be subtracted without loss, that testifies to the quality of the (moving) picture" (Fahle 2010, 167, authors translation). Meaning does not emerge separately from perception, on a lower or higher level, but together with it; perception and meaning organize themselves *in* the picture (ibid, 165). Furthermore, a picture does not consist of its content *and* its style, it can only be grasped *through* its style (cf. Merleau-Ponty 1964b, 52–57). Style is more than a mere means of representation, it grants the creation its unity, its readability, and allows an overall sense. It also opens up to us a *latent sense* (Merleau-Ponty 1973, 58) that is revealed not only through obvious shapes but also through fringes and blank spaces, "just as the painter paints as much by what he traces, by the blanks he leaves, or by the brush marks that he does not make" (ibid, 43; see also 88–89). Therefore, we have to move away from the image as the concretely visible and approach its marginal areas and immanent cracks, where the invisible shines through the visible. This invisible, which organizes itself into meaning in our perception, will be questioned in the subsequent analysis for its meaning, and potential variations of emerging meaning. With this in mind, going back to the premise of a postphenomenological analysis, we can interrogate further, how technologies and media mediate our access to the world in everyday life.

The Figure–Field–Fringe Relation

To grasp the chiasm of the visible and the invisible, we want to observe the medium through the threefold structure *Figure–Field–Fringe*—an extension of the figure-ground relation[4]—as described by Van Den Eede, Irwin, and Wellner:

> Figure is what we focus on, the target of our intentionality. However, a figure always relates to a wider ground or field within which it is situated. There is never a figure in isolation. But the focus on a figure (opaque) presupposes the invisibility of a ground (transparent). We can nevertheless try to get that environing context, the field, in view by practicing variational analysis. But in turn the field ravels out into an even wider context: the horizon or fringe, what is situated really on the border of our perceptual grasp. Here, too, we can attempt to widen our perception, and get a grasp on the fringe, but it takes great effort. (2017, xxiii)

This threefold structure—modified to include notions of Merleau-Ponty considering the visible and the invisible in moving images—is illustrated in figure 1.1 and will provide the basis for our video analysis. The schematic representation illustrates, that the Figure never stands for itself, isolated, but always appears by standing out from a background (Field). It has a certain permeability instead of a firmly defined contour[5] (see figure 1.1), yet it stands out in a certain way against the grid of the Field behind it in the momentary perception.

This background can mean both a literal visual background but also applies to social and individual self-evidence, non-conscious thought patterns, beliefs, and ideologies.[6] This relation between the known and the new is not binary; rather, something paradoxical is inherent in the enterprise of expression: expressing oneself "presupposes that there is a fund of kindred expressions, already established and thoroughly evident, and that from this fund the form used should detach itself and remain new enough to arouse attention," as Merleau-Ponty states (1973, 35). On the one hand, one can ask what emerges from a particular constellation in the current perception to attract our attention by standing out from a background of the familiar. On the other hand, we can argue, that perception, or rather the visible, is (ontologically) based on a fundamental moment of division (*déhiscence*), which leads to a chiasm of the visible and the invisible:

> Every visual something, as individual as it is, functions also as a dimension, because it gives itself as the result of a dehiscence of Being. What this ultimately means is that the proper essence [*le propre*] of the visible is to have a layer [*doublure*] of invisibility in the strict sense, which it makes present as a certain absence. (Merleau-Ponty 1964a, 187)

Images are not (only) a manifestation of the visual, but an elevation (*Aufriss*) of the visible *within* the visible; inherent in the moving images is a movement that extends beyond itself (Fahle 2010, 173). The image, as the momentarily visible, bears witness to the genesis, the arrangement of the visible, and at the same time is constantly being drawn away by it, in never-ending "weaving-processes" (Fahle 2010, 172–173; cf. Merleau-Ponty 1964a, 163). Thus, one category that the Figure–Field–Fringe pattern allows us to focus on is the visible as it momentarily builds itself up and stands in relation to other visible things (within the same image, or a sequence of images surrounding it); all within the relation Figure–Field. Within this chiasm of the (momentarily) visible and the invisible, we can find latent structures at the fringes of the visible, surrounding it and piercing through it. Taking up Merleau-Ponty, we can say that our perception, even our normal way of seeing, is imbued with a certain character, which itself is not (purely) naturalistic or essential (1964b,

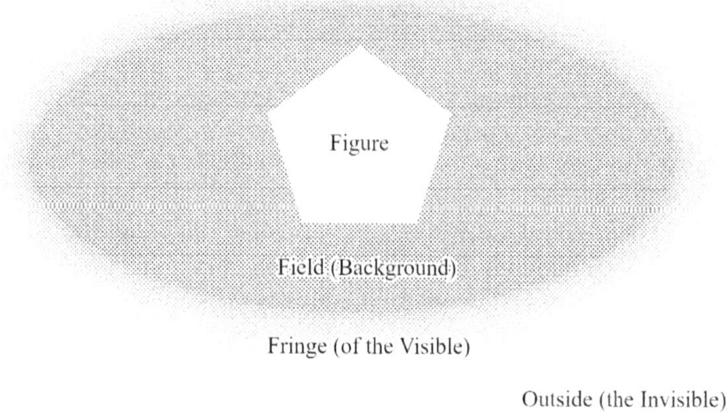

Figure 1.1 Threefold Structure Figure–Field–Fringe. *Source*: Image created by author.

53–54), but is closely tied to (cultural) socialization. In the analysis of the video, for example, we will see that certain social views, such as the ignoring of social, political, and economic contexts, appear self-evident through common latent structures and are reproduced (unconsciously) through a certain visual-esthetic style.

However, the openness at the fringes and within the visible itself offers the possibility of making these invisible structures reflexively accessible. As we will show in the following chapter, a variational analysis allows us to speculatively play with different constellations, consciously rearrange the visible, draw a distance to it, and, through this distancing, bring our perception—our seeing *with* the video—into view anew. In this way, we can explore different processes of organization of meaning in our perception of videos as media.

"WHAT ARE YOU DOING WITH YOUR LIFE?": A VIDEO ANALYSIS

The following section first describes the YouTube channel where the video is uploaded, and then the video itself. After that, the analysis of the video is unfolded.

Brief Information about the Channel

The YouTube channel *Kurzgesagt—In a Nutshell*, run by the In a Nutshell—Kurzgesagt GmbH based in Munich, was created in July 2013.

Since then, new animated short videos in English, and partly in German, of about 5–15 minutes in length on scientific and sociopolitical topics have been published monthly. In their YouTube channel description, they state that they are meant "to spark curiosity about science and the world we live in."[7] The videos are produced by professional illustrators and animators and are seen as educational videos. They can be called a popular science format. In October 2019, the format was awarded the Pedagogical Media Award.[8] Until the end of 2022, the In a Nutshell—Kurzgesagt GmbH was part of the media group of ARD and ZDF (German public television), which co-financed the German-language videos. To date, the international channel has 20.1 million subscribers[9] and the videos also have a wide reach, considering the number of views. The average age of the *Kurzgesagt—In a Nutshell* audience is 25 (Kurzgesagt 2021). In addition to YouTube, the In a Nutshell—Kurzgesagt GmbH also maintains other social media channels on which they upload videos and/or advertise their YouTube channel and their videos.

What Are You Doing with Your Life? The Tail End: A Rough Description

The video, titled *What Are You Doing With Your Life? The Tail End* has a length of 9:35 minutes and is roughly described below. Due to the focus of the analysis, the childhood and adolescent phases in particular are described in more detail. In addition, there is no separate description of the image and audio.

The first video sequence opens with a person standing in a public transport vehicle together with his or her shopping bag. Then masses of water unfold and the person drops the grocery to not drown but to swim in the opening up sea. With regard to this scene, a speaker mentions that it is difficult to keep track of one's own life because of the involvement and the variety of distractions one experiences over the lifetime so that it is challenging to think about how to shape one's own life.

This is followed by an incision to take a look at how precious a lifetime is. Based on the statement made in the video that every 20th person who watches the video will live to be 100 years old, this lifetime is quantitatively visualized in weeks by means of a time grid, and first the childhood and youth phases are depicted on it. The speaker mentions that the individual does not make a productive contribution to society during these phases of life and that he or she is equally unable to make decisions of his or her own. This statement is supported with a baby sitting under a mobile, a toddler splashing around with food, and a teenager chewing gum (see figure 1.2). It is also emphasized that adolescents are spared the consequences of adulthood because they are protected by their parents and given the freedom to develop so that they can become—as it is said—a functioning human being.

A jump to the 20th year of life follows. The time grid is used to show how many weeks have already passed. In addition, it is mentioned that people in this phase either have a profession or attend university. The working life is visualized with a person standing at a height-adjustable desk in front of a computer and wearing a headset. Next to the person are various office utensils, such as a folder and appointment book. The university setting is depicted as follows: A person is standing next to a green area, with books under his arm and waving to two other people. In the distance, a larger building can be seen. Following this, the time grid is shown again and it is pointed out that the majority of society members take up a profession between the ages of 20 and 30 and thus make a finite productive contribution to social coexistence. This period of time is also visualized in weeks with the help of the time grid and it is emphasized that this is "serious adult work." In addition, the importance of the professional phase and career-making is further emphasized, since the video producers are planning a specific video on this and it is—compared to the previous phases—the largest phase of life. The video also gives examples of what people usually do in this phase: get married, have children, build a house, get divorced, and fall in love again. In short: A so-called normal biography is illustrated.

There is another jump to retirement, which is also visualized using the time grid showing the remaining weeks—assuming that a person will live to be 100 years old. However, it is mentioned that the majority of people will not live to be 100 years old and country-specific statistics are shown that depict life expectancy. Furthermore, it is mentioned that with increasing age, the body's health decreases and diseases accumulate. This is also visually supported by statistics.

Another cut is made and it is mentioned that people often do not think about all these phases due to their involvement in their own lives. Nevertheless, according to the speaker of the video, it is helpful from time to time to think about how one wants to shape one's life so as not to regret anything later. A leap is made to the COVID-19 pandemic and it is emphasized that everyday things—such as traveling, visiting public places, going out to eat, meeting friends, and making new acquaintances—that people took for granted were no longer possible because of the social distancing.

Exemplary Analysis of a YouTube Video through Postphenomenology

We will now apply our theoretical considerations of a postphenomenology of moving images to the given video via the threefold structure

Figure 1.2 Depiction of Childhood. Screenshot from the video *What Are You Doing with Your Life? The Tail End* (Kurzgesagt 2021, 01:37).

Figure–Field–Fringe (figure 1.1). First, we will fill in the Figure–Field constellation (the Visible), shown in table 1.1. We will start with placing "life" as the Figure of this constellation, contextualized against the background (Field) of a *Time-Grid*: Time as available lifetime, objectified by a fixed, linear unit (weeks). This time grid is explicitly represented visually in the video, via the grid of points (Kurzgesagt 2021, 1:25); and is taken up visually throughout the course of the videos. However, the video not only poses the question of a mere *lifetime*, but also implicitly poses questions about a meaningfully led life—which is already indicated by the video title "What Are You Doing With Your Life?"—, thus leading us to set life as the Figure initially.

The relationality between Figure and Field can now be played with by means of variational analysis: for example, one can try to invert the constellation of Figure–Field and view the Field against the background of the Figure. Although this does not cause the figure to lose its opacity or the background to lose its transparency, it shifts the focus on the relation and the process of becoming an overall meaning. Such a variation makes clear that there can never be an unbiased, neutral view. Instead, playing with perspectives on the visible allows for an analysis of many different constellations, each with its own set and ratio of revealed and concealed dimensions and artifacts. With a phenomenology of expression, we can ask about the style: How are different figures and backgrounds contextualized and placed in relation to one another? How do the figures thereby present themselves to us through our perception? This takes place inside the Visible, within the pair-constellation Figure–Field.

In swapping Figure and Field—that is, setting life as the new Field of our observation (see table 1.2)—we could ask whether there are other concepts of time, aside from a Time-Grid in weeks, that would make sense in the horizon between time and life. One could argue that time does not have to be represented as a neutral container within which life takes place and which divides life into equal, linear time units. Other concepts of time might come to mind: experienced, subjective time, time as duration (Henri Bergson's *durée*), forms of historicity, just to name a few. If we observe timc against the background of life, we have to set dimensions like uncontrollability [*Unverfügbarkeit*] (Hartmut Rosa) and fracture lines [*Bruchlinien*] (Bernhard Waldenfels) as central moments of (experienced) time. Time cannot be controlled, anticipated, or retrospectively rewound in a linear way but creates its own historicity always anew, oftentimes arising from fractures. Beginnings and initiation points, as well as phases of transition and transformation (often into the unknown), become more relevant than a uniform, calendar-like sequence of events, or the adding up of days according to certain criteria.

Returning to the original constellation of life as Figure and the Time-Grid as Field, we will argue that the contextualization of the highly abstract, ambiguous motif (life) by a supposedly objective background (weeks as a Time-Grid) suggests a neutral objectivity that wants to observe life from the outside.[10] This implicitly introduces a form of measurability and objectification into the figure, life, itself. The style of visual representation (colorful animation), a supposed diversity of depicted lifestyles, and the direct address to recipients disguise this transformation instead of drawing attention to prevailing ambiguities. However, contextualizing life as a lifetime that one has at one's disposal in the form of 5,200 weeks is not neutral. It makes life occur in a certain way and brings with it implicit normative dimensions. What does it mean to view a passed lifetime as "used up"? What does it mean to use one's lifetime wisely? At this point, we are breaking through the pair-constellation of the visible, as the immediately perceived, extending it with its Fringe, the invisible that permeates the visible. It penetrates the visible from its borders as well as from within, makes it fundamentally intelligible, and, at the same time, points beyond the visible, transporting a latent meaning. In figure 1.1, therefore, the invisible (the outside) also penetrates the Figure–Field constellation through its punctuated depiction, reaching into it.

If we include the (immanent and outer) Fringes in our analysis, we can ask not only about constellations and depictions of the visible (depiction and

Table 1.1 Figure–Field Constellation of the Analyzed Video

Figure	*Field*
Life	Time-Grid (weeks)

arrangement of motifs in the image and image sequences) but also about the invisible that lies within the fabric of the images(s): What surrounds the constellation Figure–Field, penetrates our perception in this interplay, and manifests itself unnoticed? This invisible, which penetrates from the outside into the visible, includes the unsaid, the unshown, implicit beliefs, underlying ideologies, and a language of images and signs.[11] Within this fabric, our perception constantly performs itself anew and brings forth a constant, processual becoming of meaning of the visible.

Returning to our video, we can ask what socio-cultural dimensions and individual beliefs are involved in the form of making life readable through this hermeneutic relation. We can now extend the schematic-tabular representation to include the Fringe (see table 1.3). The Fringe column contains terms that attempt to describe the movements of the invisible into the visible. Thus, under the neoliberalism movement, one can examine moments that place individualistic choices at the center and assume an autonomous self, in part an entrepreneurial self, that is supposed to be in charge of its own fortune.[12]

The representation of childhood in the video makes some of these notions vivid:

> You spent the first 600 weeks of your life being a kid and 400 being a teenager. During this time you were pretty useless for society and also not very free to make your own decisions. As a payoff, you were somewhat spared from the consequences and responsibilities of adulthood. Mostly because well meaning adults protected you by cleaning up your mess and giving you time to find yourself. This time is necessary to make you, at least in theory, a functioning human. (Kurzgesagt 2021, 01:29–01:56, transcription)[13]

Confronted with this statement, which is exemplary for the depiction of childhood and adolescence in the video, one can ask, what does it mean to be a "functioning human"? The contextualization of childhood hints in a certain direction: one is supposed to make one's contribution by means of labor, which means to simultaneously work toward one's own retirement.[14] There is no mention of the independent significance of childhood as a period of life, bringing forth its own meaning. In addition to the narration, the visual representation of the children in the video—each shown alone, without contact to other people—supports this individualistic concept (see figure 1.1): Playing, eating, and homework are portrayed as acts of solitude and boredom.

Table 1.2 Swapped Figure–Field Constellation of the Analyzed Video

Figure	*Field*
Time	Life

Contextualized in this way, childhood and adolescence do not seem like a time in which one has made valuable experiences, and has discovered the world and oneself—in one's own way, or the way of childhood, contrary to the thesis that one could not make one's own decisions. Instead of actively living one's life, time simply passes, life happens. The rapid growth of the character in the video (the transition from childhood to adolescence and adulthood, see Kurzgesagt 2021, 1:52–2:02) illustrates a powerlessness and one-sided dependency, giving the impression not only that this is a phase of life that happens to you while you remain a passive victim, but also that this is a phase of life that needs to be put behind as quickly as possible. While the character is rapidly growing, the world around it is magically changing; suddenly the figure finds herself besieged by stacks of documents, while a pizza box sits enthroned on the side table (fast food as an expression of lacking time, even for the vital care of one's own needs) and loose sheets of documents flutter through the door slit toward the figure, who desperately tries to protect herself against them with her arms and legs; in vain, she is at the mercy of the unfolding events. One could refer to Neil Postman's thesis, that childhood is disappearing, as the boundaries between childhood and adulthood are disappearing (1982, 63–64).

Therefore, what does it mean to be "not very free to make your own decisions" in childhood? For this, one has to ask what kind of decisions are shown in the course of the whole video. These are mainly decisions of consumption and decisions of job choice; always *within* a commodified world (in which even lifetime is presented as a commodity or at least a good). Never do the world or social structures themselves appear changeable, never are collective decisions debated, and never are truly existential issues engaged with or political dimensions addressed. The act of decision itself, grounded in an isolated individual, is set as the central element. What is not addressed is that these decisions can only be made, and in part *must* be made, against the background of an existing political, social, and above all economic order. Under this assumption, children have to appear almost unfree. However, children are by no means "spared from the consequences and responsibilities of adulthood." Crises like the climate catastrophe or the COVID-19 pandemic show us that often those suffer from the consequences of socio-political and economic decisions who do not have (interpretative) power and authority.

Table 1.3 Figure–Field–Fringe Constellation of the Analyzed Video

Figure	*Field*	*Fringe*
Life	Time-Grid (weeks)	Neoliberalism, meritocracy, normal biography, positivist view of time

Especially the statement that "well meaning adults protected you by cleaning up your mess" can thus be read quite ironically.

CONCLUSION AND OUTLOOK

Recapitulating the analysis of the video, it can be stated that a view of the world and life is taken that reproduces certain social tendencies, especially individualization.[15] Decisions can only be made on an individual level, and within a certain framework, which seems to be set; social and economic changes or political demands are not addressed. Together with the contextualization of life through a fixed time-grid, the video proposes a positivistic realism, or *Capitalist Realism* in reference to Mark Fisher (2009), which reproduces a lack of alternatives regarding existing orders; change cannot be thought of outside of or beyond the individual itself. A critical postphenomenological analysis of popular (educational) videos as media offers the potential to reveal dimensions that resonate subliminally in their (moving) images, where the invisible penetrates our perception through the visible, in the process of becoming of meaning. A variational analysis, with the help of the Figure–Field–Fringe pattern, can bring different aspects into view anew. Therefore, in their transformative effect, media not only hold the potential to conceal but also to reveal certain aspects of the human world- and self-relations.

A postphenomenological analysis has the potential to open up new educational perspectives by asking how technologies and media could also present themselves to us aside from the dominant view (how they are "dominantly used," see Ihde 1993, 37), how they could be read otherwise (which "non-dominant stabilities" can be found, see Rosenberger 2020). This perspective now, in turn, confronts the individual with the challenge of coming to his or her own perspective and being able to communicate it in an argumentative manner. This approach combines creative and critical aspects by analyzing revealing and concealing processes and asking for alternative readings of the visible through variational analysis. Dealing with plurality and ambiguity can invite one to play with different perspectives but, at the same time, it challenges one to relate to different positions, to be able to orientate oneself in the midst of the fabric of plural perspectives, and to reflect on one's own position within this fabric.[16]

A postphenomenological perspective makes one aware of the relationality that also acts in and through videos as media. It can analyze particular relations and constellations of the visible and invisible in depth, and explore the openness of videos, asking for alternative readings. This interplay between focusing and opening offers many entry points for educational science, many of which remain to be explored. In the future, it will also be necessary to further illuminate the extent to which a postphenemenological perspective contributes to a more differentiated view of digitized (informal) learning environments: that is, that they not only counteract social inequality but also (re)produce it (e.g., with regard to certain life plans and the passive exposure to social frameworks).

NOTES

1. With regard to the dissolution of the so-called normal biography, there is also talk of a deinstitutionalization of coherent life course patterns (cf. Heinz 2000).
2. On imaging technologies, there is already extensive research in postphenomenology (see, e.g., Fried and Rosenberger 2021; O'Brien 2019).
3. On the relation of language and painting—and, in the digital extension, images as (creative) expression—see the chapter *Indirect Language* in Merleau-Ponty's *The Prose of the World* (1973, 47–113).
4. The figure–ground relation has been an important structure since early phenomenology and has been taken up by Gestalt psychology and phenomenological psychologies (cf. Ihde 2012, 37, 2002, 44).
5. On contour and modern painting, cf. Merleau-Ponty 1964a, 182–83.
6. See the work of Annette M. Stroß in the field of reflexive health education, where, through analysis of narratives, different mythical figurations can be elaborated via metaphors as expressive figures (cf. Stroß 2018).
7. https://www.youtube.com/@kurzgesagt
8. https://presse.funk.net/format/dinge-erklaert-kurzgesagt/
9. The German channel has to date (February 2023) 2 million subscribers (https://www.youtube.com/@KurzgesagtDE).
10. This impression is reinforced by the presentation of statistics throughout the video.
11. At this point, the analysis becomes open to many theoretical connecting points, from different theoretical fields and disciplines, especially theories of socialization and subjectivation.
12. These concepts (which need not be in any way consistent or coherent, and usually are not) could be further elaborated for a deeper analysis of the video but can only be hinted at here. Our goal is to explore the possibility of a formal framework that takes up the postphenomenological program and continue it via a phenomenology of expression to bring videos as media into view anew.
13. For a representation of childhood and adolescence, see especially 1:20–2:50.
14. See the illustration of coins jumping from the calendar section of the working life to that of retirement (Kurzgesagt 2021, 3:25–3:35); socio-economic issues, such as old-age poverty, are not addressed.
15. See Tomoki Sakata's contribution in this book, especially regarding the European ideal of individual autonomy.

16. With Jürgen Mittelstraß (1982), we could describe this skill as "orientation-knowledge" (*Orientierungswissen*) *in* the world/nature, as opposed to a "possession-knowledge" (*Verfügungswissen*) *about* (*über*) the world/nature.

REFERENCES

Albrow, Martin. 1996. *The Global Age: State and Society Beyond Modernity*. Cambridge: Polity Press.

Bach, Clemens, ed. 2019. *Pädagogik im Verborgenen. Bildung und Erziehung in der ästhetischen Gegenwart.* Wiesbaden: Springer VS.

Baumann, Leonie. 2012. "Kunstvereine: Kunstvermittlung dezentral und experimentell." In *Handbuch kulturelle Bildung*, edited by Hildegard Bockhorst, Vanessa-Isabelle Reinwand-Weiss, and Wolfgang Zacharias, 448–50. München: kopaed.

Beck, Ulrich. 1986. *Risikogesellschaft.* Frankfurt a.M.: Suhrkamp.

Beyeler, Marie, and Livia Patrizi. 2012. "Tanz—Schule—Bildung. Überlegungen auf der Erfahrungsgrundlage eines Berliner Tanz-in-Schulen-Projekts." In *Handbuch kulturelle Bildung*, edited by Hildegard Bockhorst, Vanessa-Isabelle Reinwand-Weiss, and Wolfgang Zacharias, 600–603. München: kopaed.

Bode Rainer, Bernd Hesse, and Torsten Nagel. 2012 "Kulturelle Bildung in den Soziokulturellen Zentren." In *Handbuch kulturelle Bildung*, edited by Hildegard Bockhorst, Vanessa-Isabelle Reinwand-Weiss, and Wolfgang Zacharias, 773–777. München: kopaed.

Bolwin, Rolf. 2012. "Theater als Ort Kultureller Bildung." In *Handbuch kulturelle Bildung*, edited by Hildegard Bockhorst, Vanessa-Isabelle Reinwand-Weiss, and Wolfgang Zacharias, 622–627. München: kopaed.

Deutschmann, Anna. 2021. "Institutionalisierung und Transformation: Bildungs- und Lernprozesse in und von (Protest-)Bewegungen." *ZEP: Zeitschrift für internationale Bildungsforschung und Entwicklungspädagogik* 44 (3): 4–9.

Dohmen, Günther. 2001. *Das informelle Lernen. Die internationale Erschließung einer bisher vernachlässigten Grundform menschlichen Lernens für das lebenslange Lernen aller.* Bonn: BMBF.

Düx, Wiebken, Gerald Prein, Erich Sass, and Claus J. Tully. 2009. *Kompetenzerwerb im freiwilligen Engagement. Eine empirische Studie zum informellen Lernen im Jugendalter.* Second edition. Wiesbaden: VS Verlag für Sozialwissenschaften.

Epp, André. 2011. "Methoden und Techniken der Praxisevaluation in der Kinder- und Jugendarbeit." *Deutsche Jugend* 9: 388–394.

———. 2012. ""' . . . lieber sie haben "n Mikro in der Hand anstatt 'n Klappmesser!"—Praxisevaluation, Wirkungsrekonstruktionen und Nachhaltigkeit kultureller Jugendarbeit." In *Die Kunst, über kulturelle Bildung zu forschen. Theorie- und Forschungsansätze*, edited by Tobias Fink, Burkhard Hill, Vanessa-Isabelle Reinwand-Weiss, and Alexander Wenzlik, 194–207. München: kopaed.

———. 2019. "The Mosh-Pit—An Area for Excess or a Place of Learning?" *Metal Music Studies* 5 (1): 107–114. https://doi.org/10.1386/mms.5.1.107_1.

———. 2021a. "Heavy Metal as a Profession?—The Mexican Heavy Metal Scene and Its Effects as an Informal and Non-Formal Learning Space." *Metal Music Studies* 7 (1): 149–158. https://doi.org/10.1386/mms_00039_1.

———. 2021b. "Vermittlung qualitativer Forschungsmethoden im digitalen Raum—(Didaktische) Einblicke in eine bisher wenig thematisierte Praxis der Hochschullehre." *Ludwigsburger Beiträge zur Medienpädagogik* 21: 1–13. https://doi.org/10.21240/lbzm/21/07.

Epp, André, and Laura Barth. 2021. "Lebensmittelretten als (implizites) Bildungskonzept? (Bildungs-)Wirkungen zivilgesellschaftlichen Engagements." *Voluntaris—Zeitschrift für Freiwilligendienste und zivilgesellschaftliches Engagement* 9 (2): 249–262. https://doi.org/10.5771/2196-3886-2021-2-249.

Fahle, Oliver. 2010. "Die Ästhetik der bewegten Bilder." In *Kunst. Bild. Wahrnehmung. Blick. Merleau-Ponty zum Hundertsten*, edited by Antje Kapust and Bernhard Waldenfels, 159–173. München: Wilhelm Fink.

Fisher, Mark. 2009. *Capitalist Realism. Is There No Alternative?* Ropley, Hants: Zero books.

Fried, Samantha J., and Robert Rosenberger, eds. 2021. *Postphenomenology and Imaging. How to Read Technology*. Lanham: Lexington Books.

Giddens, Anthony. 1990. *The Consequences of Modernity*. Stanford: Stanford University Press.

Hauser, Sabrina, Doenja Oogjes, Ron Wakkary, and Paul-Peter Verbeek. 2018. "An Annotated Portfolio on Doing Postphenomenology Through Research Products." *DIS'18: Designing Interactive Systems Conference*, June 9–13, 2018, Hong Kong. 459–471. https://doi.org/10.1145/3196709.3196745.

Heinz, Walter R. 2000. "Selbstsozialisation im Lebenslauf. Umrisse einer Theorie biographischen Handelns. " In *Biographische Sozialisation*, edited by Erika M. Hoerning, 165–186. Berlin, Boston: De Gruyter Oldenbourg.

Hinrichsen, Merle. 2020. *Das FSJ als biographischer Zwischenraum. (Re-)Konstruktionen von Bildungswegen junger Erwachsener*. Wiesbaden: Springer VS.

Höfer-Lück, Hanna, Malte Delere, Tatjana Vogel, and Gudrun Marci-Boehncke. 2021. "Digitale Bildungsräume im Corona-Semester. Erwartungen und Erfahrungen von Lehramtsausbildenden." *Journal für LehrerInnenbildung* 21 (2021) 2: 34–43. https://doi.org/10.25656/01:23094.

Ihde, Don. 1990. *Technology and the Lifeworld. From Garden to Earth*. Bloomington, Indianapolis: Indiana University Press.

———. 1993. *Postphenomenology: Essays in the Postmodern Context*. Evanston: Northwestern University Press.

———. 2002. *Bodies in Technology*. Minneapolis, London: University of Minnesota Press.

———. 2009. *Postphenomenology and Technoscience: The Peking University Lectures*. Albany: State University of New York Press.

———. 2012. *Experimental Phenomenology. Multistabilities*. Second edition. Albany: State University of New York Press.

———. 2016. *Husserl's Missing Technologies*. New York: Fordham University Press.

Kade, Jochen, and Wolfgang Seitter. 2007. "Wissensgesellschaft—Umgang mit Wissen—Universalisierung des Pädagogischen: Theoretischer Begrüñdungszusammenhang und projektbezogener Aufriss." In *Umgang mit Wissen. Recherchen zur Empirie des Pädagogischen*, edited by Jochen Kade, and Wolfgang Seitter, 15–42. Opladen: Verlag Barbara Budrich.

Krüger, Heinz-Hermann. 2019. *Erziehungs- und Bildungswissenschaft als Wissenschaftsdisziplin.* Opladen, Toronto: Verlag Barbara Budrich.

Kurzgesagt, In a Nutshell. "What Are You Doing With Your Life? The Tail End." *YouTube Video*, 9:35. May 25, 2021. https://youtu.be/JXeJANDKwDc.

Merleau-Ponty, Maurice. 1964a. "Eye and Mind." Translated by Carleton Dallery. In *The Primacy of Perception and Other Essays on Phenomenological Psychology, the Philosophy of Art, History and Politics*, edited by James M. Edie, 159–190. Evanston: Northwestern University Press.

———. 1964b. *Signs.* Translated by Richard C. McClearly. Evanston: Northwestern University Press.

———. 1973. *The Prose of the World.* Translated by John O'Neill. Evanston: Northwestern University Press.

———. 2019. "The Film and the New Psychology." In *Philosophers on Film. From Bergson to Badiou. A Critical Reader*, edited by Christopher Kul-Want, 97–112. New York: Columbia University Press.

Mittelstraß, Jürgen. 1982. *Wissenschaft als Lebensform.* Frankfurt a.M.: Suhrkamp.

O'Brien, Daniel Paul. 2019. "Hap-Tech Narration and the Postphenomenological Film." *Philosophies* 4 (3): 47. https://doi.org/10.3390/philosophies4030047.

OECD. 2021. *OECD Digital Education Outlook 2021.* Paris: OECD Publishing. https://doi.org/10.1787/589b283f-en.

Postman, Neil. 1982. *The Disappearance of Childhood.* New York: Delacorte Press.

Rauschenbach, Thomas. 2003. "Das Bildungsdilemma. (Un-)beabsichtigte Nebenwirkungen öffentlicher Bildungsinstanzen." *Diskurs* 13 (2): 50–58.

Richter, Ingo. 1999. *Die sieben Todsünden der Bildungspolitik.* München: Beltz.

Rosenberger, Robert. 2020. "On Variational Cross-Examination: A Method for Postphenomenological Multistability." *AI & Society.* https://doi.org/10.1007/s00146-020-01050-7.

Rosenberger, Robert, and Peter-Paul Verbeek, eds. 2015. *Postphenomenological Investigations.* Lanham: Lexington Books.

Seipold, Judith. 2014. "Mobile Learning: Structures, Concepts and Practices of the British and German Mobile Learning Discussion from a Media Education Perspective." *MedienPädagogik: Zeitschrift für Theorie und Praxis der Medienbildung* 24 (August): 30–52. https://doi.org/10.21240/mpaed/24/2014.08.15.X.

Sobchack, Vivian. 1992. *The Address of the Eye. A Phenomenology of Film Experience.* Princeton: Princeton University Press.

Socialbalde. 2022. "Top 100 YouTubers in Germany Sorted by Subscribers." Most Subscribed YouTube Channels in Germany. https://socialblade.com/youtube/top/country/de/mostsubscribed. Retrieved September 9, 2022.

Stroß, Annette M. 2018. *Gesundheit und Bildung. Reflexionsansprüche und Professionalisierungsperspektiven.* Wiesbaden: Springer VS.

Täubig, Vicki. 2018. "Informelles Lernen. Standorte bestimmen. " In *Informelles Lernen. Standortbestimmungen*, edited by Nina Kahnwald and Vicki Täubig, 3–14. Wiesbaden: Springer VS.

Van Den Eede, Yoni, Stacey O. Irwin, and Galit Wellner. 2017. "What Media Do." In *Postphenomenology and Media. Essays on Human-Media-World Relations*, edited by Yoni Van Den Eede, Stacey O. Irwin, and Galit Wellner, xvii–xxxii. Lanham: Lexington Books.

Wendt, Peter-Ulrich, ed. 2010. *mit Wirkung! Zur Nachhaltigkeit von Jugendarbeit im ländlichen Raum.* Marburg: Schüren.

Wiltse, Heather. 2017. "Mediating (Infra)structures: Technology, Media, Environment." In *Postphenomenology and Media. Essays on Human-Media-World Relations*, edited by Yoni Van Den Eede, Stacey O. Irwin, and Galit Wellner, 3–25. Lanham: Lexington Books.

Chapter 2

A Note on the Materiality of Educational Frog Dissection

Robert Rosenberger

The dissection of frogs in grade-school biology classrooms remains a widespread practice across the United States and Canada.[1] And it also remains fraught. There is a history of resistance, and some regions have even enacted policies in which students must be provided an alternative assignment if requested. The intensity of the debates over these issues waxed and waned over the decades, and the arena is varied, including newspaper and magazine editorials, occasional public protests, and local legislatures considering policy proposals (e.g., Huntsberry, 2015; Gregory and Berger, 2018; Averett, 2020). Academics argue their cases (e.g., Keiser and Hamm, 1991; Orlans, 1991; Kline, 1995; Sapontzis, 1995; Balcombe, 2000; Friesen, 2011; Oakley, 3013), and do empirical research into learning outcomes (e.g., Zemanova and Knight, 2021; Ormandy et al., 2022). I have been a participant in the criticism of classroom frog dissection in both academic and public settings (Rosenberger, 2011, 2013),[2] and I would like to continue this work here. In addition, I suggest that the practice of frog dissection and the computer-simulated alternatives that have been developed provide a useful arena for thinking through issues of technology and pedagogy.

One reason that issues of frog dissection and computer-simulated alternatives may be useful to revisit at this moment in time is that they in some ways take on precisely the opposite configuration of much of today's conversation on educational technologies. A lot of the contemporary discussion on educational computer technologies is constituted by hype. While there are of course many thoughtful advocates for cutting-edge digital and online tech in education, the zone is flooded by advertising and naïve and shallow cheerleading. I suggest that the character of the discussion on frog dissection and its alternatives is very much the reverse of all of this; the propaganda in general is *against* computer-simulated alternatives. As such, reflection on issues of frog dissection has the potential to help carve out some core values that cut through the hype, ones that exist separate from the immediate discourse.

At least in the U.S. and Canadian context, the activity of dissecting a frog corpse in the classroom (what I'll refer to here as "corpse dissection") is pervasive. It is a staple of public school education for older children. It is strongly set within the culture of this part of the world, long-established as a normal and expected part of childhood educational upbringing. It has the cultural cache of a rite of passage. Researcher and critic on this topic Jan Oakley reports that the literature reveals 78–94% of biology and science instructors teach frog dissection, and 88–95% of students say they've done it at some point during their education, amounting in the United States to 10–12 million dissected animals per year (2013, 4). She summarizes that "The current culture of North American school science continues to operate within a dissection paradigm: that is, dissection remains commonplace" (Oakley, 2013, 4). Arguments against the use of frog corpses for classroom dissection amount to an attempt to uproot the status quo. Where in many areas across these countries computers have encroached upon multiple aspects of childhood education (especially in wealthier regions), frog dissection has become established as a conspicuous holdout against the digitalization of classroom experience.

The dominance of the practice of classroom corpse dissection appears to at least largely be the result of the views held by teachers themselves, what they take to be student expectations, as well as wider cultural momentum. Arguments in support of the practice often point to the vivid visceral hands-on quality of the experience as an important part of the learning process. For example, a survey was conducted by Pamela Osenkowski and colleagues to determine the views of more than 2500 American biology teachers (2021). They found that while the majority would grant an individual student's request for an alternative assignment, that majority also "believe dissection to be the best way to teach anatomy and/or physiology" and that "the use of dissection in biology is an important tradition," as well as that the hands-on experience of this activity is irreplaceable by alternatives (Osenkowski et al., 2021, 239). These views are reflected in position statements maintained by both the National Association of Biology Teachers (NABT) and the National Science Teachers Association (NSTA). The NABT official statement on "The Use of Animals in Biology Education" acknowledges that dissenting students should be provided with alternatives, but maintains that:

> While the increased quality and accessibility of dissection simulation software has helped address concerns from students and parents opposed to dissection, these alternatives are not without limitations. Utilizing a software-only approach may constitute a disservice to many students and does not acknowledge the well-documented educational benefits of hands-on dissection. (NABT, 2019)

The NSTA statement on "Responsible Use of Live Animals and Dissection in the Classroom" similarly "opposes regulations or legislation that would eliminate an educator's decision-making role regarding dissection or would deny students the opportunity to learn through actual animal dissection" (NSTA, 2008). They support this practice which can help students develop observational skills, learn about organisms' structures, and "develop a greater appreciation for the complexity of life" (NTSA, 2008). Or as Terry D. Keiser and Roger W. Hamm put it, "The more one has to involve a variety of our sensory abilities in exploring an animal's anatomy, the greater will be our appreciation and understanding of its structure and function" (1991, 15). Whatever else it does, the act of dissecting a frog corpse in the classroom undeniably constitutes a vivid and memorable experience.

Critics take issue with many aspects of classroom corpse dissection. For example, some point to the growing evidence that alternatives to corpse dissection produce equal, or perhaps even better, learning outcomes. For example, in their recent systematic review of 20 studies published between 2005 and 2020, Elizabeth Ormandy and colleagues found that "Our data show that in 95% of studies, students do at least as well—and in most cases better—when they use NAMs (non-animal teaching methods) compared to animal dissection" (2022, 402). However, most of all, the criticisms tend to center on the unethical treatment of the frogs themselves, the importance of student freedom of choice on an important ethical matter, and to the environmental damage done as frogs for dissection are snatched from their habitats.[3] The quantity of frogs used for school dissections every year number in the millions, with the majority taken from the wild. And we cannot assume the conditions are humane. As F. Barbara Orlans writes, "frogs typically suffer considerably when they are captured from the wild for sale in schools. Often they are held for days in overcrowded collection sacks where they may suffer dehydration and disease" (1991, 14). Laws that require schools to provide alternatives to frog dissection are cast as providing respect to students' individual ethical positions. The resistance to the widespread practice of educational corpse dissection can be understood as a form of animal and environmental advocacy. Resistance efforts thus involve multiple strategies, including the development of argumentation and consciousness-raising efforts, the push for policy changes such as student choice laws that allow individuals to opt out (with lessons imparted in an alternative manner), and the creation and refinement of dissection alternatives, such as computer and physical simulations.

I suggest that the dissection debate can be largely seen as a conflict between two values. On the one side, defenders of the status quo of the pervasiveness of corpse dissection tout the value of hands-on science education and steadfastly maintain that real-life animal dissection is an irreplaceable learning experience. On the other side, critics of corpse dissection—and advocates for alternatives—push the value of respect and care for animals and the environment.

It has been my observation that, due to this established framework for this debate, *the deck is stacked against critics of corpse dissection.* That is, the structure of the debate itself sets up a losing argument for advocates for alternatives to the dissection of frog corpses in the classroom. My observation is that *this framework has the effect of casting classroom corpse dissection as the educational ideal.* Alternatives are thus set up with the task of attempting to prove that they are as close to that ideal as possible. Since computer screens and mouse clicks cannot ever be expected to recreate the experience of the hands-on vivisection of an actual frog corpse (and seemingly may never even come close), critics are set up to fail. Even as some simulated dissections improve, and as animal and environmental advocates refine their arguments, they remain saddled with the burden of endorsing alternatives that explicitly pale in comparison to the established educational ideal.

In my view, it is essential to tear down this whole framework. Corpse dissection cannot be permitted to maintain its status as the implicit ideal for teaching these lessons in biology education. We must find a way to explore the structure of the debate in which any abandonment of corpse dissection is understood as an inherent educational sacrifice.

I suggest that it can be useful to turn to the philosophy of technology, and in particular to the postphenomenological perspective. Ideas from postphenomenology can be used to draw out the details of the experience of our relationships with technology. This could help to emphasize the importance of approaching *both* computer-simulated dissection *and* corpse dissection as forms of technology usage. It could help to emphasize how *both* computer-simulated dissection *and* corpse dissection transform and shape student experience in specific ways. This sets up a better context for comparing the two, one that reveals (rather than obscures) the tradeoffs that come with the educational practice of cutting up a frog corpse in the classroom, and one that prompts new thinking on the possibilities for computer-simulated alternatives.

TECHNOLOGICAL MEDIATION

Phenomenology may seem at first to be an odd perspective from which to launch a criticism of classroom corpse dissection. This is because, in one sense, phenomenological tools are almost ideally positioned to reveal the deep differences between the experience of cutting open an actual frog corpse and that of using a computer simulation. I agree that these two experiences are radically different. However, this is not the only point that can be made in this comparison. I suggest that it is also essential to explore how both of these practices—corpse dissection and computer-simulated alternatives—are instances of technological mediation.

The postphenomenological philosophical perspective is a useful place to start. Postphenomenology is a contemporary international and interdisciplinary school of thought that brings phenomenological insights together with commitments from American pragmatists and other postmodern perspectives to conduct investigations into the details of people's experiences with technologies (e.g., Verbeek, 2011; Rosenberger and Verbeek, 2015; Ihde, 2016; Rosenberger, 2017; Irwin, 2016; Wellner, 2016; Hasse, 2020; de Boer, 2021; Kudina, 2024). A central idea in this perspective is the notion of "technological mediation." Don Ihde, the founding figure in postphenomenology, explains that technologies are a "mediational means of experiencing an environment or world" (2016, 127). Technologies are "not primarily experienced—except in certain non-use contexts—to be objects or things" (Ihde, 2016, 127). That is, a technology can be understood to be not merely a thing in the world that we experience, but also as something that comes in between the user and the world and changes the way that we experience the world.

In this way, technological mediation is transformational. As Jesper Aagaard puts it, "technologies do not just carry meaning from A to B but influence or 'mediate' how the world is present to us and how we are present in the world" (Aagaard, 2022, 1882). This changed user-and-world-transformed-together-by-technological-mediation is referred to within postphenomenology as "human–technology relations." And such relations are non-innocent. This conception of technology usage prompts us to consider the concrete details of specific relations, and to consider the various tradeoffs that are incurred.

This conception of human–technology relations puts the postphenomenological perspective in a kin group with what is sometimes called "posthumanism" (e.g., Barad, 2007; Braidotti, 2013; Lewis, 2021; Wakkery, 2021). As postphenomenological anthropologist of education Cathrine Hasse puts it,

> Posthuman theory wants us to focus on how materials co-shape our situated creative engagements with surroundings . . . The individual subject . . . is not an individual as an a priori entity but an individual that comes into a being-with-materials. (Hasse, 2020, 55)

This perspective emphasizes how human users should not be conceived in isolation from their devices and larger world. Peter-Paul Verbeek has put special focus on the ethical implications of posthumanism, writing that, "technologies are morally active. They help to shape human actions, interpretations, and decisions that would have been different without these technologies"

(Verbeek, 2011, 57). In this way, Verbeek has redirected postphenomenological work toward issues of the moral transformations of human-technology relations. This makes the project of postphenomenology not merely one of description, but of prescription. As Verbeek puts it, "The central technological question in a technological culture, to conclude, is 'what kind of mediated subjects do we want to be?'" (2011, 84).

What does all of this mean for questions of frog dissection in education, and for educational technologies and computing more generally? My suggestion is that it could be helpful to turn to a postphenomenological conception of technological mediation for addressing several things. By conceiving of both frog corpse dissection and its alternatives as *mediators* of human experience, we can better highlight the special tradeoffs involved in each. If corpse dissection were to be conceived as a form of technological mediation, rather than as an educational ideal, then this sets up a different context for comparison. Rather than understanding any alternative to corpse dissection as an imitator, and evaluating it in terms of how well or poorly it imitates, we are prompted to contrast one specific form of technologically mediated education including its distinctive advantages and problems against an alternative form of technologically mediated education complete with its own distinctive advantages and limitations. I suggest that this new context puts a spotlight on two important things: (1) the underemphasized distinctive educational possibilities for dissection simulation; and (2) the specific problematic educational tradeoffs of corpse dissection.

EXPANDING DISSECTION SIMULATION

Most computer simulations attempt to capture some elements of the experience of corpse dissection, often through the use of images onscreen of a frog opened up and pinned down, with simulated scalpels and clamps controlled with a mouse or touchscreen interface. Norm Friesen describes the experience of the use of the "FrogGuts" online dissection simulation as follows:

> instead of first seeing the animal "lying in the sink" or "flattened" in a box or jar, the first experience of the frog occurs while waiting for it to load in the browser window. Handling and even cutting into the animal, furthermore, is a question of clicking on the correct virtual instrument (the scalpel) and gliding it, in effect, across part of the computer screen. (2011, 92)

In these frog dissection simulations, often the user is directed to make a series of slices, and then they follow these directions by dragging the mouse across a particular space on the screen, or similarly engaging the touchscreen interface (e.g., figure 2.1). Sometimes this results in a line drawn onscreen where the "cut" has been made. If the lines are drawn correctly onscreen by the scalpel icon controlled by the mouse or fingertip, then the simulation moves on to the next screen in which the frog is revealed to be opened up. The simulation may include instructional text or video with each new opened-up layer or removed bit. Then onto the next cut and next screen.

For example, figure 2.1 shows a moment early in the usage of the Digital Frog 2.5 computer simulation. Upon making the incision lines on the simulated frog onscreen with the mouse, a video of frog dissection plays showing those cuts, and then the screen switches to a view of the digital frog opened up along those lines. From there, we are directed to perform further virtual incisions to explore the interior parts of the frog.

My suggestion is that we set aside, at least for the moment, the charge of recreating as best as possible the experience of corpse dissection within our simulations. And let's brainstorm possibilities for biology education that these kinds of digital technologies make possible when dissection simulations are decoupled from corpse dissection. Of course, contemporary simulation developers are exploring some of these possibilities already.[4] We can build further on this work.

One dimension on which computer simulations of frog anatomy could be expanded is in terms of their spatiality. The digital frog could potentially be

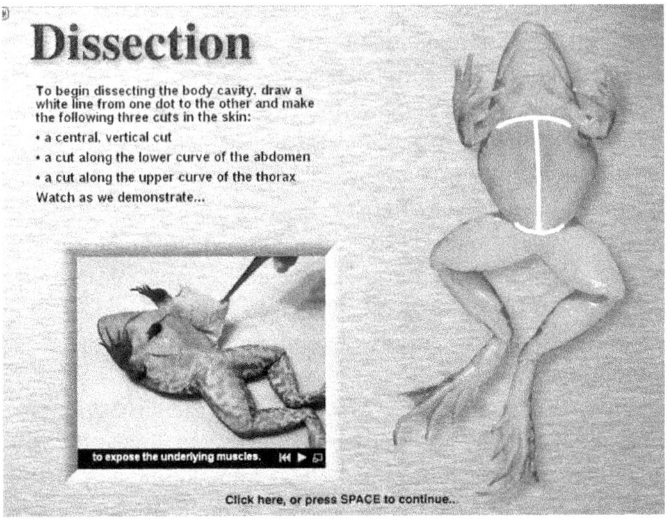

Figure 2.1 Screen Capture of the Virtual Frog 2.5 Simulation. *Source:* Permission received from Celia Clark, President of Digital Frog International.

explored within its relationship to its digital environment. This opens up the potential for lessons about frog habitat to be incorporated into the activity. In this way, the learning activity could be expanded from the exploration of anatomy to include consideration of frog ecology, with connections made across these branches of study.

These spatial considerations could also be expanded to include dimensions of frog motility. Simulated frogs could specifically be investigated with respect to their bodily movements, the kind engaged by a living frog. These could include the frog's various iconic characteristics: its hop, its ribbit sound, its amphibious swim, and its out-stretching tongue. In its simulated way, the digital frog could be made to move on its own in a manner more akin to a living frog than the ready-for-dissection corpse. This could be extended to include the dynamics of the frog's internal systems. Views inside the simulated frog could display the movement of the digestive system, respiratory system, or circulatory system as they would move in a living frog were we able to see inside one while it remained alive.

Another way that simulations could be augmented in terms of a spatial dimension would be changes to the size of the digital sample frog's internal systems. The simulation could potentially involve zooming in to explore smaller parts of the frog's body, say, exploring parts of the frog's eye, heart valves, or inner ear. We could imagine a simulation that includes a zooming-in capacity that enlarges the sample down to the cellular level.

Simulations of frog dissection could also be expanded in terms of their temporality. Where a frog corpse comparatively provides an exploration of one moment of a frog's life—that is, its status at the moment of its death—a simulation could potentially include further aspects of a frog's life, life cycle, and biological history. The simulated frog's life could include exploration of its different phases, for example, including lessons about its hibernation process, burying itself under the mud for the winter, and the accompanying changes to the dynamics of its internal anatomy. The simulation could be expanded as well to include aspects of the frog's distinctive life cycle, growing across its lifespan from a tadpole, sprouting legs, maturing, and participating in reproduction. (See figure 2.2 for an example of a simulation that includes some interactive lessons on this issue.)

This temporal expansion could even be imagined to include aspects of the species' evolutionary history.

One distinctive educational advantage of corpse dissection is the consequentiality of the dissection activity; all cuts are final. Slices cannot be undone, and there may be some pedagogical value to maintaining irreversible consequences to the choices and steadiness of hand of one's cutting. In his phenomenologically-informed contrast of the experiences of corpse dissection and simulated dissection, Norm Friesen notes that

Figure 2.2 A Screen Capture of Digital Frog 2.5's Coverage of the Life Cycle of the Frog, Including a Moment of Its Video of a Frog Morphing Through Its Various Stages Featured in the Middle of the Screen. *Source:* Permission received from Celia Clark, President of Digital Frog International.

> parts of the dissected body can be cut or removed in error, rendering further steps in the dissection process impossible to perform . . . there is no way to simply undo a particular action of decision: In keeping with its natural origins, the body of the rat, pig, or other animal cannot be the subject of any sort of "undo" command—or be "refreshed," "reset," or "rebooted." (2011, 99–100)

Any irreversible consequentiality to cut decisions within a simulation would be merely programmed in, and thus seemingly somewhat arbitrary. For Friesen, this stresses the intracorporeal nature of the dissection of animal corpses, one of the many things that can never be captured within the simulation. Douglass Allchin has been similarly reflective about these differences, suggesting that "A proper aim is discovery, not destruction. One should separate and clarify: Trace pathways, find boundaries, encounter connections—quite impossible if things are pre-cut and disappear as preformed units in a single mouse-click" (2005, 370). These comments reflect some pedagogical values that some defenders may maintain are exclusive to the material interaction with the frog corpse as a natural object. There may be a value in the student's encounter with risk, with the possibility of unfixable error, with a demand for carefulness in these terms.

And yet there is a corresponding and opposite pedagogical value to be noted in the case of simulations: repeatability and variability. Simulated dissections could be performed over and over. Mistakes can be identified, and then the activity could be retried in a way not possible in corpse dissection (without enlisting additional corpses). While this repeatability of simulation is not the same educational advantage as the finality of incisions made to corpses, it is its own advantage nevertheless. And this corresponds to a separate educational value: learning from mistakes, and appreciation for the work of correcting errors, practice, and improvement.

It is also possible to imagine simulations in which this repeatability is offered in terms of varied simulated frog samples (e.g., size, frog species, male and female morphological differences, variations in stomach contents, and so on). Such variability in the particularities of individual frogs offered up for dissection would only be possible in terms of corpse dissection again through the enlisting of additional corpses.

Another distinctive advantage of dissection simulation, and, in particular, computer-simulated dissection, is that it can potentially be integrated with other media. It is not difficult to imagine a simulation built to include sound, photography, video, text, other forms of media, and links to further resources on the internet to learn more about frog biology and ecology. Of course, contemporary frog dissection simulations already incorporate some of these possibilities. It is also possible to imagine particular additional activities added to the dissection experience made possible by computer instantiation, such as typing and writing tasks, automated texts that prompt users to identify internal body parts, or tasks that set up the user to do internet research. An additional area of potential media expansion is scientific imaging. Ihde himself notes this when he writes, "Today's biology education, as with today's scientific practices, takes up a wide set of variant instrumentally mediated perspectives, each producing different ranges of knowledge, which, when synthesized, produces a much richer and robust result than the limited dissection/simulation example given" (2011, 204). In this way, the simulated dissection of the simulated specimen could include instruction about the actual ways biologists conduct scientific study, including through high-tech imaging work, rather than the comparatively artificial version of scientific investigation we get in the dissection of a frog corpse presented as if the activity is like an experiment.

Simulated dissection can thus potentially be expanded in many directions, and in distinctive ways with educational possibilities that may far exceed what is possible with corpse dissection, at least in terms of certain specific things. It may not even always be necessary to simulate the dissection activity itself. While corpse dissection allegedly retains its own distinctive educational advantages, we see that simulation retains its own—albeit different—educational advantages, and ones that have not yet met their pedagogical potential.

These different educational advantages offered by expanded simulation can only be realized if we shed the obligation for simulated dissection to merely mimic the activity of corpse dissection.

CORPSE MEDIATION

Once again, to be perfectly clear, I am absolutely not arguing that simulation—even in the expanded forms reviewed and brainstormed in the previous section—comes close to capturing the experience of what it is like to physically cut open an actual frog corpse. My suggestion is that there are other questions to ask about corpse dissection that go beyond whether simulations can adequately mimic them, questions that are in addition to ethical and environmental concerns. The important move here is to explicitly conceive of corpse dissection as a form of technological mediation.

Because the central material object in this learning activity is the body of an actual animal, it might be tempting to conceive of it merely as an encounter with an aspect of the "real" world. However, when we approach this educational activity as a form of mediation, we can highlight just how much of it is artifactual, and we can evaluate the educational tradeoffs involved. In addition, because this educational activity takes on a scientistic posture—that is, because the classroom dissection experience is often presented as if the child is conducting something like a scientific experiment—it might be tempting to assume that it all takes place exclusively within a context of facts and objectivity, and thus that it would be somehow above subjective judgment or opinion, perhaps open only to questions of scientific standards and validity. However, when we consider corpse dissection in terms of technological mediation, it is clear that these educational practices are deeply co-constituted by values and social decision-making, all of which make their way into the learning experience. Let's then proceed by conceiving the frog corpse not as an educational encounter with the real world. Let's conceive of *the frog corpse itself as a mediator* of educational experience, a pedagogical tool, an object that transforms our relationship with the lessons to be learned.

Reconceiving corpse dissection in terms of technological mediation sets up a particular context of questioning: How does the frog corpse—as a thing, as an artifact, as a classroom technology—mediate the learning experience? And what educational limitations are introduced by the specificities of this learning tool?

Perhaps the most salient artifactual feature of this experience is that the frog presented to students for dissection is *already dead*. The already dead frog is, of course, a natural object that could be encountered in the wild. However, this is distinctly not the case for these frogs presented for dissection.

These frogs, as they are presented to students, were specifically killed, and killed for the sake of this educational activity. This basic feature of corpse dissection has a number of consequences for what can be taught with them, what cannot, and what additional lessons may be imparted implicitly.

One consequence of this status is that, as already dead, the frog corpses are immobile. Frog corpses are, of course, transportable and often bendable or posable, and are changeable in that they are available for the central activity of being cut open. However, in juxtaposition to a living frog, the corpse-for-dissection is *comparably inert.* Frogs have many distinctive motions, from their hop to their out-stretching tongue. The interior of the living animal—the ostensible object of investigation of the dissection exercise—is also in motion, from blood circulation, to digestion, to respiration. *Corpse dissection offers at best a snapshot of these dynamics.* (For example, when students root around the corpse's stomach to inspect its contents, they get a look at those contents from only one moment, the moment that it was killed.) That is, the frog corpse's status as "already dead" renders it inherently unable to display aspects of frog motion, both in terms of the full frog's bodily motion and in terms of the dynamics of its internal systems. Here, it is worth noting again that this inherent limitation of corpse dissection happens to be the opposite for dissection simulation; as noted above, computer simulations are open to being expanded into exactly these and other issues of frogs' bodily dynamics. Simulations could make possible detailed explorations of bodily motion. This includes the possibility of exploring the dynamics of interior bodily systems.

In addition, the frog corpse presented to students for discussion has *already been removed from its environment.* Rather than an uncomplicated encounter with the real frog as it is, the frog-corpse-for-dissection has been snatched from its natural habitat, killed (as mentioned), prepared for transportation (such as through treatment with formaldehyde), and shipped. The corpse-for-dissection is necessarily one extracted from its surroundings. As noted in the previous section, this is another limitation of corpse dissection that happens to be a comparable educational advantage of computer simulation: virtual frogs could potentially be presented within, and in connection with, their virtual environments. However, it is important to not merely consider the inherent educational limitations that come with corpse dissection; we must also consider the actual educational situation it creates, and how it necessarily contains specific problematic elements.

The already-dead and already-removed-from-its-habitat frog corpse creates a particular learning environment for the student. In the posthumanist language of postphenomenology, the technological mediation of the frog corpse co-shapes a particular learner and a particular kind of classroom. The mediation of the corpse carries along implicit commitments and lessons about the value of animal life. It co-constitutes the ethical decision-making situation as one in which students opt out as conscientious objectors, or remain willingly complicit.

Lastly, it is important to put a spotlight on a few telling details of the materiality of corpse dissection. They help to reveal a problematic implication of the corpse dissection activity.

Let's start with an early moment of corpse dissection: the unpacking of the dead frogs. One of the things I remember best about my own experience of grade-school biology frog dissection was something that was not an official part of the activity but instead something I just happened to notice. The frogs for my class had been shipped in a large, transparent, vacuum-sealed bag. Perhaps I was looking at a bag of corpses, still sealed, waiting for use for another class. I don't exactly recall. But in any case, it was striking. For me, one of the most memorable aspects of the frog dissection experience wasn't the individual dissection of my own frog (although that was also memorable), it was the sight of the large number of dead frogs, splayed and overlapping, draped over one another in various positions (and not in the standard, ready-to-go, tummy-up, sitting-there-before-you-on-the-desk arrangement), pressed together by the vacuum seal, all leaving it often unclear where one corpse ended and the next began.

This status of the dead-frogs-ready-for-dissection is pedagogically non-innocent. An implicit lesson is communicated by this treatment. Jan Oakley puts it this way:

> Dissection's hidden curriculum teaches lessons about the ethics of animal use: specifically that there is little value in animal life. Although its stated intent is to teach students about biology, dissection also teaches that it is ethically acceptable to kill animals in the name of school. (2013, 7)

It is not difficult to infer from the dissection activity that animal life is worth less than an afternoon classroom lesson that could have been learned otherwise. One reason this lesson is communicated implicitly (rather than explicitly to the student) is because of the specificity of the technological mediation of the activity: the frog isn't killed in front of the student. The already-dead frog is presented as a learning object.

These implicit lessons about the negligible value of the life of animals are carried through from the opening of the bag or jar or frogs, through the dissection activity itself, and then even potentially through the moments after the activity has come to a close. A focus on the materiality of the frog corpse as a mediating technology next raises this question: What happens to the frog corpse after the dissection activity is complete? This is of course dependent on the classroom management of the teacher. Is the corpse tossed into the trash? It is true that such a decision may be incidental to the explicit lessons about anatomy. But they are central to the implicit lessons on the value of animal life.[5] Are students themselves tasked with discarding the spent corpse? Or does a teacher decide to omit this chore from the activity? This decision-making scenario, an ethically loaded one, is co-constituted by the specificity of the technological mediation of the frog corpse.

CONCLUSIONS

One thing that's interesting about the case of the conventional debate over frog dissection and its simulated alternatives is that it tends to obscure issues of technological mediation. My observation is that this debate tends to take place as one in which the hands-on dissection of an actual frog corpse is cast as an educational ideal (albeit one saddled for some people with ethical problems), and simulated alternatives are thus charged with the task of mimicking corpse dissection as best they can. In this way, I have tried to show that reconceiving this debate in terms of technological mediation helps to highlight both computer simulation and corpse dissection as material practices with distinct pedagogical tradeoffs.

Stepping back, the re-conception of this debate in terms of technological mediation has the effect of emphasizing a few general points. One thing that is highlighted is the fact that both sides of this debate should be understood as specific material practices. It is clear that frog dissection simulations are technologically mediated; the dissection experience within computer simulations is of course entirely artifactual. In addition, although often cast as an encounter with the "real" world, classroom corpse dissection is also deeply technologically mediated. This emphasizes the fact that educational practices should be evaluated in terms of the implications and effects of the mediation of their associated technologies, even when those practices are not mediated by computers. We should thus evaluate the educational tradeoffs of the technological mediations of all of our teaching practices, and this goes for those mediated by computers and those that are not. Honestly, I share the same reservations about educational computing as many critics. However, any hesitancy about educational computing—or resistance to the pervasive pro-digital hype—should not be based on anything inherent to computing. The specificity of the tradeoffs depends on the case.

The topic of educational frog dissection usefully puts these issues into relief. This is because the example of corpse dissection is one that is at once non-digital—it's not a computer-mediated practice—and is at the same time a practice that introduces problematic pedagogical tradeoffs as a result of its technological mediation. The conventional discussion on frog dissection explicitly casts simulation as encumbered with pedagogical tradeoffs resulting from its computer mediation; computers cannot recapture the experience of cutting up an actual animal corpse. The conventional discussion simultaneously casts corpse dissection as an innocent encounter with the "real." However, in this chapter, we have seen that by reframing the discussion in terms of technological mediation, both simulated dissection and corpse dissection are revealed to have their own distinctive pedagogical tradeoffs. In particular, we have seen above that simulation has the potential to be expanded in new ways that could enable new pedagogical possibilities. And we have seen that corpse dissection is correspondingly beset with its own pedagogical limitations.

As a final comment on the case of educational frog dissection, it seems clear to me that the conventional setup of the debate must be overturned. And when set on the proper stage of comparison—that is, one of technological mediation in which the tradeoffs of both sides are considered—dissection simulation emerges as the preferable option not only morally and environmentally, but also pedagogically.

NOTES

1. My favorite review of these issues is (Oakley, 2013) now available online at: www.animalsandsociety.org/research/digital-archive/.
2. See also the 2011, Vol. 15(3), special issue of the journal *Techné: Research in Philosophy and Technology* on this issue.
3. There are many online resources posted by societies specifically focused on issues of animal dissection, as well as animal welfare advocacy groups. Some examples include:

    - https://navs.org/
    - https://aavs.org/
    - https://www.animalearn.org/index.html
    - https://www.peta.org/issues/animals-used-for-experimentation/animals-used-experimentation-factsheets/dissection-lessons-cruelty/
    - https://awionline.org/content/dissection-alternatives

4. For some examples of dissection simulations, see:

    - "Frog Dissection" app, see: https://www.frogvirtualdissection.com/
    - Gizmos' "Frog Dissection" see: https://gizmos.explorelearning.com/find-gizmos/lesson-info?resourceId=1097
    - "Froguts" see: https://thesciencebank.org/pages/froguts?_pos=1&_sid=7f8604f41&_ss=r
    - VictoryXR's "Dissection Simulator: Frog Edition" see: vxr.direct/dissection-simulator-frog-edition/

- Digital Frog International's "Digital Frog 2.5" see: https://www.learningservicesus.com/the-digital-frog-2.5
- For 3D models of internal frog systems, see, for example: Biosphera's "3D Frog Anatomy," see: https://biosphera3d.com/product/3d-frog-anatomy-software/
- Visibile Biology's interactive 3D frog model: https://www.visiblebody.com/blog/no-frog-no-prob-a-3d-visual-guide-to-frog-anatomy
- For physical simulated frog dissection units, see, for example: Syndaver's "SynFrog Dissection Model": https://syndaver.com/product/synfrog-frog-dissection-model/
- VetEffect's "Frog Dissection Training Manikin": http://veteffects.com/products.htm

5. For other examples of this argument on the implicit problematic lessons about animal life taught by frog dissection, see: Orlans (1991); Sapontzis (1995); Rosenberger (2011).

REFERENCES

Aagaard, Jesper. 2022. "On the Dynamics of Zoom Fatigue." *Convergence: The International Journal of Research Into New Media Technologies* 26(8): 1878–1891.

Allchin, Douglas. 2005. "'Hands-Off Dissection?' What Do We Seek in Alternatives to Examining Real Organisms?" *The American Biology Teacher* 67(6): 369–374.

Averett, Nancy. 2020. "High School Dissections Are A Science Class Tradition. But Are They Doing More Harm Than Good?" *Discover*. January 27, 2020. www.discovermagazine.com/the-sciences/the-argument-against-high-school-animal-dissections.

Balcombe, Jonathan. 2000. *The Use of Animals in Higher Education: Problems, Alternatives, & Recommendations*. Washington: Humane Society Press. https://www. humanesociety.org/sites/default/files/docs/use-of-animals-highereducation.pdf.

Barad, Karen. 2007. *Meeting the Universe Halfway: Quantum Physics and the Entanglement of Matter and Meaning.* Second edition. Durham: Duke University Press.

Braidotti, Rosi. 2013. *The Posthuman.* Cambridge: Polity Press.

de Boer, Bas. 2021. *How Scientific Instruments Speak: Postphenomenology and Technology Mediations in Neuroscientific Practice.* Lanham: Lexington Books.

Friesen, Norm. 2011. *The Place of the Classroom and the Space of the Screen: Relational Pedagogy and Internet Technology.* New York: Peter Lang.

Gregory, Ted, and Susan Berger. 2018. "Is Dissecting a Frog in Science Class Ethical? Protesters Challenge the Long-Standing but Controversial Practice." *Chicago Tribune*. https://www.chicagotribune.com/news/ct-met-animal-dissect-peta-protest-highland-park-20180530-story.html.

Hasse, Cathrine. 2020. *Posthuman Learning: What Robots and Cyborgs Teach Us About Being Ultra-Social.* London: Routledge.

Huntsberry, Will. 2015. "Dissecting a Frog: A Middle School Rite of Passage." *NPR. All Things Considered.* https://www.npr.org/sections/ed/2015/02/25/384348150/dissecting-a-frog-a-middle-school-rite-of-passage

Ihde, Don. 2011. "Dissection and Simulation: A Postphenomenological Critique." *Techné* 15(3): 203–205.

———. 2016. *Husserl's Missing Technologies.* New York: Fordham.

Irwin, Stacey. 2016. *Digital Media: Human-Technology Relations.* Lanham: Lexington Books.

Keiser, Terry D., and Roger W. Hamm. 1991. "Dissection—The Case For." *The Science Teacher* 51(1): 13.

Kline, A. David. 1995. "We Should Allow Dissection of Animals." *Journal of Agricultural and Environmental Ethics* 8(2): 190–197.

Kudina, Olya. 2024. *Moral Hermeneutics and Technology: Making Moral Sense through Human-Technology-World Relations.* Lanham: Lexington Books.

Lewis, Richard S. 2021. *Technology, Media Literacy, and the Human Subject: A Posthumanist Approach.* Cambridge: Open Book Publishers.

National Association of Biology Teachers. 2019. "The Use of Animals in Biology Education." *Position Statement.* Revised and Adopted by the NSTA Board of Directors. https://nabt.org/Position-Statements-The-Use-of-Animals-in-Biology-Education.

National Science Teaching Association. 2008. "Responsible Use of Live Animals and Dissection in the Science Classroom." *Position Statement.* Adopted by the NTSA Board of Directors, June 2005. Revised 2008. https://www.nsta.org/nstas-official-positions/responsible-use-live-animals-and-dissection-science-classroom.

Oakley, Jan. 2013. *Animal Dissection in Schools: Life Lessons, Alternatives and Humane Education.* [Policy paper.]. Ann Arbor, MI: Animal and Society Institute. Available online at: www.animalsandsociety.org/research/digital-archive/.

Orlans, F. Barbara. 1991. "Dissection—The Case Against." *The Science Teacher* 58(1): 12.

Ormandy, Elizabeth, Janella C. Schwab, Samantha Suiter, Nicole Green, Janice Oakley, Pamela Osenkowski, and Christine Sumner. 2022. "Animal Dissection vs. Non-Animal Teaching Methods: A Systematic Review of Pedagogical Value." *The American Biology Teacher* 84(7): 399–404.

Osenkowski, Pamela, Ignas Karaliunas, and Merari Diorio. 2022. "Educators' Views on the Use of Dissection Alternatives in American Biology Classrooms." *Alternatives to Laboratory Animals* 50(3): 235–243.

Rosenberger, Robert. 2011. "A Phenomenological Defense of Computer-Simulated Frog Dissection." *Techné* 15(3): 215–228.

———. 2013. "Do Dissections Teach Animal Cruelty?" *Atlanta Journal-Constitution.* July 22, 2013: A10.

Rosenberger, Robert, and Peter-Paul Verbeek, eds. 2015. *Postphenomenological Investigations: Essays in Human-Technology Relations.* Lanham: Lexington Books.

———. 2017. *Callous Objects: Designs Against the Homeless.* Minneapolis: Minnesota University Press.

Sapontzis, Steve F. 1995. "We Should Not Allow the Dissection of Animals." *Journal of Agricultural and Environmental Ethics* 8(2): 181–189.

Verbeek, Peter-Paul. 2011. *Moralizing Technology*. Chicago: Chicago University Press.

Wakkery, Ron. 2021. *Things We Could Design: For More than Human-Centered Worlds*. Cambridge: MIT Press.

Wellner, Galit. 2016. *A Postphenomenological Inquiry of Cell Phones*. Lanham: Lexington Books.

Zemanova, Miriam A., and Andrew Knight. 2021. "The Educational Efficacy of Humane Teaching Methods: A Systematic Review of the Evidence." *Animals* 11(1): https:// doi.org/10.3390/ani11010114.

Chapter 3

Affordances as Solicitations

The Role of Technical Mediation in Academic Care

Kristy Forrest

A central epistemological stance in the postphenomenological approach to the philosophy of technology, the concept of technical mediation offers an analytical framework to interpret the role technology plays in human existence and society. Requiring a re-orientation of traditional dichotomies which view the human–technology relation as an external one between pregiven, distinct entities, technical mediation instead views the relation as mutually co-constitutive. Within other disciplines that engage with questions of technology (such as science or educational studies), there has traditionally been a less explicit focus on mediation, due to the penchant for an instrumentalist as opposed to a relational lens. Yet, given the new pressures of the digital age, whereby conceptualizations of the human and the technical are at the centre of public discussion, it is now imperative that educators engage with the philosophical elements of technology as part of their professional practice of academic care. Indeed, as emerging policies (UNESCO 2023) increasingly expand the educators' role beyond merely training children for technical proficiency and focus on more expansive concepts of generational care and impact, there exists the potential for philosophy to inform pedagogical practice in novel ways. Therefore, in the need to evaluate the affordances of technology, the projects of educators and philosophers align. To this end, through a discussion of ChatGPT in relation to the Australian context, this paper will offer a teacher mediation-care theorem, positing that educating young people now necessitates *care for the quality of their technical interactions,* thus entailing the inclusion of technical mediation as part of pedagogical content knowledge.

TECHNICAL MEDIATION THEORY IN POSTPHENOMENOLOGY

Technical mediation theory emerges from the ongoing engagement with *technics* in the philosophy of technology, which concerns the interrelationship between technology, society, and the human. The translator of Gilbert Simondon's *Culture and Technics* provides a useful definition:

> The 'technics' is a rendering of the French noun la technique. The latter is used to denote the general domain of technologies, techniques, methods, arts and practices, at once material and cognitive, through which humans engage and reshape their environment and psycho-social milieu. For Simondon, technology per se must be understood as only a part, albeit an extremely important one, of this broader modal sphere of technical activity and relationality. (Fraser 2015)

Following Simondon, postphenomenological philosophers sought to establish the technical milieu by adopting the 'mediating approach' (Verbeek 2015). Mediation theory argues that humans and digital technology are not two separate entities where an interaction takes place, but rather they are entwined and are the *result of* this interaction (Dorrestjin 2017). In this manner, the concept of technical mediation is the central epistemological position of postphenomenology (Kraalingen 2022).

Mediation theory is inspired by Lev Vygotsky's theory of mediated action, which posits that human action and psychological functions are mediated by tools (technical and psychological) that shape the mind and environment (Vygotsky 1978). Tool mediation explains the appropriation of socially developed forms of acting in the world as well as (via the process of internalization) an individual's psychological functioning. For example, a person learns of the structural properties of a car and how to use it via the previous experience of others, (as well as their own attempts). In addition, a person's cognitive map of a city is influenced by whether or not they are a car driver (Ding 2021).

In seeking to explain how technologies shape the human–world relation, postphenomenology endorses the co-constitution of people and their environment (De Boer 2018). Under this reading, the human–technics relationship is forged through co-existence. Simondon describes man as 'being among the machines that operate with him' (1958, p. 18). This concept of *being-with* negates the tendency to examine technology from a purely technophilic or technophobic perspective, as tools with the capacity to enhance or destroy our *essential nature*. Instead, through a process of mediation, we think, act, perceive, and dwell with and through technology (Hayles 2010).

Second, the pragmatic commitment of postphenomenology to an anti-essentialist perspective is reflected in its view of technologies as multistable (Rosenberger 2020, 377). Multistability argues that a technological artefact has no predetermined or single essence but rather has various partially determined trajectories or possibilities, depending on the context. This echoes Gibson's concept of affordances, which initially emerged from ecology psychology as a means of describing what the environment affords or offers the individual (Gibson 1966). When applied to technology, affordances are often translated as possibilities for action, understood as 'technical features . . . that enable and constrain certain actions' (Aagaard 2018).

Therefore, multiple relations to a technology are possible, while the potential of these relations is limited by the materiality of the technology (Rosenberger 2020; Aagaard 2018). Thus, by replacing a uni-directional account of interaction with mediation, and challenging a foundationalist view of technology through multistability, postphenomenology rejects the notion that technology offers predetermined pathways of action to a predetermined subject. Indeed, as Verbeek argues, 'when you stop seeing technologies as *part of the world* and start seeing them as *part of our relations with the world*, then suddenly the whole perspective changes and technology becomes foundational' (2015). Such an orientation has significant theoretical and practical implications for educators, as will be outlined subsequently.

YOUTH UNDER DIGITALIZATION

The transformation of home and educational spaces into ecosystems of digital devices, platforms, and networks has resulted in the first generation of youth who occupy truly hybrid lives. This is where the traditional rites of passage and patterns of childhood are not *wholly* facilitated by adult mentors in traditional social institutions but instead through immersion in a media-saturated online world. Yet, it is still the home and school that are charged with the responsibility, as Bernard Stiegler frames it, of 'taking care of youth and the generations' (2010). For parents, and educators working with the first generation to 'grow up digital', there is an oft-stated sense of 'confusion' and lack of consensus regarding the specifics of the interrelationship between digital technology and society, in particular the impact on the development of children (Livingstone 2017; Gonski 2020).

Indeed, the divergent positions tend to coalesce around contrasting views on the relative value of a digitally ascetic past or radically opposed determinist positions of alarmist or excitement regarding the future (Selwyn 2012; Feenberg 1992). This approach is evidenced in literature since 2000 which highlights the impact of technology on the learning, socialization, and subjectivity of an entire generation of youth. This entailed an important conceptual shift from the generic 'learner' (under an instrumentalist view of learning) to the possibility of a *new type of child* that emerges alongside technology. This was initially provoked by Mark Prensky's seminal 2001 article 'On the Horizon' which offered a new contention of how *growing up* immersed in a digital world had the potential to alter dispositions and capabilities. For Prensky, this was the arrival of the 'digital native', whose childhood immersion in games and the internet equipped them with new, exciting capabilities that were shamefully neglected by their 'digital immigrant' educators.

However, after 2010, this techno-progressivist thesis was largely rejected, replaced instead by a growing sense of caution that nonetheless retained the focus on generational essentialism, where a singular relationship between technology and new types of personhood are proposed (Carrigan 2020). Using language laced with urgency and lament, psychologists and social theorists argue that the fundamental changes to the environment in the digital realm (such as intense and varied integration streams, hyper-connectivity, and the behavioural conditioning embedded in social media) have nefarious neurological consequences that are most evident in young people. Under this view, information flow and digital media are causally responsible for shaping subjectivity (Petland 2011) while participation in social media is purposefully curated to destroy the attentional capacity of the young, as well as modifying behaviour for the purposes of profit, all of which has potentially devastating implications for youth self-identity (Steigler 2011; Zuboff 2020). This sociological view cohered with growing psychological research, which, accompanied by alarmist titles such as 'Alone Together' (Turkle 2011) and 'How Smartphones Destroyed an Entire Generation' (Twenge 2017) suggested correlations between digital technology and the degradation of children's cognition, maturation, communication skills, and an increase in mental health issues. Conversely, these technologies also created positive opportunities for young people in terms of increased access to information and connection to global knowledge sources beyond the confines of their own communities.

The current focus on the potential harms of technology has led to an increased intensity in the moral response to technology in schools. The rising sense of panic has resulted (in the Australian context) in the banning of mobile phones and the new ChatGPT in schools in several states. In addition, beyond the bans, the policy response has identified the need for 'new understandings' (Gonski 2020) for parents and teachers to direct children safely around the dangers of digital technology while simultaneously equipping them for a future characterized by its ubiquity. Far from being a tokenistic cry, this requirement has been codified in recent curriculum iterations, where a new strand of *Practising Digital Safety and Wellbeing* has been included as part of the General Capability of Digital Literacy (ACARA 2021). The new curriculum requires that students are educated in concepts of digital self-care, well-being, as well as privacy and the curation and protection of online identity.

Thus, it is now apparent that a rapid shift in our techno-social conditions has in turn resulted in an increased interest in the public media discourse and the educational policy level in the potential of technology to *structure and shape the lives of children*, rather than simply be employed as a tool for learning.

Yet, the philosophical concept of mediation that is necessary for making sense of the 'structuring' and 'shaping' is yet to be translated into an educational context. Instead, educators are still grappling with the *impact of* technology on their human students rather than considering their *relations with* it. In addition, as noted above, the educational analysis tends towards uni-directionalism and essentialism rather than accounting for multi-stability. Indeed, Krallingen (2022) notes that the specific type of teacher reflection needed for technical mediation is a 'neglected aspect of pedagogical practice', as such knowledges are not accounted for in the pre-existing teacher competencies in relation to technology.

For example, under dominant models, such as the TPACK framework offered by Mishra and Koehler (2006), a teacher's understanding of technology is articulated in terms of an integration between functional knowledge of technology (TK), content knowledge of their discipline (CK), and pedagogical knowledge (PK), including the combinations TCK and TPK. Thus, teachers must have technical proficiency while also being able to ensure the same proficiency in their students, while utilizing technology for distinct, pedagogical outcomes. Yet, the new concepts of digital literacy mandate knowledge of the *mediating* aspects of technology. For example, e-safety requires further skills in evaluating issues of privacy, safety, and ethical use of technology, while the emphasis on well-being and identity requires knowledge of social practices and the use of digital texts in everyday lives (Pangrazio et al. 2020) and how they further shape our subjectivity.

THE CORRELATION OF MEDIATION AND CARE: THE CHANGING ROLE OF TEACHER

Therefore, such curriculum requirements clearly demand a more nuanced understanding of the affordances of technology that go beyond an impact analysis (while not abandoning it entirely) to consider relational and constitutive elements. This further entails a novel form of academic care that requires technical mediation as a part of teacher thinking and pedagogical practice.

Broadly speaking, an analysis of impact manifests in the general wariness of or celebration of technology that is indicative of what Kiran and Verbeek (2010) identify as the central features of ethical assessments of technology: trust and distrust. When considering the value of technology, we employ criteria such as reliability, safety, and risk, often supported by strategies for mitigating harm. In this manner, educators assume a role similar to that of technical assessors, charged with the 'prediction, anticipation, explication and avoidance of possibly harmful impacts of new and emerging technologies' (De Boer et al. 2018). On the trust side, educators are simultaneously tasked with predicting the potential positive outcomes of technological use and thus look to enhance such opportunities. An example of this is the tension surrounding the use of cell phones in schools and subsequent bans, with one affordance view pointing to their capacity to destroy learning and connection (Lemov 2022) and the opposing view pointing to the need for students to develop their digital self-regulation skills (Orlando 2022).

A different approach to affordances moves away from the mere consideration of harmful or beneficial action possibilities and focuses instead on the process of human becoming. With its basis in the Greek concept of noesis and articulated by Bernard Stiegler in 'Taking Care of Youth and Generations', it conjugates the question of technics with that of time, focusing specifically on the intergenerational flows which cultivate the formation of subjects via desire, memory, and knowledge (2010). Stiegler offers an account of the institutional obligations that follow from such relations, 'involving a creation of educative milieus which care for and pay attention to the future' (Heaney 2020). Thus, given the role of technics in constituting the human future, there is now a necessity for educators to care *both about and through* technical mediation.

Indeed, the emerging forms of academic care shift the locus of concern from a moral subject that is considered separate from technology to an acknowledgement of ongoing hybridization, where our very being is entwined with technology (Dorrestjin 2017; Hayles 2010). Thus, caring for young people within the current milieu demands a capacity to both articulate and cope with the technical conditions of our existence as part of savoir-vivre, or knowing how to live (Stiegler; 2010), and caring for 'the quality of interactions and fusions with technology' (Dorrestjin 2017). This is supported by the aforementioned inclusion of social and ethical dimensions in national curricula which implies a position that digital literacy is both a functional, cognitive, and social practice (Pangrazio et al. 2020).

From these accounts, I have identified three central themes to define the pedagogical responsibilities of academic care in relation to technical mediation:

1. Concern for the inter-generational transfer of cultural memory and the cognitive skills and attentional capacity required for such a task.

2. Concern for the socialization and subjectivity of youth as constituted through technics.
3. The need to prepare students to cope within the conditions of a hybridized existence.

This further justifies the need for a new kind of thinking about technology as Kiran and Verbeek (2010) argue:

- An approach to technology as *mediating* human existence, rather than threatening it or merely facilitating it.
- An approach to humanity as being fundamentally interwoven with technology, rather than being authentic and autonomous.

Thus, the shift from regarding technology as merely functional to something with the mediating power to shape our subjectivity and interactions dramatically reframes the scope of educators' pedagogical and affective labour. Where previously an educator was tasked with facilitating a student's instrumental proficiency in using technology, their task now is epistemological, ontological, and ethical, in that they are charged with determining *responsible forms of technical mediation* as an act of care for young people and the community at large.

What does this entail? Following postphenomenology, this requires an educator to develop a deep understanding of the types of *interactions* and *relations* that occur between students and digital technology (which is the mediating factor) and the practices and experiences which emerge from such relations (Verbeek 2015). This involves clarifying and conceptualizing our lived experience and interrelationship with digital technology in our educative spaces or learning 'lifeworlds', or 'reading the mediating roles of the technology' (Kiran and Verbeek 2010, p. 244). Importantly, this requires more than simply identifying affordances as possibilities for action, but reframing them as *solicitations*, which invite or repel us to act in particular ways depending on our context. Indeed, we must accept that technologies do not afford action possibilities to preexisting subjects with fixed goals, but subtly invite (Verbeek 2005) and facilitate (Rosenberger 2014) certain comportments while inhibiting and foreclosing others. As stated above, the analysis is grounded in the lived experience and habitual use of technological artifacts by teachers and students which 'guarantees a much needed fidelity to the state of the actual' (Aagaard 2018). For a teacher, mapping these invitations using postphenomenology to focus on the routines of the classroom allows for a more 'exacting capture of the mediating impact of technologies' (Kraalingen 2022) that goes beyond a phenomenological approach which focuses primarily on experience to focus instead on the *how of mediation*.

Based on this knowledge of interaction gained through mediation analysis, an educator then moves to skillfully 'appropriating it' in specific ways, *evaluating* and selecting such forms of mediation as part of the practice of academic care. This task is normative, as it requires educators to assume responsibility and an 'ethos of care' for the psychic and social wellbeing of learners alongside technology, by guiding them through the challenges of current conditions, while avoiding overly dichotomous or essentialist thinking (Biesta 2015).

Therefore, the expanded view of digital literacy requires expertise in both the *skills and social* tradition, whereby literacies cannot be reduced to functional skills alone. Likewise, teachers cannot focus purely on pedagogy or technology use, but rather embrace an 'entangled' view, which encompasses the dynamic, social elements of technology (Fawns 2022). I will now consider what this might look like in practice.

TECHNICAL MEDIATION AS CARE: A CONCRETE EDUCATIONAL EXAMPLE

In focusing on articulating the mediating effects of technology and its affordances, the methodology of postphenomenology is primarily descriptive. Frameworks for analysis include the foundational work of Don Ihde, who offers three pathways to understand the process of mediation: types of relation, points of contact, and types of influence. Alternatively, Kiran (2015) offers a framework based on four dimensions of technological mediation under a single postpheomenological perspective: ontological, epistemological, practical, and ethical.

Building on Ihde's conceptual repertoire of concepts (outlined above) that describe how technologies both shape and reshape humans and the world through mediating effects, Kiran argues that each dimension contains a shaping movement and *corresponding* and *contradictory* downplaying movement. Labelling this as the 'two-sidedness' of dimensions of technological mediation, he argues this permits the comprehension of a 'multiplicity of possible mediating movements' (Kraalingen 2022), which then informs how we exist in our technological lifeworlds.

This paper will adopt Kiran's approach (while still incorporating some of Ihde's relational terminology) in considering how postphenomenological concepts can assist educators in a Senior English Literature class in articulating the mediating impact of the new technology of ChatGPT for the purposes of academic care. I will posit that Kiran's dimensions of mediation provide a means of mapping affordances as solicitations and thus provide a basis from which educators can expand the criteria on which they assess the mediating elements of technology through a pedagogical lens.

Similarly to the humanities and in contrast to the sciences and social sciences, the discipline of Literature is primarily print-based and thus has been at the epicentre of the significant changes resulting from digitalization. Like all disciplines, the study of Literature is characterized by a series of practices and values. In terms of practice, Literature sanctifies the processes of close reading of primary texts, analytical writing, and dialogue with peers in a close circle (hence the term 'literature circle') and views these as the core techniques through which students learn to form meaning and disciplinary knowledge. In terms of values, Literature is elevated as one of the primary means by which the transfer of cultural ancestry occurs through the study of canonical texts, while it also operates as a means of ethical training, through the cultivation of a sympathetic imagination (Nussbaum 1998). Finally, student writing of long-form responses is the key mechanism by which Literature teachers make inferences regarding the quality of student thinking (Christolodou 2023).

While the rise of non-print texts in addition to the plethora of information now available online has already provoked discussions regarding the changing role of reading in schools, writing has largely retained its privileged status, until now.

ChatGPT is a generative language model which operates in practice as an artificial intelligence (AI) chatbot that is trained and designed to hold natural conversations and produce human-level text in response to prompts. Its release online in December 2022 generated a frenzy of attention in the education sector, including think pieces with doomsday predictions such 'The End of High School English' (Herman 2023) and those adopting a more progressive approach by urging districts to not 'Ban Chat GPT in Schools, but Teach With It' (Roose 2023). For the latter, potential affordances include personalized tutoring, unlocking student creativity, and better preparation for an adult life alongside AI.

Yet, both perspectives fall into the default position of viewing technology as something that either threatens or enhances human existence (Verbeek and Kiran 2010), creating tensions in schools where teachers on the ground are having to make evaluative decisions about how to guide their students. So the question is, how can Literature educators understand it in terms of its *mediating* impact?

THE ONTOLOGICAL DIMENSION

The ontological level considers the capacity of technologies to shape the world in which we find ourselves, as 'humans, individuals and societies' (Kiran p. 125). The movement (borrowing from Heidegger) is characterized as 'revealing-concealing'. Revealing refers to reality as it presents itself to the actors, or what is enhanced or reinforced through the way we employ tools functionally, 'in order to do' something, and also the ways of *being* that the technology permits through its use in context. In contrast, in an opposite movement, technology minimizes and weakens other aspects of the same reality. In terms of care, this dimension links directly to the concern of Literature educators to attend to the socialization and subjectivity of their students, which occurs through close reading and dialogue. Analytical-reflective questions to ascertain the ontological could include:

- What elements of human existence (particularly in relation to teaching and learning identities) are revealed or concealed by a technology?
- What are the implications of the technology for our being in the world, and thus processes of socialiszation?

When a Literature student initially prompts ChatGPT with a question such as 'how does The Great Gatsby critique the American Dream?' and receives a fluent answer, the relation is arguably one of *alterity*, similarly to using an ATM (Idhe 1990) where the student interacts with the machine while the world remains in the background. Yet, as the student engages in further dialogue with the program, prompting it for further detail or precision and perhaps even writing with it, the relation arguably shifts to one of *embodimen*t, 'where the technology forms a unity with a human being, and this unity is *directed* at the world' a relationship that Ihde schematizes as *(human – technology) → world* (Ihde 1990). Hayles (2010) makes the case that embodiment 'then takes the form of extended cognition, in which human agency and thought are enmeshed within larger networks that extend beyond the desktop computer into the environment' (p. 3).

Ontologically speaking, the highly collaborative, embodied relation of humans and technology within ChatGPT reinforces our identity as hybrid-beings and works to further conceal (or even make redundant) the human–technology binary. Yet, in its capacity to produce writing (such as poems and essay) in a manner similar to a human literary scholar challenges the assumption that there is something unique about human writing and creativity that is distinct from machine learning. In addition, given the traditional role of writing in shaping student academic identities (Ivancic 1998), we can question if using text generation tools will impede this process of socialization into the language and thinking of a scholarly community which is integral to academic care.

Finally, it poses several challenges to the identity and purpose of the teacher. The ability of ChatGPT to produce knowledge and explain concepts conceals the role of the teacher as a person, replacing it instead with a novel concept of the teacher as an optimal source of information output. However, it also reveals a new mode of instruction, which is to guide students in how to best utilize the technology, such as designing increasingly precise prompts for the best outcome. It further introduces an additional relation, that of (student–technology–teacher) —> *world.* Within this relation, given the onus is still on the human to ensure the accuracy and validity of the output, an emergent role as the 'human loop' (Lifschitz 2023) between ChatGPT and its product emerges for all students and teachers.

EPISTEMOLOGICAL DIMENSION

The epistemological dimension examines the magnification–reduction structure of technology, which relates to Ihde's argument that 'any technology that augments or strengthens our perceptual capacities will simultaneously weaken or reduce other aspects of our experiential presence' (quoted in Kiran 2015, p. 128). In terms of academic care, this element of mediation is arguably the most important dimension for educators, when considering how technologies impact the attentional and perceptive capacities of their students, in addition to their ways of knowing. Key analytical–reflective questions include:

- What perceptive or cognitive capacities does the technology augment/magnify or reduce?
- What kinds of knowledge does the technology require for successful interaction?

Most importantly for Literature educators, ChatGPT contributes to the ever-changing shift away from close reading to 'distance' reading, which is the process of 'understanding literature not by studying particular texts, but by aggregating and analysing massive amounts of data' (Moretti 2000). Through its capacity to synthesize and evaluate critical readings on literary texts, the tool negates the need for us to engage in such processes. Thus, a key affordance is a fundamental change to what constitutes 'reading', with close textual analysis being replaced by machine algorithms.

In this way, it forms the next iteration of a long process by which knowledge has been exteriorized via technics, through memory substitutes or externalizations such as writing, machines, or apparatus. Stiegler views this as a kind of reduction (which he labels grammatization) and a dangerous form of proletarianization, whereby the individual or collective knowledge of previous generations 'escapes' the student, so that knowledge is no longer theirs, and belongs to the apparatus itself (Stiegler 2011).

Yet conversely, the apparatus itself cannot think contextually and thus is prone to error, which places the onus back on the human user to verify facts and critique using expertise (Lifschitz 2023). In a similar vein, to successfully interact with the chat, a user requires knowledge of effective grammatical structure for prompt design, as well as disciplinary knowledge in order to evaluate its output. Thus, as a sub-section of digital literacy, AI literacy emerges as a new domain of knowledge.

PRACTICAL DIMENSION

The practical dimension considers the practical implications of technology for our daily lives through an examination of its enabling-constraining abilities. With its focus on functionality and possibilities for action, this dimension concurrently relies on the concept of affordances and aligns most closely with the educator's need to determine potential risks. The analytical–reflective questions for educators are:

1. What possibilities or activities does the technology enable?
2. What possibilities or activities does the technology constrain?

The capacity for ChatGPT to easily produce academic work affords several possibilities, such as student dependency on the technology (including a myopic focus on the writing outcome instead of the writing process) and the possibility of academic integrity being compromised. It thus enhances the need for more rigorous authentication mechanisms. In contrast, through collaborating with a machine to produce written work, different modes of collective endeavour between human and AI become possible. Yet, in having a machine do the searching, collating, and communication of information on our behalf, our capacity to do these things for ourselves is arguably constrained, which has implications for our education system as a whole and particularly our understanding of what academic literary 'qualification' (Biesta 2015) now looks like.

ETHICAL DIMENSION

Finally, the ethical dimension relates to the manner in which we attach value to the behaviours that emerge from the practical dimension (Kiran 2015). Yet, technical mediation in and of itself 'does not question the desirability of a technology as such' (de Boer et al. 2018). In order for us to ascertain the value assigned to an affordance requires external standards of ethical assessment, which for a teacher are pedagogical. The primary role of postphenomenology then is to build a 'preliminary orientation by which different value deliberations might proceed' (de Boer et al. 2018, p. 305). For Kiran, the 'two-sidedness' that emerges from the other dimensions leads to the attribution of both positive and negative values, which he labels the 'involving-alienating' structure.

For example, the use of ChatGPT to cheat on assessments can be viewed as damaging to academic integrity and an impediment to teachers who use writing as a mechanism to draw inferences about student thinking. Beyond this obvious ethical implication, however, are further concerns such as algorithmic bias, which relates to the concerning capacity of the technology to perpetuate harmful stereotypes or discrimination (Kiran 2023). In addition is environmental concerns, due to the significant carbon footprint of both training and running large language models (Furze 2023). Legal concerns for educators include the potential breaches of copyright infringement, in addition to the potential for privacy violations or the misuse of student data (Kiran 2023; Furze 2023).

Yet, its capacity as an assistive tool for those who have difficulty with language generation is both positive and involving. Likewise, a novice learner using ChatGPT can arguably find entry points into the foundational elements of Literature. Yet, if students are not genuinely engaged in the close reading and writing process, they are arguably alienated from gaining important disciplinary knowledge and mastering cognitive skills such as analysis, synthesis, and evaluation, which are the precursors for being able to think critically and solve problems (Ellerton 2015). Furthermore, the technological efficiency argument which sees value in outsourcing our knowledge acquisition to search engines ('just google/chatGPT it') ignores the importance of fluency and automaticity to learning, which requires rich academic background knowledge (Christolodou 2023). Indeed, possessing knowledge in long-term memory is a much more efficient process than having to consult machines first.

BRINGING IT TOGETHER: MEDIATION AS CARE

At this point, the 'two sidedness' of technology that results from postphenomenological analysis arguably provides us with a complex picture of the mediating elements of ChatGPT for both students and teachers across four domains. Indeed, understanding the affordances of the technology as solicitations helps us to recognize how the practices of reading, thinking, and writing essential to the study of Literature are mediated in very different ways by ChatGPT as opposed to the use of older technologies such as primary text, pen, and paper.

Yet, simultaneously, the same 'two-sidedness' could potentially result in a feeling of moral ambivalence and uncertainty, seemingly leaving the 'educational confusion' noted at the outset unresolved. Indeed, nine months into the release of the technology, schools and universities are still 'scrambling' to cope with its implications (Nakano 2023). Under such circumstances, erring on the side of caution (through actions such as bans) *appears as* the safest option if we consider care purely in terms of harm-mitigation. However, this leads us back towards dystopianism while ignoring the nature of our technically mediated existence.

Instead, as I have argued above, the knowledge gleaned from technical mediation can form the basis on which pedagogical decisions as academic care can be enacted. This involves following the philosophy of technology in its urging of a return to a Socratic and Foucauldian concept of care, which, as signposted in the opening section, entails caring for the (now hybrid) self by through 'knowing how to live' alongside technology (Dorrestjin 2017). This neatly aligns with the role of teachers, as within the educative relationship, teachers are concerned with the formation of attention that allows a student to come to being, and orientate them in a temporal sense to the knowledge of past generations, the immediate present and possibility of a future (Stiegler 2010).

Therefore, for these new forms of care, knowledge of technical mediation (and its multiplicity of affordances) does not provide clear moral pathways but rather provides a 'pro-ethics . . . .which builds a preliminary orientation by which different value deliberations might proceed' (de Boer et al. 2018 p. 305). Technical mediation attunes us in different ways to the pedagogical *potentialities* in technology so that we can avoid the perils of ignorant surrender or defending ourselves on the basis of protecting an assumed frontier of autonomous human freedom. Instead, new forms of freedom emerge in the space between the enablers and constraints of technological mediation, as 'a person's ability to relate [and thus negotiate] what determines and influences him or her' (Verbeek 2011, p. 156). Importantly, this requires an active engagement with technical mediation in order to equip ourselves with the necessary understandings to cope.

For teachers, this means assisting students in developing what Kiran and Verbeek (2010) label as 'confidence', which is framed as new form of trust predicated on the *knowledge of mediating affects.* This approach entails a recognition and understanding of technical mediations (rather than avoidance or submission) and requires a 'secondary form of distance' where one is both with technology while simultaneously taking an analytical stance towards this involvement. In a classroom context, this means centralizing the affordances of technology explicitly for students, based on how it is habitually used. Indeed, Saracas (2017) argues that virtue ethics, with its 'concern for habits, practices and communities of moral formation', provides the best alignment of ethical theory with the postphenomenological view of technical mediation.

This approach also correlates with the proviso of academic care to ensure a student, guided by a teacher is habituated alongside technology into states of character and dispositions that enable them to flourish.

Therefore, in our current milieu, where our hybridization is a given, our attention is turned to the quality of our technological interactions and ensuring that the technologically mediated self has the necessary tools for self-care, or retaining their freedom. For our Literature student, this might mean prioritizing the skill of close reading primary texts *alongside* the skill of prompting the chatbot to assist with argumentative structure once foundational knowledge has been acquired. In this way, the student's cognition and cultural memory are not compromised through over-dependency on a machine, yet they are nonetheless improving by thinking with and through the technology, harnessing the co-constitutive potentiality.

CONCLUSION

Thus, while postphenomenology cannot provide teachers with unequivocal answers on how best to live with technology, it does provide the conceptual tools for teachers to adopt the necessary analytical stance to begin this deliberation. Indeed, the analytical approach of postphenomenology has the potential to orientate teacher attitudes so they align with new concepts of digital literacy, where an 'entangled' view of technology is the norm. Second, in offering a framework for 'reading' technical mediation, it can arguably expand on a teacher's cognitive map of the relations between students and technology in addition to the technical milieu in which they live and learn. This, in turn, provides the knowledge base for new practices of freedom and habitual methods of 'coping' characterize the new forms of academic care.

REFERENCES

Aagaard, Jesper. 2018. "Magnetic and multistable: reinterpreting the affordances of educational technology." *Int J Educ Technol High Educ* 15 (4). https://doi.org/10.1186/s41239-017-0088-4.

Christodoulou, Daisy. 2023. *'Six Things Schools Need to Know About Chat GPT'*. https://substack.nomoremarking.com/p/six-things-schools-need-to-know-about.

de Boer, Bas. 2021. "Explaining multistability: postphenomenology and affordances of technologies." *AI & Society*. https://doi.org/10.1007/s00146-021-01272-3.

Ding, Oliver 2021. *On Mediation Theory*. https://www.activityanalysis.net/the-concept-of-mediation/.

Dorrestijn, Stephen. 2017. "The care of our hybrid selves: ethics in times of technical mediation." *Found Sci* 22: 311–321.

Ellerton, Peter. 2015. "The skills and values of inquiry." *Proceedings of the International Conference of the Philosophy of Education Society of Australasia.* Melbourne, VIC, Australia: Philosophy of Education Society of Australasia.

Fawns, Tim. 2022. "An entangled pedagogy: looking beyond the pedagogy—technology dichotomy." *Postdigit Sci Educ* 4: 711–728. https://doi.org/10.1007/s42438-022-00302-7.

Feenberg, A. 1992. "Subversive rationalization: technology, power, and democracy," *Inquiry* 35(3–4): 301–322. https://doi.org/10.1080/00201749208602296.

Fraser, Olivia. 2015. *Culture and Technics.* https://www.radicalphilosophy.com/article/culture-and-technics-1965.

Furze, Leon. 2023. *Chat GPT in Education: Back to Basics.* https://leonfurze.com/2023/01/17/chatgpt-in-education-back-to-basics/comment-page-1/#:~:text=ChatGPT%20can%20be%20used%20by,around%20privacy%20and%20data%20security.

Gibson, J. J. 1966. *The Senses Considered as Perceptual Systems.* Allen and Unwin, London.

Gonski Institute for Education. 2020. *Growing Up Digital Australia: Phase 1 Technical Report.* Gonski Institute for Education. UNSW, Sydney.

Hayles, N. Katherine. 2012. *How We Think: Digital Media and Contemporary Technogenesis.* Chicago: UC Press.

Heaney, C. 2020. "Rhythmic nootechnics: Stiegler, Whitehead, and noetic life." *Educ Philos Theory* 52(4): 397–408.

Herman, Daniel. 2023. "The end of high school english." *The Atlantic.* https://www.theatlantic.com/technology/archive/2022/12/openai-chatgpt-writing-high-school-english-essay/672412/.

Idhe, Don. 1990. *Technology and the Lifeworld. From Garden to Earth.* Bloomington: Indiana University Press.

Ivancic, Roz. 1998. *Writing and Identity.* Amsterdam: John Benjamins Publishing.

Kiran, Asle H. 2015. "Four dimensions of technological mediation." In *Postphenomenological Investigations: Essays on Human-Technology Relations*, edited by Robert Rosenberger and Peter-Paul Verbeek, 123–140. Lanham: Lexington Books.

Kiran, Asle H., and Peter-Paul Verbeek. 2010. "Trusting our selves to technology." *Know Techn Pol* 23: 409–427. https://doi.org/10.1007/s12130-010-9123-7.

Lifschitz, Jodi. 2023. *10 Commandments for Chat GPT Success.* https://dataloop.ai/blog/10-commandments-for-chat-gpt-success/.

Lemov, D. 2022. "Take away their phones so we can rewire schools for belonging and achievement." *Education Next.* https://www.educationnext.org/take-away-their-cellphones-rewire-schools-belonging-achievement/.

Livingstone, Sonia., Gerison Landsdown, and Amanda Third. 2017. *The Case for a UNCRC General Comment on Children's Rights and Digital Media. A report prepared for Children's Commissioner for England,* LSE Consulting.

Mishra, Punya, and Matthew J. Koehler. 2006. "Technological pedagogical content knowledge: a framework for integrating technology in teacher knowledge." *Teach Coll Rec* 8(6): 1017–1054.

Moretti, Franco. 2000. "Conjectures on world literature." *New Left Review* 1.

Nakano, Eari. 2023. *ChatGPT-Wary Universities Scramble to Prepare for New School Year.* https://www.bloomberg.com/news/articles/2023-08-15/how-will-colleges-handle-chatgpt-in-new-school-year-it-s-a-work-in-progress#xj4y7vzkg.

Nussbaum, Martha. 1997. *Cultivating Humanity.* Harvard University Press.

Pangrazio, Luci., Anna-Lena Godhe, and Alejo Gonzáles López Ledesma. 2020. "What is digital literacy? A comparative review of publications across three language contexts." *E-Learning and Digital Media* 17(6): 442–459.

Rosenberger, R. 2014. "Multistability and the agency of mundane artifacts: from speed bumps to subway benches." *Hum Stud* 37(3): 369–392.

———. 2020. "On variational cross-examination: a method for postphenomenological multistability." *AI & Society.* https://doi.org/10.1007/s00146-020-01050-7.

Sacasas, L. M. 2017. *The Ethics of Technological Mediation.* https://thefrailestthing.com/2017/11/18/the-ethics-of-technological-mediation/.

Selwyn, N. 2012. "Making sense of young people, education and digital technology: the role of sociological theory." *Oxf Rev Edu* 38(1): 81–96. Institute of Education, University of London, UK.

Simondon, Gilbert. 1958. *On the Mode of Existence of Technical Objects.* Minneapolis: Univocal Publishing.

Turkle, Sherry. 2011. *Alone Together: Why We Expect More from Technology and Less from Each Other.* New York:Basic Books.

Twenge, Jean M. 2017. Have smartphones destroyed a generation? *The Atlantic.* https://www.theatlantic.com/magazine/archive/2017/09/has-the-smartphone-destroyed-a-generation/534198/.

UNESCO. 2023. *Global Education Monitoring Report* 2023: *Technology in Education—A Tool on Whose Terms?* Paris: UNESCO.

van Kraalingen, Imre. 2022. "Theorizing technological mediation in the outdoor classroom." *Postdigit Sci Educ* 5: 754–776. http://doi.org/10.1007/s42438-022-00315-2.

Verbeek, Peter-Paul. 2005. *What Things Do: Philosophical Reflections on Technology, Agency, and Design.* Penn State: Penn State University Press.

———. 2015. "Cover story: beyond interaction: a short introduction to mediation theory." *Interactions (ACM)* 22(3): 26–31. https://doi.org/10.1145/2751314.

Vygotsky, Lev. 1978. *Mind in Society: The Development of Higher Psychological Processes.* Cambridge, MA: Harvard University Press.

Chapter 4

Mediated (Mis)Conduct

Turnitin as an Audience for Academic Work

Eliott Rooke

This chapter offers a postphenomenological account of academic integrity and conduct in the higher education sector focusing on plagiarism in student essays. The chapter begins by setting out the context of academic integrity, reflecting on why concerns have grown of late and why they primarily revolve around student conduct. I then introduce the technology at the heart of this discussion – *Turnitin* – and provide a postphenomenological interpretation of how it mediates the relationship between students, staff, assessments, and working practices. This discussion is built across three sections. The first provides a general overview of Turnitin's mechanics, the second explores the role of hermeneutic strategies in the act of interpreting Turnitin's reports, and the third explores the possibility that Turnitin is an intended audience for students' work which therefore guides their writing. Finally, I end this chapter by drawing on these points to propose three recommendations for reducing plagiarism in higher education.

WHAT IS 'ACADEMIC INTEGRITY'?

The idea of academic integrity underpins much of the work done in the higher education sector. It suggests an assurance of authenticity and honesty without which two of the main products of universities – research and qualifications – would have little value. Employers would not favour applicants with degrees conferred by uncredible institutions, and the widespread impact of falsified academic research has been seen before (Hilgartner, 1997). Yet, academic integrity remains a poorly defined concept. Unlike our colleagues working in medical fields, there is no oath to which we swear to abide. Instead, academic integrity more loosely refers to a set of procedurally accepted ethics, values, and working practices – some, but not all, of which may be individually governed or regulated by committee or policy. What these practices are, however, varies between institutions and countries depending on local concerns (Bretag, 2016). Although students and staff alike are expected to work to these protocols, the phrase 'academic integrity' is often used as a byword for specifically student (mis)conduct in academic literature and institutional policies (Macfarlane, Zhang, and Pun, 2014). This focus is not entirely undeserved. Academic staff, as employees, are governed by contractual and legal obligations. Student behaviour is – quite literally in most countries (Draper and Newton, 2017; Awdry, Dawson, and Sutherland-Smith, 2022) – a more lawless area. Left largely unchecked by legislation, academic integrity has instead become the centre of not one but two flourishing industries in what Roe and Perkins have described as a technological arms race (2022). On one side of the market, essay mills, AI text generators, and exam sharing websites seek to help desperate students take shortcuts – or just exploit them (Draper et al., 2021). On the other side, companies sell proctoring services and analytical tools to equally desperate institutions needing to protect their reputation and commercial viability.

Concerns about misconduct – the breaching of institutional rules relating to academic integrity – are by no means new. However, the ever-growing prevalence of the internet has increased the perceived risk by creating additional opportunities for students to engage in unauthorized working practices. Pivoting to online teaching and assessment during the COVID-19 pandemic, while necessary, is widely regarded to have undermined standards of academic integrity (Gamage, de Silva, and Gunawardhana, 2020). Assessment formats designed to reduce the potential for misconduct, such as proctored, closed-book exams, were rendered infeasible and the abruptness of the transition left little opportunity to keep pace. These new opportunities to cheat coupled with the heightened pressures on students faced living through a pandemic to create a 'perfect storm' for certain kinds of misconduct (Brown and Hammond, 2022) such as contract cheating and plagiarism. In turn, this has prompted a renewed interest in the tools and methods institutions use to identify and respond to misconduct. Much of the higher education sector's ability to enforce principles of academic integrity revolves around these technological tools.

Of these tools, Turnitin, the technology I focus on in this chapter, is a clear market leader with eighty-one of the world's top one hundred universities and 98% of all UK universities subscribing to the service (Turnitin, 2023). Academic misconduct can take many forms – Weber-Wulff (2014) provides an extensive typology including fourteen classifications – but Turnitin primarily focuses on issues of authorship. As a subcategory of misconduct, authorship concerns are themselves diverse. Turnitin lists twelve forms of "unoriginality" (Turnitin, 2021), including eight types of plagiarism differentiated by what is taken and from whom. The success of Turnitin is testament to the level of concern surrounding students engaging in unauthorized working practices (which is notably not the same as the level to which such practices are present) and the extent to which the technology is engrained in higher education means it plays a central role in determining what 'authorized' working practices are in the first place. However, staff are often less clear on precisely what tools like Turnitin are doing and how they are doing it. The centrality of these technologies to our relationship with students and their work means it is pivotal we unpack their mediating influence. It is here that a postphenomenological perspective can provide insight.

TURNITIN: A POSTPHENOMENOLOGICAL INTRODUCTION

The postphenomenological unit of analysis is not the Husserlian 'thing itself' but the relationships within which it is embedded (Rosenberger and Verbeek, 2015; Greenwood and Cox, 2022). As such, to view Turnitin through a postphenomenological lens is to see the technology as being woven into the processes and practices of higher education, and to ask what role it plays in their implementation. Ihde (1990) divides human–technology relations into four categories: embodiment, background, hermeneutic, and alterity. It is the penultimate of these, hermeneutic relations, that most aptly describes users' interactions with Turnitin. Technologies in hermeneutic relations enable new perceptive capabilities by providing the user with a representation of some aspect of the world, often one that was previously unseeable (Ihde, 1990). The exact process for using Turnitin varies between institutions depending on which features are purchased and how they are integrated into wider procedures, but Turnitin is generally framed as a post-submission analytic tool. Once a student has submitted their coursework through the Turnitin system, a 'report' is generated. This report is a composite of text (the student's essay) and image (a visual overlay). The overlay highlights areas of text that the system has matched to another source – which I shall call a *comparator*. These comparators can be intra-corpal or extra-corpal (Lancaster and Culwin, 2005). Intra-corpal matches occur when the comparator is within the same submission cohort. This may indicate collusion but equally could result from students including the assignment question or title in their submission or relying on a similar selection of readings as their classmates. Extra-corpal matches arise when the comparators are external, such as published articles, and therefore may indicate other kinds of unoriginality.

At this stage, it is necessary to make a distinction. Turnitin is not a *plagiarism* detection program but rather a *similarity* detection program. The software cannot determine unoriginality, only the similarity that might indicate it. This distinction is crucial for two reasons. First, similarity is no guarantee of misconduct. Misconduct is not a textual artefact, but a process relating to the production of the work. This means that the relationship Turnitin mediates is not a staff–student relation (although there are elements of that) nor a staff–essay relation. What Turnitin mediates is the relationship between the user and the working practice behind the essay's construction. When viewing a student's submission, it would not normally be possible to see how that piece of work was created. However, by detecting similarities *and* identifying the comparators, Turnitin reveals what Wiltse (2014) calls *traces*. Traces are 'some part of the world being "caught in the act"' (Wiltse, 2014, 155), which provides the user with insight into which sources were used, the extent to which they were relied upon, and whether the student engaged with them sufficiently to re-state the argument in their own words. Turnitin functions somewhat like an x-ray machine, revealing the underlying structure of an essay. However, because we are only seeing traces which need to be deciphered rather than, say, a recording of the student completing the assignment, Turnitin is a non-isomorphic technology. Ihde uses the terms isomorphic and non-isomorphic to distinguish between technology which produce images directly representing their subject, such as a camera, and those which transform it into a non-analogous readable format, such as a watch (Ihde, 2009).

This leads to the second reason: the readouts from non-isomorphic technologies require interpretation. They are not neutral representations which faithfully convey an objective truth about the world (Ihde, 1993) meaning Turnitin does not itself answer the question of whether the student is the author of the essay. Instead, an academic member of staff must interpret the information provided in the report and reach a conclusion themselves. But the transformative process behind generating these reports is reductive. Students (hopefully) spend a considerable amount of time preparing their work prior to submission and the report can only represent a fraction of that. Nothing is conveyed about background reading not directly incorporated, previous drafts, conversations with staff or other students, or prior knowledge. Ihde describes the 'flattening' process of imaging technologies (1993, 50) as they translate three dimensions into two; this is something Rosenberger has since expanded on in his work studying photography of the Martian surface (2008, 2011, 2013). As these images are all taken from directly above, Rosenberger notes how the heights of mountains must be deduced indirectly using other factors, such as the lengths of the shadows they cast. In the case of Turnitin, the dimension being lost is not height but time. The user's encounter with the student's working practice is limited to a perspective that only sees the final submission, anything beyond this snapshot must be inferred by other means. As I discuss later, this is an important limitation.

Beyond the report, there are two other features to Turnitin which I must note despite not being the focus of the chapter. The first of these is the *score*. The Turnitin score indicates the percentage of the submission included in a match. This calculation includes the full text of the submission, not just the main body of the assessment, and so is inflated by matches in the bibliography as well as the title. Along with the report, the score is the part of Turnitin users directly interact with. I have chosen not to focus on it in too much detail here because the primary purpose of the score is to identify which reports need to be reviewed, with higher scores suggesting a higher likelihood of misconduct. However, the score is reductive to the point of nigh-irrelevance and it is generally good practice to review every report. The second aspect, and truthfully the real value of Turnitin, is the *crawler*. The crawler is the search algorithm that checks submissions against other sources on the internet and in the Turnitin database. While one could manually enter key terms into a search engine for a somewhat similar result, as described by Weber-Wulff (2014), this would quickly become infeasible with long or numerous submissions. Despite the crawler being the heart of the program, it is essentially a black box with some aspects of the repository hidden entirely. For example, if the comparator is an essay submitted to another institution, the overlay will only reveal the name of the institution and the match score. The score and the crawler could both be said to function in background relations with the user (Ihde, 1990) as they are not directly interacted with but do shape the user's experience of the system and interpretation of the report. Equally, one could argue that the crawler is engaged in a hermeneutic relationship with the user, albeit one which is rarely fully actualized. Deep within its black box, the crawler is theoretically readable. Yet, most students and staff who interact with it would never do so in a way that would require them to decipher it.

INTERPRETING TURNITIN REPORTS

As Turnitin cannot make the decision to accuse work of impropriety itself – at least, not for the time being – the process of interpreting the report usually falls to a member of academic staff. What Turnitin does offer is an indicative likelihood that a piece of work is the product of plagiarism (the score) and direction towards areas of concern and their potential comparators (in the overlay) which can be used to help make this decision. The fact that it is usually an *academic* member of staff is an important part of the process; these are, after all, matters of *academic* integrity and working practices. There are specific conventions and contextual nuances which need to be considered prior to reaching a conclusion (Bretag and Mahmud, 2009). In this section, I articulate these through the postphenomenological language of variational analysis and hermeneutic strategies (Rosenberger, 2020, 2014), highlighting the need for training – covering both a broad knowledge of academic working practices and the specific features of Turnitin – prior to attempting interpretation. This not only has specific connotations for staffing and the design of academic conduct processes, but also implications for students should they be given access to review their own Turnitin reports.

Turnitin reports require interpretation because they are multi-stable. At any given point, a report both is and is not indicative that the work is the product of plagiarism. That a report can be read in multiple ways says nothing of the work itself, nor the working practices entailed in its production. Rather, as Rosenberger (2021, 52) notes, the multistability of the image reflects the extent to which the technology is mediating our perception of that which it reveals. In other words, it is a consequence of the transformative process in which something unobservable (the student's working practice) is made visible through a hermeneutic relation (Ihde, 2009). These multiple stabilities all revolve around a constant factor without which they could not retain meaning or relation. Whyte (2015) calls this point a *pivot* and argues that it is through the pivot that there is a consistent sense of identity between stabilities. We might say that the multiple interpretations of a Turnitin report are connected by a *textual similarity pivot* as, regardless of the reason for the match, each stability entails a level of textual similarity being identified. Rosenberger combines Whyte's idea of a pivot with his own work on *axes* (2008, 2011). Where the pivot is the point around which variations are determined, the axis refers to the feature which allows for these multiple interpretations. In the cases of Turnitin, we can call this an *origination axis* in recognition of the fact that the differing stabilities stem from different interpretations of how the work was created.

The interpretation of the origination of work is fundamental for deciding whether it is a product of plagiarism. One would always expect that a student's work would contain traces of the work of others, this is a natural consequence of drawing on existing literature. The question is therefore not whether traces of literature are present, but whether their use was legitimate. To consider this in the context of an example, let us imagine the following sentence was taken from a student's work and was not accompanied by any attribution to another author: '*Postphenomenology can be viewed as an offshoot of phenomenology*'. An innocuous enough sentence. But, if this chapter were submitted to the program, Turnitin would flag this sentence as a match. And rightly so! For I have lifted this sentence verbatim from Verbeek's (2005, 133) work with only minor grammatical changes to fit the syntax of its new context. That this text matches Verbeek's own work – that there is *similarity* – cannot be denied. Determining whether it constitutes *plagiarism* is far more complex, in no small part because definitions of plagiarism are themselves highly contested (Carroll, 2016). The process of determining which stabilities can be interpreted from the Turnitin report is an act of variational analysis (Ihde, 2009) but Rosenberger takes this a step further with the proposal of variational cross-examination (Rosenberger, 2020, 2017, 2014) as a means of contrasting different stabilities. For the interpreter of a Turnitin report, variational analysis is fairly simple. There are a finite number of stabilities that matter, and these are likely to be consistent between assessments. We could identify three groupings of possible stabilities for any given essay viewed through Turnitin: that the essay was produced through legitimate means and the student is the true and sole author; that there is evidence of misunderstandings of academic convention but insufficient evidence of intent to deceive the marker or gain advantage; and that there is sufficient evidence to suggest wilful misconduct. These could be broken down for finer granularity, for example, distinguishing between degrees of severity or the kind of misconduct (for example, poor paraphrasing versus contract cheating), but the likelihood is that those decisions would be made in subsequent processes once further evidence has been collected and the student given an opportunity to make their case. There are also, of course, theoretically feasible interpretations that likely appear nonsensical to those even vaguely familiar with the content (Rosenberger, 2020, 71). Someone who has never encountered Turnitin and lacks any knowledge of the context may not distinguish between the overlay and the essay, thus believing that the colour coding is part of the original submission, perhaps indicating emphasis. I disregard these here; any interpreter in higher education would likely have some familiarity with the context of the assessment and Turnitin's interface. It is the interpreter's role to determine which of these stabilities is the most accurate representation of the student's working practice. Whichever stability appears most 'valid' for them is the dominant stability.

Which stability appears as dominant is dependent on the hermeneutic strategy of the interpreter (Rosenberger, 2008). These strategies emerge from the situatedness of the observer, which brings some elements of the image to the fore while diminishing others (Friis, 2017). Returning to our hypothetical example, one interpreter may focus on the source of the information and determine that, because it is a readily available text which directly links to the topic, coincidence is unlikely. Thus, they may view a *similarity-as-plagiarism* stability as dominant. Another may focus on the argument and phraseology and reach the equally plausible conclusion that the claim is, in a manner of speaking, established knowledge that can only be expressed in so many ways and therefore may be (or may have already been) stated in the same way by multiple authors acting independently. Thus, the second interpreter's dominant stability is *similarity-as-coincidence* or *similarity-as-commonplace*. Yet a third may consider the sentence in the context of the rest of the report and decide that, lacking evidence of plagiarism elsewhere but agreeing coincidence is unlikely, it is indicative of problems in working practices that are insufficient to constitute malfeasance but may instead result from laziness or misapprehension. In other words, they may view a *similarity-as-mistake* or *similarity-as-misunderstanding* stability as dominant. Finally, a fourth reader – the reader of this chapter – will hopefully perceive another stability: *similarity-as-intention*. Guided by the broader context of its use, this reader sees the similarity as a deliberate decision, used for illustrative or evocative purposes. This leaves a total of six stabilities: similarity-as-plagiarism, similarity-as-coincidence, similarity-as-commonplace, similarity-as-mistake, similarity-as-misunderstanding, and similarity-as-intention. Once again, that a stability appears dominant does not inherently mean that it is 'correct'. The fact that a similarity-as-plagiarism stability is perceived as dominant would not mean a student actually engaged in misconduct, merely that the work is perceived in such a way in that moment and by that reader.

The interpreter must, however, recognize that in drawing attention to textual similarity, Turnitin is shifting the focus away from other aspects of the essay which may speak to whether it is the product of academic misconduct. How it compares to the student's prior work, how long they spent writing the essay, and the consistency of the author's voice all may indicate whether or not evidence of misconduct is present but are pushed into the relative background by Turnitin's overlay. This is part of the magnification/reduction structure that Ihde (1990) has claimed is typical of human–technology relations. Although a binary magnification/reduction structure can itself be reductive, simplifying the technology to 'good' or 'bad' (Kiran, 2015), here we can think about how it makes the student's working practices not just present but present 'in a specific way' (Verbeek, 2008, 15; de Boer, 2021, 10). That is to say, the Turnitin interface partially determines the kinds of questions that are likely to be asked about the essay. As a mediating technology, it can also invite the possibility of new and different types of questions (Verbeek, 2008). The second interpreter in the above example may have never questioned the sentence had Turnitin not flagged it. In this way, Turnitin re-orients the question away from 'is there evidence of misconduct?' towards 'is there a legitimate explanation for this textual similarity?' and, in doing so, creates a default position of suspicion, potentially undermining the relationship between students and staff (Macfarlane, 2021).

Once a dominant stability is established by the interpreter, the process continues. If their conclusion was that there was no concern, then the marking process can resume. If there are issues requiring further action, then recommendations must be made. This will normally comprise a referral to an institutional misconduct procedure, requiring discussion with non-academic staff and eventually feedback to the student. Both avenues require the interpreter to convey their hermeneutic strategy, something Friis (2015) has noted is prone to miscommunication. Hermeneutic strategies evolve from the perceiver's situatedness and thus may resist communication. When liaising with non-academic staff, it would be expected that they would possess sufficient familiarity with Turnitin and the academic context to facilitate their own variational analysis. They may view a different stability as dominant, which would be a valid interpretation (Rosenberger, 2021), and may prove a fruitful avenue for continuing the investigation (Friis, 2017) through variational cross-examination. However, even if not, one would anticipate the shared level of skill and understanding to allow both parties to recognize the potential validity of each interpretation, through their own variational analysis, even if they do not see it as the dominant stability. Communication with students, by comparison, may be much less straightforward. Although student's may be able to view their Turnitin report, their relationship with Turnitin is markedly different to that of the other interpreters. Turnitin does not mediate students' relationship with their own working practices in the same way – they know how they produced their work without the need to decipher it by interpreting traces. However, Turnitin can mediate students' understandings of academic (mis)conduct and practice more broadly precisely because of how it shapes staff members' perception.

TURNITIN AS AN AUDIENCE

The ability to interpret images requires a specific set of skills (Rosenberger, 2021; Friis, 2017). For members of staff, academic-facing or not, one would anticipate that, if their job entails reading Turnitin reports, they would have received training which would then be supplemented by the experience of reviewing dozens of reports each year. This experience forms a core part of their hermeneutic strategy, allowing them to make sense of the reports in the ways I have described. Students, however, will likely only review their own reports and may, at best, be given a cursory overview of Turnitin. Students' hermeneutic strategies are therefore derived from an assortment of other experiences and are more likely to comprise a combination of common sense, limited training and experience, available online resources, and assumptions. Together, these elements may limit students' ability to derive actionable guidance from the report. While certainly a concern, the possibility that Turnitin reports are *illegible* to students is not, I think, the most significant risk. Rather, the potential for students to *misinterpret* reports, pursuing warped ideas of correct practice, is much more troubling.

Considering the interpretation of a Turnitin report from the postphenomenological position that technologies do not simply alter a priori experiences of the world, but actively co-constitute human–world relations (Verbeek, 2005), it is necessary to think how Turnitin is partially defining what students perceive plagiarism to be. By taking Turnitin reports at face value, students would provide disproportionate weighting to those features that Turnitin foregrounds, namely, textual similarity, without the technical knowhow required to situate these in the broader context of academic integrity. Rolfe (2011) found that when students were given access to their Turnitin report prior to final submission, therefore allowing them time to re-edit their essays using the report as a form of feedback, citation practices actually diminished. Rolfe attributes this to the fact that while Turnitin can identify areas of potential concern, it is unable to speak to whether those areas contain evidence of misconduct nor what action should be taken to resolve any problems that do exist. In Wiltse's language of traces, this is because 'the *means* of producing traces are separated from the *ends* of the traces themselves' (2014, 173). An unskilled interpretation would see only the visible end of the trace – the match – without following it back to the cause – the working practice – and so conflate similarity with plagiarism.

This is where the focus on Turnitin's role as post-submission analytic tool falls short of recognizing its true importance to academic integrity. When we write, we adapt our style and approach with our audience in mind (Magnifico, 2010). Students, knowing that their work will be viewed by Turnitin, would view it as an intentional audience – one that we *write for* – rather than an incidental audience – one who has an indirect relationship with the work, such as a moderator. If students anticipate how their work will be framed by Turnitin, and adjust their approach to writing accordingly, then Turnitin's influence is also present *prior* to submission. However, if students are misinterpreting Turnitin reports in such a way as to view similarity, rather than plagiarism, as the concern, then their writing styles will endeavour to remove similarity even when this might further embed plagiarism. For example, if a student wanted to reduce the number and extent of matches, they could remove references or use a word substitution approach to paraphrase a quote. Both would achieve a lower Turnitin score, but would be detrimental to the integrity of the work. They need not have access to their Turnitin reports prior to submission for this to be the case. It would gradually become learnt practice from reviewing the reports of previous submissions.

Remediating this is no small task. Students entering higher education would be expected to possess neither familiarity with the specific technologies at hand, such as Turnitin, nor the broader notion of 'good academic practice' from the outset. Indeed, the practice of using citations in any form is not usually present in UK education until university. Thus, there is little foundation on which to build. This is not to say that students should not be given access to Turnitin reports, but rather that it is necessary for considered, extensive training to be provided. Chris Ireland and John English (2011) argue that the pedagogic approaches used to teach students about plagiarism and academic integrity may well require more consideration than other aspects of teaching given the potential severity of the consequences should they misunderstand. Perkins, Gezgin, and Roe (2020) found success in reducing cases of plagiarism through a dedicated 'Academic English Masterclass' consisting of twenty-four hours of teaching spread across a semester and assessed by examination. However, the investment in staff time and resources this would require may render it infeasible for most institutions. Furthermore, given that students report the difficulty of balancing multiple commitments as a factor driving them towards misconduct (Comas-Forgas and Sureda-Negre, 2010), a time-intensive course could prove detrimental. Instead, shorter courses could focus on providing students with the tools to develop their own, informed hermeneutic strategies through which they could interpret Turnitin reports, thus facilitating a degree of self-led learning. By acknowledging the role their informal experiences already play in directing writing styles, this would be more a matter of 'course correction' than a substantive new workload.

CONCLUDING RECOMMENDATIONS

Throughout this chapter, I have made the case that similarity detection programs like Turnitin can be valuable to higher education, but only if used appropriately. The improper implementation of similarity detection can prove detrimental to students' working practices and sector-wide understandings of academic integrity. Using Turnitin effectively requires a clear understanding of how it mediates our encounters with students' working practice, something that postphenomenology helps us understand. By way of conclusion, in this section, I offer three recommendations for implementing similarity detection programs into university procedures. First, however, it is important to recognize that the nature of the claims postphenomenological work allows us to make are inherently situated within a specific investigatory context and cannot be assumed to hold true beyond it (Rosenberger, 2017). Combined with the continuing advances seen with technologies like Turnitin – and newly emerging opportunities for evading them – there are limits to the kinds of recommendations that can be made. Thus, those I offer here seek to work within the inevitable dynamism of their context by speaking to the broader relationship between academic integrity, technology, and the higher education sector.

The first recommendation I propose is simply that we openly acknowledge the limitations of these technologies in our policies, amongst staff, and in conversations with students. Turnitin, as it is currently available, is not a fully packaged solution. Rather, it is a tool that should form part of a suite of approaches for identifying and remedying concerns of academic misconduct. A postphenomenological lens helps us understand that Turnitin produces a magnification/reduction structure (Ihde, 1990) which de-emphasizes features of work, but these features must not be ignored. Similarity is neither always indicative of, nor inherent in, acts of plagiarism. Using the work of Verbeek (2005, 2008), I have argued that Turnitin actively co-constitutes understandings of misconduct, working to conflate plagiarism with similarity. Building it into a more diverse suite of strategies would lessen this and aid in the recognition of other principles of good practice. A further part of this limitation entails acknowledging that, even if these technologies provided a solution to misconduct, that would not in itself constitute a guarantee of integrity. Academic integrity must instead be understood to be a much broader phenomenon that is contextually determined. Although the absence of misconduct would likely be a common factor to any definition of integrity, as would be gaining ethical approval, abiding by institutional regulations, and engaging in non-exploitative working practices. Technologies like Turnitin make no appraisal of these factors. So, while academic conduct must remain a part of the conversation around academic integrity, it should never become the full extent of it.

This leads into my second recommendation, which is that we must recognize the structural and systemic influences of misconduct and address these. Foltýnek et al. assert that attempts to identify plagiarism come 'far too late in the writing process' (2020, 28). I am inclined to agree. A perhaps cynical take is that using similarity detection software is comparatively easy, despite what I have discussed throughout this chapter. But the promise of a technological solution for misconduct is a myth which must be dispelled. There will never be a technological panacea for academic misconduct. Solutions must instead recognize the structural issues which underpin students' motivations to cheat or result in inadvertent misconduct. Part of this means considering the points above and developing a clearer sense of what 'academic integrity' truly means. Importantly, this is likely to be different for staff and students, given that their work's reach, influence, and capacity to cause harm will all differ too. Absent a consistent set of rules or principles (Bretag, 2016), students are left to navigate a complex and changing environment which compounds other issues. These include but are by no means limited to: low engagement, high workloads, demanding expectations, incomplete understandings of academic convention, a lack of standard practice, limited resources, misplaced understandings of the value of their qualification (i.e., valuing the final grade over the skills and knowledges they gain), or poor assessment design. Similarity detection software is integral to facilitating this, but it can only ever be part of the solution and must be properly embedded within a wider infrastructure.

My final recommendation is that students should receive training to interpret Turnitin reports. This would provide students with the resources to develop hermeneutic strategies which allow them to engage in variational analysis, making Turnitin report viable resources for self-development. I offer this recommendation with Macfarlane's (2021) critique of the default use of similarity detection software as undermining student-staff relationships in mind. Providing students with training can recast Turnitin as an educational tool aiding them develop good working practice, rather than something accusative. Recognizing that Turnitin may already influence how students write is an important first step for this. Ideally, this would be coupled with providing them access to review their reports prior to submission. I do not share Rolfe's (2011) concern that this would allow students to edit their work solely to evade Turnitin. For those determined to cheat, there will always be a path forward. But making reports available would help students proactively reflect on their working practices and better understand academic integrity. Providing training that covers the points raised in my first recommendation would therefore be essential to ensure that students were aware of not just what Turnitin reports could tell them, but what they cannot show.

These three recommendations are by no means a template to resolving concerns about misconduct, but rather offer prompts to reflect on how we engage with it. The postphenomenological lens has aided in identifying them by shifting the focus away from the technologies or students themselves and onto the relationship between the various actors involved in producing academic work. By acknowledging that definitions of academic integrity, misconduct, and good/bad working practices are emergent and relational, we can recognize that tools like Turnitin are not implicated solely 'after the fact' but work to actively construct what it means to write with integrity in the higher education sector. Moreover, these recommendations would allow existing technologies and resources to provide a firmer foundation for future growth. This growth is doubtlessly going to be necessary, and I fear the sector is ill-prepared for it. As I write this chapter, speculation about what impact large language models like OpenAI's ChatGPT will have on higher education is rife, with some suggesting that we may need to fundamentally rethink the use of written assessment (Susnak, 2022). I suspect this hype will fade in time, but the underlying questions about what academic integrity is, how we can safeguard it, and what role technologies play in this will linger. We must be prepared to answer.

REFERENCES

Awdry, Rebecca, Phillip Dawson, and Wendy Sutherland-Smith. 2022. "Contract cheating: To legislate or not to legislate – is that the question?" *Assessment and Evaluation in Higher Education*, 47(5): 712–726. https://doi.org/10.1080/02602938.2021.1957773.

Bretag, Tracey. 2016. "Defining academic integrity: International perspectives – introduction." In *Handbook of Academic Integrity*, edited by Tracey Bretag, 3–5. Singapore: Springer.

Bretag, Tracey, and Saadia Mahud. 2009. "A model for determining student plagiarism: Electronic detection and academic judgement." *Journal of University Teaching and Learning Practice*, 6(1): 49–60. Retrieved from http://ro.uow.edu.au/jutlp/vol6/iss1/6.

Brown, Stephen, and Kay Hammond. 2022. "Plagiarism in higher education: Navigating a perfect storm." *European Journal of Education and Pedagogy*, 3(5): 100–103. https://doi.org/10.24018/ejedu.2022.3.5.452.

Carroll, Jude. 2016. "Making decisions on management of plagiarism cases where there is a deliberate attempt to cheat." In *Handbook of Academic Integrity*, edited by Tracey Bretag, 219–299. Singapore: Springer.

Comas-Forgas, Rubén, and Jaume Sureda-Negre. 2010. "Academic plagiarism: Explanatory factors from students' perspective." *Journal of Academic Ethics*, 8(3): 217–232. https://doi.org/10.1007/s10805-010-9121-0.

de Boer, Bas. 2021. "Explaining multistability: postphenomenology and affordances of technologies." *AI & Society*. https://doi.org/10.1007/s00146-021-01272-3.

Draper, Michael, and Philip Newton. 2017. "A legal approach to tackling contract cheating?" *International Journal for Educational Integrity*, 13(11): 1–16. https://doi.org/10.1007/s40979-017-0022-5.

Draper, Michael, Thomas Lancaster, Sandie Dann, Robin Crockett, and Irene Glendinning. 2021. "Essay mills and other contract cheating services: To buy or not to buy and the consequences of students changing their mind." *International Journal for Educational Integrity*, 17(13): 1–13. https://doi.org/10.1007/s40979-021-00081-x.

Foltýnek, Tomáš, Dita Dlabolová, Alla Anohina-Naumeca, Salim Razı, Július Kravjar, Laima Kamzola, Jean Guerrero-Dib, Özgür Çelik, and Debora Weber-Wulff. 2020. "Testing of support tools for plagiarism detection." *International Journal of Educational Technology in Higher Education*, 17(46): 1–31. https://doi.org/10.1186/s41239-020-00192-4.

Friis, Jan K.B. 2015. "Towards a hermeneutic of unveiling." In *Postphenomenological Investigations: Essays on Human-Technology Relations,* edited by Robert Rosenberger and Peter-Paul Verbeek, 215–226. London: Lexington Books.

———. 2017. "Gestalt descriptions embodiments and medical image interpretation." *AI & Society*, 32(2): 209–218. https://doi.org/10.1007/s00146-015-0615-6.

Gamage, Kelum, Erandika de Silva, and Nanda Gunawardhana. 2020. "Online delivery and assessment during COVID-19: Safeguarding academic integrity." *Education Science*, 10(11): 301–325. https://doi.org/10.3390/educsci10110301.

Greenwood, Michelle, and Julie Cox. 2022. "Seduced by technology? How moral agency is mediated by the invisibility of everyday technologies?" *Organization Studies*, 44(4): 523–543. https://doi.org/10.1177/01708406221107455.

Hilgartner, Stephen. 1997. "The Sokal affair in context." *Science, Technology, and Human Values*, 22(4): 506–522. https://doi.org/10.1177/016224399702200404.

Ihde, Don. 1990. *Technology and the Lifeworld.* Bloomington: Indiana University Press.

———. 1993. *Postphenomenology: Essays in the Postmodern Context.* Evanston: Northwestern University Press.

———. 2009. *Postphenomenology and Technoscience: The Peking University Lectures.* Albany: SUNY Press.

Ireland, Chris, and John English. 2011. "Let them plagiarise: Developing academic writing in a safe environment." *Journal of Academic Writing*, 1(1): 165–172. https://doi.org/10.18552/joaw.v1i1.10.

Kiran, Asle. 2015. "Four dimensions of technological mediation." In *Postphenomenological Investigations: Essays on Human-Technology Relations* edited by Robert Rosenberger and Peter-Paul Verbeek, 123–140. London: Lexington Books.

Lancaster, Thomas, and Fintan Culwin. 2005. "Classification of plagiarism detection engines." *Innovation in Teaching and Learning in Information and Computer Sciences*, 4(2): 1–16. https://doi.org/10.11120/ital.2005.04020006.

Macfarlane, Bruce. 2021. "The distrust of students as learners: Myths and realities." In *Trusting in Higher Education,* edited by Paul Gibbs and Peter Maassen, 89–100. Switzerland: Springer Nature.

Macfarlane, Bruce, Jingjing Zhang, and Annie Pun. 2014. "Academic integrity: A review of the literature." *Studies in Higher Education*, 39(2): 339–358. https://doi.org/10.1080/03075079.2012.709495.

Magnifico, Alecia. 2010. "Writing for whom? Cognition, motivation, and a writer's audience." *Educational Psychologist*, 45(3): 167–184. https://doi.org/10.1080/00461520.2010.493470.

Perkins, Mike, Ulas Gezgin, and Japser Roe. 2020. "Reducing plagiarism through academic misconduct education." *International Journal of Educational Integrity*, 16(3): 1–15. https://doi.org/10.1007/s40979-020-00052-8.

Roe, Jasper, and Mike Perkins. 2022. "What are automated paraphrasing tools and how do we address them? A review of a growing threat to academic integrity." *International Journal of Educational Integrity*, 18(15): 1–10 https://doi.org/10.1007/s40979-022-00109-w.

Rolfe, Vivien. 2011. "Can Turnitin be used to provide instant formative feedback?" *British Journal of Educational Technology*, 42(4): 701–710. https://doi.org/10.1111/j.1467-8535.2010.01091.x.

Rosenberger, Robert. 2008. "Perceiving other planets: Bodily experience, interpretation, and the Mars Orbiter Camera." *Human Studies*, 31(1): 63–75. https://doi.org/10.1007/s10746-007-9078-1.

———. 2011. "A phenomenology of image use in science: Multistability and the debate over Martian gully deposits." *Techné: Research in Philosophy and Technology*, 15(2): 156–169. https://doi.org/10.5840/techne201115214.

———. 2013. "Mediating Mars: Perceptual experience and scientific imaging technologies." *Foundations of Science*, 18(1): 75–91. https://doi.org/10.1007/s10699-012-9286-7.

———. 2014. "Multstability and the agency of mundane artifacts: From speed bumps to subway benches." *Human Studies*, 37(3): 369–372. https://doi.org/10.1007/s10746-014-9317-1.

———. 2017. "On the hermeneutics of everyday things: Or, the philosophy of fire hydrants." *AI & Society*, 32(2): 233–241. https://doi.org/10.1007/s00146-016-0674-3.

———. 2020. "On variational cross-examination: A method for postphenomenological multistability." *AI & Society*, 38: 2229–2242. https://doi.org/10.1007/s00146-020-01050-7.

———. 2021. "A primer on postphenomenology and image reading." In *Postphenomenology and Imaging: How to Read Technology*, edited by Samantha Fried and Robert Rosenberger, 3–95. Lanham: Lexington Books.

Rosenberger, Robert, and Peter-Paul Verbeek. 2015. *Postphenomenological Investigations*. London: Lexington Books.

Susnjak, Teo. 2022. "ChatGPT: The end of online exam integrity?" arXiv:2212.09292v1. https://doi.org/10.48550/arXiv.2212.09292. Last accessed February 18, 2023.

Turnitin. 2021. "The plagiarism spectrum 2.0." Accessed February 18, 2023. https://www.turnitin.com/resources/plagiarism-spectrum-2-0.

———. 2023. "Similarity." Accessed February 18, 2023. https://www.turnitin.com/products/similarity.

Verbeek, Peter-Paul. 2005. *What Things Do*. Pennsylvania: Pennsylvania State University Press.

———. 2008. "Obstetric ultrasound and the technological mediation of morality: A postphenomenological analysis." *Human Studies*, 31(1) 11–26. https://doi.org/10.1007/s10746-007-9079-0.

Weber-Wulff, Debora. 2014. *False Feathers: A Perspective on Academic Plagiarism*. Berlin: Springer.

Whyte, Kyle. 2015. "What is multistability? A theory of the keystone concept of postphenomenological research." In *Technoscience and Postphenomenology: The Manhattan Papers*, edited by Jan K.B. Friis and Robert Crease, 69–81. London: Lexington Books.

Wiltse, Heathers. 2014. "Unpacking digital material mediation." *Techné: Research in Philosophy and Technology*, 18(3): 154–182. https://doi.org/10.5840/techne201411322.

Chapter 5

Equipping Tablets

In-Depth Interviews with Early Adopters on the Sedimentation of Human–Technology Relations in Schools

Markus Bohlmann and Martin Wilmer

ON THE POSTPHENOMENOLOGY OF A KEY TECHNOLOGY IN THE EDUCATION SYSTEM

The digitalization of schools in Germany has received a major boost in recent years with the widespread introduction of a large numbers of tablets. Thanks to the funds of the federally financed "Digitalpakt Schule" (Digital Pact for Schools), a funding package worth billions, and the financial contributions of the Länder and cities in and after the lockdowns of the COVID-19 pandemic, tablets are now available for widespread use in many schools. Actors in the education system hope this will bring about the "digital revolution" hoped for by the Conference of Education Ministers (KMK 2017, 8, see also 2021). So far, the digitalization of the education system has been rather slow in the years from 2011 to 2021. It was carried by individual committed teachers and class-wide pilot projects. But a key technology here was already the tablet (Bohlmann 2022, chap. 11). There are several prescriptive didactic proposals for tablet use and a few studies of empirical educational research on the learning benefits of tablet use. But there is still no philosophy of technology analysis of the human–technology relations in schools that are changing with this widespread digital key technology.

This study focuses on a group of early adopters (see on the term: Dedehayir et al. 2017; the classic approach by: Rogers 1962). There was a group of tablet users who voluntarily and independently used tablets at school in the previous years in the sense of bring-your-own-device. This took place where tablet use was neither piloted nor introduced class-wide but was also no longer prohibited as in previous years. It was a relatively short period of time when not only tablets no longer fell under the no-mobile-phones-in-class rules in German schools, but also could not yet be provided in class size by the school side.

In the tradition of the social construction of technology (SCOT) (Pinch and Bijker 1987), in which postphenomenology also stands in a certain way, such early use is crucial for the development of technology and—one can add from a postphenomenological perspective—also for the establishment of human-technology relations (Ihde 2009, 8). Postphenomenological terminology often subsumes this process under the term "sedimentation" (Rosenberger and Verbeek 2015, 25; Rosenberger 2012). Here, a technological practice settles down, is still plastic at the beginning, but then becomes a stone-like solid form. This form henceforth not only structures the perceptions and possibilities of action of early users but also structures human–technology relations for the entire later technological practice. In this study, we conducted in-depth interviews with early adopters of tablet use in school. We recruited four early adopters as interviewees shortly before they started their university studies at the University of Münster in Germany. Each of them had regularly used a tablet in school lessons for at least half a year. Our aim was to gain exploratory insights into the phenomenological dimensions of their technological practices to draw conclusions about the educational system-wide process of sedimentation of this human–technology relation that is underway. To this end, we conducted interviews using a specific trigger and questioning technique so that our early adopters could relive particular situations of their tablet use at school and tell us about their human–technology relations.

DESIGN OF THE STUDY

How can people come to understand the human–technology relations in which they themselves are involved? Answering this question seems particularly difficult with tablet use in schools for four reasons.

First, the tablet as a platform technology is per se not the focus of attention. As a product of the "cultural technique of flattening " (Krämer 2022, 83), it may be only the world, or some section of it, in the educational case especially the instruction, that we perceive "through" the tablet. So, there was a danger that the interviewees would only describe the phenomena of the classroom situation, which were not specific to the *use* of the tablet.

Second, there is a common bias not only among students to consider technology as a tool first. This is instrumentalism, a frequently observed naïve position in the philosophy of technology. Andrew Feenberg calls it "the most widely accepted view of technology" (Feenberg 1991, 5). It is very widespread among students as well. A description of a phenomenon against the background of this common student conception is in danger of drifting into an evaluation of good or bad functioning.

Third, it should be difficult—at least as postphenomenological theory suggests (Ihde 1990, 36–37)—for students to leave the Galilean or Cartesian separation that corresponds to our technical-scientific worldview. We learn this worldview at school and it might be difficult, especially for students, to think in terms of relations instead of subject-object distinctions. Here it could happen that students tend to subjectively describe themselves as technology users or the technology as an object, but precisely not their human-technology relations, because they lack the conceptual means to do so.

Fourth and finally, the use of tablets in schools is accompanied by hopes for the modernization of the educational system. Stefan Aufenanger mentions location and time-independent learning, self-directed learning, better communication, increased cooperation, and possibilities for class internal differentiation as some of the hopes that typically accompany the introduction of tablets in the educational system (Aufenanger 2020, 31). Thus, from the inside view of individual actors, human–technology relations are often shaped by the outside view that carries with it all the expectations of the system-wide function of tablets.

So, in designing this study, for these four reasons, it was clear that we could not simply *ask* the students in our study about their human–technology relations in tablet use. So we have done two things.

First, we did a small preliminary study with 23 tenthgrade students to develop an exploratory learning material that, based on their level of learning, provided them with three basic concepts to understand human–technology relations at a fundamental level. We then gave this learning material to the four participants in the main study as a trigger before we conducted in-depth interviews with them.

Second, we used a certain questioning technique in which the relation in action is relived through the description of life-world situations and the recollection of one's own experience.

We will now describe in turn first the development of the learning material, then the questioning technique and finally the process of conducting in-depth interviews.

Preliminary Study: Development of a Learning Material That Makes Human–Technology Relations Understandable

Instructional designs for the philosophy of technology that make use of postphenomenology usually presuppose a fundamental knowledge of the philosophy of technology tradition (Bohlmann 2022, chap. 9 and 10). This knowledge is currently rarely taught in German schools (Bohlmann 2022, chap. 8). But in North Rhine-Westphalia, where our study was carried out, there is epistemology as a standard topic in philosophy classes. Russell´s classical epistemological position from *The Problems of Philosophy*, including Cartesian scepticism, is common in German philosophy education. Russell explains scepticism using the iconic example of his desk (Russell 1912, 10).

This is where Heidegger's concept of "equipment" from paragraphs 15 and 16 of *Being and Time* can come in as a basic philosophy of technology concept (Heidegger 1962, 95–107[66–76]).[1] Heidegger's concept of "equipment" is—literally—a rather easy-to-grasp concept of a human–technology relation and in North Rhine-Westphalia it can be connected to student's prior knowledge from epistemology. Heidegger refers directly to the being of a desk—possibly Russell's prominent table—in *Being and Time* (Heidegger 1962, 81[55]). So, we developed a material, a short explanatory text on Heidegger's concept of equipment as a trigger for our main study and pre-tested this learning material.

Heidegger's original text from *Being and Time* is difficult to access even in German because of the fundamental-ontological references, so we developed a text that introduces Heidegger's distinction between ready-to-hand (ger: "zuhanden") (Heidegger 1962, 98[69]) and present-at-hand (ger.: "vorhanden") (Heidegger 1962, 67[42]) to conceptualize equipment (ger.: "Zeug") (Heidegger 1962, 98[69]). Due to its broader theoretical function, the concept of "equipment" falls short of the concepts of postphenomenology when it comes to describing human–technology relations. But in some respects "equipment" can offer a first access to these relations in school; we will come back to this shortly.

The preliminary study to develop the material was conducted at a Gymnasium in North Rhine-Westphalia, Germany, with 23 tenth-grade students (age approx. 16 years), 12 male, 11 female, who had just previously read Russell's text in class. In the preliminary study, the following short explanatory text on the core passages of §§ 15 and 16 from Heidegger's *Being and Time* was developed in multiple development cycles using the tools of design research (Bakker 2018). Here is an English translation of the explanatory text developed:

> M1: On Heidegger's Understanding of the Being of Things
>
> Martin Heidegger is one of the most influential German philosophers of the 20th century. And he is one of the first to have thought systematically about technology. He believes that previous philosophy was far too much influenced by traditional, handed-down ways of thinking. In doing so, it loses sight of human existence as it really is.

> Above all, Heidegger considers it wrong in the philosophical tradition that things in the world are understood as merely "existing" things that are none of our concern. Bertrand Russell, for example, stands in this tradition when he describes the "table." For he describes it as a thing with four legs on which a plate is fixed, which has a certain colour, etc. . . . . This description takes place on a very abstract level. The table is "present-at-hand" and we describe our optical impression of it in a very specific way.
>
> Heidegger finds this too abstract and theoretical. He wants to start with his philosophy in everyday life. He wants to describe how we "really" perceive things. In Heidegger's view, things in the world do not simply exist without us having anything to do with them. On the contrary, when we live, we act. We must therefore understand the world and the things in the world from action, from practice. The table is not simply present for us. Above all, it offers us meaningful possibilities for action: we can sit down at it; we can put our tablet on it; we can put it aside when it is in the way and much more. We can even, like Bertrand Russell, just look at the table and describe it in all its components. But for Heidegger, this is also only one possibility among many of how we can relate to the table in an active way. Heidegger therefore thinks that things are not primarily present for us, but "ready-to-hand." They are familiar to us because we deal with them and use them. This is how we have to understand things. This is why Heidegger also calls the things we use "equipment" (ger.: "Zeug"), as in "vehicle" (ger.: "Fahrzeug") or "tool" (ger.: "Werkzeug"). (based on: Heidegger 1962, paras. 15, 16)

Finally, in the preliminary study, we exploratively determined whether learners, after reading this text, are able to complete a task in which they can transfer three aspects of the human–technology relation, using Heidegger's "hammer" as an example (Heidegger 1962, 98[69]), to a technology in their lifeworld. Here, appropriately, it was the use of tablets in school.

These are the three basic aspects of the human–technology relation of Heidegger's "equipment," as developed in §15 and §16 of *Being and Time,* that were tested in the preliminary study with the translated original text we used. In the endnotes, you will find the extent to which postphenomenology today offers more complex views in these aspects than the easy-to-learn concept of "equipment" here:

1. Stability: For Heidegger "equipment" always has a purpose or function. We use something to. . . . The hammer, for example, is there to hammer.[2]
2. Transparency: When we use "equipment," we are often not aware that we are using it. For example, when we hammer, we are not permanently aware that we are holding a hammer. After all, we need our attention for hammering.[3]

3. Hermeneutics of Being: According to Heidegger, being must be understood from praxis, that is, from being-in-the-world. We can only use "equipment" in connection with other "equipment" in a certain praxis. To hammer, we need not only the hammer, but, for example, a nail, a picture, and a wall with some good lighting. This is arguably something like what Heidegger later called a framework (ger.: "Gestell") and what Albert Borgmann then called a device paradigm, as the "compound of commodity and machinery" is the device, and the "distinctive pattern of the division and connection of its components" is its paradigm (Borgmann 2003, 18).[4]

In our preliminary study students could mostly correctly apply concepts 1–3 including the hammer example to the tablet as equipment after reading the explanatory text M1. Table 5.1 shows the percentage of correct[5] examples, the percentage of given answers[6] and the most frequently mentioned example for each of the three aspects.

The results indicate that the students understood the three aspects of the equipment quite confidently, with transparency being the most difficult aspect. The number of examples given per person on the stability aspect suggests that most students here actually understood the tablet even in terms of *multi*-stability, a concept that arguably goes beyond Heidegger (see footnote 2). The same is true of the "hermeneutics of being." For example, one learner answered that it needs a "teacher" as additional equipment for educational tablet use, one said "additional people," four said that it needs "a hand/finger," and five said "media competence/experience in operation." The students in the preliminary study thus already extended the hermeneutic analysis to the practical context of use by human beings, which also already goes beyond the simple context analysis in the sense of technical additional accessories, as suggested by "equipment."

With the M1-learning-material, we could be quite sure, based on the preliminary study, to be able to evoke conceptual thinking about human–technology

Table 5.1 Results of the Preliminary Study

Aspect of "Equipment"	*% of Correct Examples*	*% of Given Answers*	*Most Frequently Mentioned Example*
Stability	74/74 (100%)	23/23 (100%)	Saving and retrieving documents, books, working material (15)
Transparency	21/26 (80.7%)	19/23 (82.6%)	When editing tasks/notetaking (8)
Hermeneutics of Being	68/69 (98.5%)	23/23 (100%)	Digital pen (18)

relations. The material was to be used in our main study as a trigger at the beginning of the interviews with our four early adopters.

The Questioning Technique: Remembering Involvement

As already mentioned in the introduction, the time of early use of tablets is now largely over at schools in Germany. The 21-year-old female Interviewee D, the oldest in the study, describes this transition in the third and validating part of the interview:

> I think that only happened when I had long since left school, that the students all got such a device from the school. Of course, those who already had a tablet could use their own, but there were also tablets from the school that could be distributed. (D, 164–168)[7]

She also talks about how the software on the tablets was then managed by the school (160–164) and how teachers received training (172–178).

Our interviewees have thus just experienced the decisive years in which the technological practices of tablet use in schools were consolidated and the human–technology relations sedimented. Because, in such cases, the use of the tablet in general is always path-dependent, it was clear to us that we can learn the most about the human–technology relations of tablet use for educational purposes in general from the experiences of early adopters. But how can we facilitate the reliving of human–technology relations in the interview when the real experience was years ago? As a solution to this problem, we developed an interview guideline in which always the same questions about certain situations of tablet use in class were included, with the possibility of asking more in-depth questions about concrete experiences and perceptions in the situation. Technological practice in early phases of use not only sediments in the technology, but it also sediments in human memory. Table 5.2 shows all the situations asked about in the second part of the interview.

We needed the text on Heidegger's concept of "equipment" in particular as a trigger for question 1. While tablet technology is generally seen as a haptic technology and as a platform and as such should be per-se "ready-to-hand," it is actually not at all clear whether the students saw their own early use this way and in what dimensions they interpret Heidegger's concept here. Whether this works along the three aspects from the preliminary study or otherwise was completely open. The first question on the "present-at-hand"/"ready-to-hand" distinction was, therefore, the most important question of the in-depth interview. We assumed that the answer here would also have an impact on the further course of the interview.

In questions 2 and 3, we took the idea of questioning situations of functioning as well as non-functioning from the work of Tafdrup, Andersen, and

Table 5.2 Interview Guide to Access Situations of Memorable Human–Technology Relation in Tablet Use

Introduced by the Interviewer with: "Please Describe a Situation in Which . . .
. . . the tablet was "ready-to-hand" or "present-at-hand" to you in class (1)
. . . you have used your tablet according to its intended purpose (2)
. . . your tablet did not function or did not function as it should have (3)
. . . you used your tablet for a purpose other than that for which it was intended (4)
. . . you were caught using your tablet by someone else (5)
. . . you were alone when using the tablet (6)
. . . you were in a community when using the tablet (7)
Continued by questions aiming at the depth of the experience: How did it make you feel? What was it like for you?

Hasse (2019). It is very likely that the "purpose" of tablets had yet to be found by the early adopters of our main study. Only this then determined what good functioning or malfunctions are in the first place. The question about the use for *another* purpose (question 4) stems from technological practices of bricolage as described by the anthropologist Claude Lévi-Strauss and taken up by Don Ihde (f.e. Ihde 1990, 128). Such practices are typical of the early use of technological artefacts. Questions 5 to 7 are aimed at social constellations: being caught, being alone, being together. While digital technologies are often feared as socially disruptive in the school system, postphenomenological theory suggests that they simply restructure social relations between humans through technology. While postphenomenology itself does not go so far as to understand technology itself as a social actor, Peter-Paul Verbeek's engagement with Bruno Latour's actor-network theory at least results in understanding technological mediation as socially constitutive. In a nutshell: "technologies are social entities" (Verbeek 2011, 4). It is very likely that these social constitutive processes were particularly interesting in early use. User groups will not have been fixed yet but had to find themselves in and through practice.

The Main Study: In-Depth Interviews with Four Early Adopters

We recruited the interviewees through a request during orientation week at the start of the winter semester at Münster University. Our request was directed at all first-semester students in philosophy who had used a tablet for at least half a year in at least one subject during their time at school. It could be assumed that the prospective philosophy students already had a special affinity for reflecting on their own human–technology relations. A total of four prospective students came forward, and we conducted interviews with each of them.

The in-depth interviews lasted about half an hour each and consisted of three parts:

1. The trigger with the explanatory text M1 on Heidegger's "equipment," a passage from Russell's *Problems of Philosophy*, to look up the school material referred to here again if necessary, as well as a comprehension question: "Heidegger makes a distinction between 'ready-to-hand' and 'present-at-hand'—please describe what he means by this."
2. The seven questions from table 5.2 with follow-up questions about concrete experiences and feelings in the situation—that was the essential in-depth part of the interviews.
3. Some questions about the context, such as: "Did you have your own tablet?," "Was the software installed on the tablet by the school?," in order to be able to better interpret certain situations in the later evaluation. These passages will then also only be dealt with as additional context in the following analysis.

The four interviews were conducted at the end of October to the beginning of November 2022 in a room at the University of Münster and were audio-recorded. In the following, the interviewees are referred to by synonyms. The short form just refers to the chronological order of the interviews, A–D, the longer in addition to the gender and age of the interviewees: A-M-19, B-M-19, C-F-18 and D-F-21. The interviewer in each case was one of the authors of this study, bound to the interview guide prepared beforehand. The interviews were anonymized and transcribed, following simple transcription rules of empirical social research.[8] We subsequently categorized the transcripts using Philipp Mayring's Qualitative Content Analysis (Mayring and Fenzl 2019).

The categories were formed inductively from the material after we first openly sifted through all the material. Anchor examples were found, and codes were applied to the whole material. The coding units were semantic units, that is, one or more words with meaning. The context unit was the respective individual answer. The evaluation unit was the whole material. As a rule, a passage could not be coded more than once but it was possible to code several passages, even within a single interview, with the same code. We did not code completely, that is, answers in the interview could remain without code—in this case, they were simply not interesting for interpretation. We checked reliability by applying the coding scheme to the material a second time independently of the first coding by a different coder. Inter-rater reliability, when rigorously checked[9] across all codes and the entire sample, was at 55.9% direct match with $\kappa = .49$, which is a moderate agreement by coders according to Landis and Koch (1977).

MAIN RESULTS

Table 5.3 shows the inductively developed codes with definitions, anchor examples, and number of applications per interview.

As can be seen in table 5.3, except for "state-of-mind/mood," all the codes appear in all four interviews. In most cases, the codes could also be set several times per interview. This means that the interviewees came back to these points in other questions or mentioned them again in the third part of the interview on the context of tablet use.

This leads us to the interpretation that these codes related to an impressive re-experiencing of the human–technology relations that our interviewees had established with their tablets. This is also the case for the first two codes, which resulted from the different interpretations of the text on Heidegger's "equipment" M1. The students also spoke about these perspectives in the following main part of the interview, so we assume that our trigger worked as such and that we were able to evoke a reflection on their own human–technology relations.

We will now interpret the codes group by group, focusing on specific coded passages from the interviews.

Dealings and State-of-Mind/Mood

All interviewees interpreted the tablet's being "ready-to-hand" and its status as "equipment" in a way, which Heidegger describes as a human–technology relation resulting from practical use, that is, "our 'dealings' in the world and *with* entities within-the-world" (ger.: "Umgang", Heidegger 1962, 95[66f]). D-F-21 uses the word "utility item" (D, 32) (ger.: "Gebrauchsgegenstand"). A-M-19 describes situations where the tablet was off or not in use as "present-at-hand" and any active use as "ready-to-hand":

> When the teacher explained something on the blackboard, or on the PowerPoint presentation or on the overhead project, then my tablet was present-at-hand, then I didn't use it directly *ahem and it was ready-to-hand when we did any work assignments, read any texts *ahem when we did research on the internet and *ahem wrote our own texts. (A, 33–38)

B-M-19 and D-F-21, on the other hand, also describe a relational level of being more deeply connected to the device as equipment. This existential constitution can be described with Heidegger's "state-of-mind" (ger.: "Befindlichkeit") or "mood" (ger.: "Stimmung") (Heidegger 1962, para. 29). B calls it "simply a different approach to looking at the world" (B, 36). Very revealing is D's account of how she realized that she was "more of an

Table 5.3 Codes of the Qualitative Content Analysis in the Main Study

Code	*Definition*	*Anchor Example*	*Applic.*
Dealings	A practical distinction is made between "ready-to-hand" and "present-at-hand"	"That is, he [Heidegger] describes these as things that offer possibilities for action here" (A23-24)	A: 2 B: 3 C: 2 D: 2
State-of-Mind/Mood	A felt difference between "ready-to-hand" and "pesent-at-hand," Heidegger's "state-of-mind" especially "mood"	"Because things are at hand, because one is somehow in relation to things and that is simply a different approach to looking at the world" (B34-36).	A: 0 B: 2 C: 0 D: 4
Airdrop-Community	A community of tablet users is established by sharing texts via airdrop or similar	"That means if you forgot your book, for example, the other person could airdrop or send the text" (A40-42).	A: 5 B: 2 C: 3 D: 3
Purpose: Productivity	The purpose of the tablet is identified as classroom productivity	"Well, you can use a tablet exclusively to *ahem promote productivity" (A48-51).	A: 3 B: 5 C: 3 D: 2
Practice: Pen and Paper	The tablet is used for drawing, writing, and painting	"There is a drawing app on my tablet and when I was bored in class, I used it to draw" (A89-90).	A: 1 B: 2 C: 4 D: 3
Practice: Keeping Order	With the tablet it is easier to keep things in order and not to forget anything	"It was so much easier to create different folders" (A63-64)	A: 3 B: 2 C: 1 D: 2

analogue person" (D, 41–42). When D-F-21 did not use the tablet for writing or organization (see below) but tried to use it to study for the final exams, the German Abitur, she did not feel in the "mood" to use the tablet. She eventually put the tablet aside and went back to learning with Post-its and index cards. She switched technologies because she could not relate to the tablet while learning actively.

> I can touch the index card, I find that just using your senses makes a lot of difference, so I can, yes, everything is the same on a tablet, I used GoodNotes and everything looks the same and if I have the index cards or then use Post-its in

> another compartment, I don't know, then I can assign the Post-its directly in my head in one compartment and the index cards in the other and that's not possible with the tablet. (D, 60–66)

D was the only one of our interviewees who at least described *trying* to learn actively with the help of the tablet. She failed, but the rest of the interviewees did not even think of the practice of active learning as a human–technology relation with the tablet. It is also very interesting that D-F-21 describes the display in this context as one-dimensional, "everything looks the same." The variety of human–technology relations of a display, its multistabilities as desktop, paper, window into a virtual (game) world, etc., as Ihde describes them (Ihde 2012, 146f), are not experienced here.

The Airdrop-Community

All four interviewees reported that a community of tablet users was established through sharing scanned documents with each other. At the beginning of tablet use in school, worksheets were often only physically brought into the classroom as printouts by teachers. So typically, you had to scan the worksheet with an app (via photo function). This has often led to a ritual of sharing, as at the beginning of a meal when table fellowship is formed by passing the food around: "You can do that [being part of a community of tablet users] especially when not everyone got a worksheet and not everyone had to scan it in, but often we had someone scan it in and then send it around by airdrop" (C 89–91). The function of the iPad to share data via Bluetooth is called "airdrop," our participants reported the practice just described as "airdropping," which was sometimes also used to share work results, sometimes even without the teacher noticing: "even if you didn't understand things, you could just ask others to send you the solution without having to interact so that the teacher noticed or that you [short break] so you could generally solve things much more easily" (A66-69). B-19-M und D-21-M used the exact same word to describe the group segregation of students through these practices, they called it: "two-class society" (ger.: "Zweiklassengesellschaft," B, 132 and D, 124). D describes in her interview that this also had to do with economic divisions,[10] because early tablet users often financed their tablet themselves or with the help of their parents:

> You could already tell who had the tablet, a kind of, yes, this sounds stupid now, but a two-class society, I'd say, *ahem, those who had the tablet were, those were the ones who had the money, *ahem, I don't know, so I'd say they were just a bit more aloof than the others. So that was always the thing of the people who still wrote on paper, their words, so to speak. (D, 123–129)

Keeping Order, Tidying, Writing, and Drawing

The last three codings—"Purpose: Productivity", "Practice: Pen and Paper," and "Practice: Keeping Order"—were close to each other, which was noticeable to us especially when calculating the inter-rater reliability. Often, when one coder set a code here, another set another of this group of codes. This suggests that these codes can probably be seen as one—that is, for the interviewees, the purposeful use of the tablet appears as a single practice rather than several different stabilities.

As already described, active learning is precisely *not* one of the tablet practices described by early adopters. Instead, all four interviewees reported practices of keeping order in administrative activities, and practices that could also be done with a physical pen and paper. The latter, by the way, corresponds with a result from the preliminary study, namely that the "digital pen" is seen as the most important additional "equipment" of the tablet (see table 2).

The tidying practices of tablet use were accompanied by a feeling of relief:

> So it [the tablet use] was actually quite pleasant. *Ahem I just liked the fact that I didn't always have to lug around a lot of books and a whole bunch of paper, and that I had everything on one device, so to speak. That made a lot of things easier for me. (D, 75–77)

Our interviewees often described the function of the tablet as analogous to the function of a college notebook, which is considered a particularly messy way of storing worksheets in school—you don't file them but leave them stuck between the pages of the notebook. Here, the digital tablet solves the problem and you feel better: "It was very practical, so I didn't have to take individual pieces of paper with me and I couldn't lose my papers. Before, I was a college notebook person who had different worksheets in a college notebook from different subjects" (A, 60–65).

The substitute character for the college notebook is particularly evident in the fact that all interviewees understand the practices of recording, writing, text-marking, and drawing as essential functions of the tablet: "Notes written down, transcripts, blackboard pictures and stuff like that" (D, 70–71). This function of the tablet is clearly already embodied in a postphenomenological understanding, which can be seen particularly in the fact that several interviewees described how they also used the drawing functions to pass time in class (A, 89–93; B, 112–113). A particularly impressive example has D-F-21:

> *D:* . . . I once coloured a giraffe in philosophy class!
>
> *Exp:* Could you explain that again. . ..

> *D:* yes . . .
>
> *Exp:* Why a giraffe? [laughs]
>
> *D:* Yes, giraffes are my absolute favourite animals, yes, and I had, we had a trainee teacher who was supposed to do the lessons and it was really really awful and boring and that's why I coloured in the giraffe. (D, 104–111)

The physically felt boredom and displeasure with teaching led directly into the technological practice of doodling here. This embodied use of the digital tablet as a writing device extends even to the physical objectivation of the tablet, that is, it is now and then even used for writing when it is not even switched on, namely as a pad by C-F-18:

> *C:* . . . what comes to my mind is that you use the iPad as a writing surface when the surface is bad.
>
> *Exp:* Can you describe that in more detail?
>
> *C:* Well, for example, there are different desks that are not as smooth as here, or you are somewhere in the schoolyard or something. Then you need a reasonably smooth surface to write on.
>
> *Exp:* Okay, so like a clipboard?
>
> *C:* Yes [laughs]. (C, 60–65)

At the end of the interview, C even starts talking about the "tablets" when she means the students using tablets: "towards the end of school time it was actually the rule that almost the whole class consisted of tablets" (C, 110–112). This embodiment is particularly impressive given the short time tablets had generally been used in schools at the time of these accounts and the potential multi-functionality of the devices in the classroom.

DISCUSSION: DROPPING IN, DRAWING OUT

In this study, we explored the practice of using tablets in schools through in-depth interviews with four early adopters. Our aim was to predict how this practice is now sedimenting into widespread use. First of all, there is a loss of technological practices compared to an even earlier, teacher-driven phase of tablet use. In his case study on the digitalization of the education system in Germany from 2011 to 2021, Markus Bohlmann describes a phase of tablet use by early teacher pioneers (Bohlmann 2022, 192f). A variety of apps were used here, "Garageband" in music lessons, "VideoDelay" in sports lessons, and virtual memory sites in history lessons. These had a subject-matter-specific function. The tablet use of the early adopters from this study, on the other hand, must already be understood in the transition to widespread use. Here, the tablet must fulfil a function in every subject and be able to be used substitutively even where the teacher has not at all adjusted the lesson planning for its use. Our interviewees described this in the context section of the interview:

> I also had a Latin teacher who is a bit older and he was always the type who printed it out for everyone and didn't really care. He didn't make any additional effort or provide any links or videos or anything like that, because he wasn't interested in it. (B, 62–68)

However, it would be wrong to say that the teachers were not attuned to the use here and that it was therefore deficient in some way. Instead, the tablet use described here is a technological practice in transition, while other technological practices still exist. In *Technology and the Lifeworld* Don Ihde introduces path dependency of technologies with the example of medieval European buildings, often perceived as particularly authentic but actually built from reused antique stones (Ihde 1990, 192). Thus, the human–technology relations established through tablet use here also seem very path dependent and oriented towards already existing technologies in school, especially the college notebook. It was particularly impressive for us to see how our interviewees described the management functions and "tidying up" with their tablets as distinct from the messiness of the college notebook. On the other hand, they used the digital tablet as a substitute for the college notebook to such an extent that they embodied this practice. They even did not play digital games, surf the internet, or scroll social media to pass time in class—no, they were doodling and drawing. It is too early to say that the practice of tablet use as a mere writing device is deficient, but at first glance they seem to fall short of the didactic potential of all the different apps used by tablet pioneers in the very first years of tablet use in school (Bohlmann 2022, 192f).

However, there are indications from this small exploratory study that another widespread hope in the German educational system, the hope for active and self-directed learning with tablets, will probably not be fulfilled (see Aufenanger 2020, 31; KMK 2021, 11). No such practice has been established by early adopters; significant is our interviewee D-F-21's description of how she even put the tablet away and went back to flashcards and Post-its when she was studying for the Abitur, the final exams of the Gymnasium in Germany. No human–technology relation is established in active learning here. Perhaps practices of memorizing are rather spatial in nature and therefore not suitable for the flat screen. But that is only a guess.

A further and, from a normative point of view, more hopeful indication is the positive community building of a culture of "commons," as described by sociologist Felix Stalder (2016, 245). Through sharing in class and the practice of "airdropping," tablet use is community-building from a postphenomenological perspective. Now that the social problem of tablet financing has been solved in many places in Germany, the "two-class society" described by our interviewees should probably no longer be of any significance; the community-building function of sharing materials, tasks, and solutions remains. Here, the educational system will have to come up with solutions for open forms of sharing worksheets, papers etc. in class, especially in view of the historically given data protection sensitivity and copyright legislation in Germany.

In conclusion, we would like to summarize the results of our study in an image that echoes the two movements we described. Community building through "airdropping" involves tablet users in a new and shared technological practice (dropping in). However, the practice of taking notes, recording, drawing, and organizing as with the college notebook, which is also described by our interviewees, points out of the tablet practice to other important knowledge practices of school learning (drawing out). A very brief postphenomenological description of the tablet use of early adopters and the sedimentation of this human–technology relation might then be this image: dropping in, drawing out.

NOTES

1. We quote here first the Blackwell edition translated by John Macquarrie and Edward Robinson, but in brackets the German pagination of the seventh edition, which is laterally identical to the first edition of 1927.

2. According to Don Ihde, Heidegger sees "stability" through assignment, but not "multistability," the many possible contexts of use of each technology (Ihde 1999, 46f). For Ihde, display screens are an example of multi-stability in demarcation to Heidegger (Ihde 2010, 133).

3. A common postphenomenological criticism is that Heidegger conceives "equipment" not only as a single stability, but this stability if functional always as also *transparent*. Rosenberger and Verbeek say in their *Field Guide*: "Heidegger holds that a hammer can only ever be experienced as transparently embodied or as broken" (Rosenberger and Verbeek 2015, 28). Following Whyte (2015), many postphenomenologists today see the "pivot point" at which an investigation starts as decisive for the number of stabilities and their degree of transparency. Rosenberger, however, assumes that there is at least one "dominant stability," which then has to be freed from its transparency through "critically contrasting stabilities" (Rosenberger 2021, 383) found by postphenomenological investigations.

4. Peter-Paul Verbeek says "Heidegger does not fully succeed in developing an adequate hermeneutical perspective on technology, for he reduces technology to its conditions of possibility and thereby fails to connect with specific technologies" (Verbeek 2005, 95). Verbeek sees this hermeneutic narrowing especially in Heidegger's later philosophy after the *Kehre*. However, in a hermeneutic analysis that moves away from the human-technology relation, as in the example here, this important point of criticism from a postphenomenological perspective can also be applied to the early Heidegger.

5. In this case, one of us assessed whether an answer was correct or incorrect and the other proved the rating just by counterchecking. Since this was only a preliminary study, we did not use a more objective method of assessment as the test of inter-rater reliability in the main study.

6. An answer was counted as missing if one of the 23 students in the pre-study could not give any example of an aspect at all.

7. In the case of quotations from interview passages, the respective line number of the transcript is given. We conducted and transcribed the interviews in German and then translated the transcription into English.

8. In the spirit of a culture of research data sharing, the anonymized transcripts are freely available for download by researchers from the authors' *ResearchGate* profiles.

9. Inter-rater reliability was tested per individual code. If in one of the rows where rater A set a code, a code of the same type could be found by rater B, the match was scored 1, if no code or another type was found, 0 was scored.

10. This is sometimes referred to as a first-order digital divide, the unequal access to digital technology, as opposed to the second-order digital divide, the unequal use of digital technologies (Riggins and Dewan 2005).

REFERENCES

Aufenanger, Stefan. 2020. "Tablets in Schule und Unterricht—Pädagogische Potenziale und Herausforderungen." In *Mobile Medien im Schulkontext*, edited by Dorothee M. Meister and Ilka Mindt, 29–45. Wiesbaden: Springer Fachmedien Wiesbaden. https://doi.org/10.1007/978-3-658-29039-9_3.

Bakker, Arthur. 2018. *Design Research in Education: A Practical Guide for Early Career Researchers*. London: Routledge.

Bohlmann, Markus. 2022. *Bildung—Philosophie—Digitalisierung. Eine Curriculumtheorie. Reihe Digitalitätsforschung Band 3*. Edited by Jörg Noller, Malte Rehbein, and Sybille Krämer. Berlin, Heidelberg: J.B. Metzler. https://doi.org/10.1007/978-3-662-65792-8.

Borgmann, Albert. 2003. "The Invisibility of the Culture of Technology." In *Power Failure. Christianity and the Culture of Technology*, edited by Albert Borgmann, 11–24. Grand Rapids: Baker Publishing Group.

Dedehayir, Ozgur, Roland J. Ortt, Carla Riverola, and Francesca Miralles. 2017. "Innovators and Early Adopters in the Diffusion of Innovations: A Literature Review." *International Journal of Innovation Management* 21 (08): 1740010. https://doi.org/10.1142/S1363919617400102.

Feenberg, Andrew. 1991. *Critical Theory of Technology*. New York: Oxford University Press.

Heidegger, Martin. 1962. *Being and Time*. Translated by John Macquarrie and Edward Robinson. New York: Harper & Row.

Ihde, Don. 1990. *Technology and the Lifeworld. From Garden to Earth*. Bloomington: Indiana University Press.

———. 1999. "Technology and Prognostic Predicaments." *AI & Society* 13 (1): 44–51. https://doi.org/10.1007/BF01205256.

———. 2009. *Postphenomenology and Technoscience. The Peking University Lectures*. Albany: State University of New York Press.

———. 2010. *Heidegger's Technologies: Postphenomenological Perspectives*. New York: Fordham University Press.

———. 2012. *Experimental Phenomenology, Second Edition: Multistabilities*. Albany: State University of New York Press.

KMK. 2017. "Strategie der Kultusministerkonferenz 'Bildung in der digitalen Welt'. Beschluss der Kultusministerkonferenz vom 08.12.2016 in der Fassung vom 07.12.2017." https://www.kmk.org/fileadmin/Dateien/pdf/PresseUndAktuelles/2018/Digitalstrategie_2017_mit_Weiterbildung.pdf.

———. 2021. *Lehren und Lernen in der digitalen Welt. Die ergänzende Empfehlung zur Strategie "Bildung in der digitalen Welt."* Bonn: KMK. https://www.kmk.org/fileadmin/Dateien/veroeffentlichungen_beschluesse/2021/2021_12_09-Lehren-und-Lernen-Digi.pdf.

Krämer, Sybille. 2022. "From Dissemination to Digitality: How to Reflect on Media." *Media Theory* 5 (2): 79–98. https://journalcontent.mediatheoryjournal.org/index.php/mt/article/view/147.

Landis, J. Richard, and Gary G. Koch. 1977. "The Measurement of Observer Agreement for Categorical Data." *Biometrics* 33 (1): 159–74. https://doi.org/10.2307/2529310.

Mayring, Philipp, and Thomas Fenzl. 2019. "Qualitative Inhaltsanalyse." In *Handbuch Methoden der empirischen Sozialforschung*, edited by Nina Baur and Jörg Blasius, 633–48. Wiesbaden: Springer Fachmedien Wiesbaden. https://doi.org/10.1007/978-3-658-21308-4_42.

Pinch, Trevor J., and Wiebe E. Bijker. 1987. "The Social Construction of Facts and Artifacts: Or How the Sociology of Science and the Sociology of Technology Might Benefit Each Other." In *The Social Construction of Technological Systems. New Directions in the Sociology and History of Technology*, edited by Wiebe E. Bijker, Thomas P. Hughes, and Trevor J. Pinch, 17–50. Cambridge: MIT.

Riggins, Frederick J., and Sanjeev Dewan. 2005. "The Digital Divide: Current and Future Research Directions." *Journal of the Association for Information Systems* 6 (12): 298–337.

Rogers, Everett M. 1962. *Diffusion of Innovations*. New York: Free Press.

Rosenberger, Robert. 2012. "Embodied Technology and the Problem of Using the Phone While Driving." *Phenomenology & the Cognitive Sciences* 11 (1): 79–94. https://doi.org/10.1007/s11097-011-9230-2.

———. 2021. "Technological Multistability and the Trouble with the Things Themselves." In *The Oxford Handbook of Philosophy of Technology*, edited by Shannon Vallor, 374–92. Oxford: Oxford University Press. https://doi.org/10.1093/oxfordhb/9780190851187.013.42.

Rosenberger, Robert, and Peter-Paul Verbeek. 2015. "A Field Guide to Postphenomenology." In *Postphenomenological Investigations: Essays on Human–Technology Relations*, edited by Robert Rosenberger and Peter-Paul Verbeek, 9–42. Lanham, Md.: Lexington Books.

Russell, Bertrand. 1912. *The Problems of Philosophy*. London: Thornton Butterworth.

Stalder, Felix. 2016. *Kultur Der Digitalität*. Frankfurt a.M.: Suhrkamp.

Tafdrup, Oliver Alexander, Bjarke Lindsø Andersen, and Cathrine Hasse. 2019. "Learning to Interpret Technological Breakdowns: A Path to Technological Literacy." In *Human-Centered Digitalization and Services*, edited by Marja Toivonen and Eveliina Saari, 139–52. Singapore: Springer Singapore. https://doi.org/10.1007/978-981-13-7725-9_8.

Verbeek, Peter-Paul. 2005. *What Things Do. Philosophical Reflections on Technology, Agency, and Design*. University Park: Pennsylvania State University Press.

———. 2011. *Moralizing Technology: Understanding and Designing the Morality of Things*. Chicago: University of Chicago Press.

Whyte, Kyle Powys. 2015. "What Is Multistability? A Theory of the Keystone Concept of Postphenomenological Research." In *Technoscience and Postphenomenology*, edited by Jan Kyrre Berg, Olsen Friis, and Robert P. Crease, 69–81. Lanham: Lexington Books.

Chapter 6

Fugitive Pathways

Sensors, Lines, and Knots in the Academic Library

Lesley Gourlay

The genealogy of the concept of the *postdigital* as presented in Jandric et al. (2022) can be traced back to Negroponte (1998). As I have explored elsewhere (Gourlay 2023), there are three main ideas inherent in his notion which are in tension; that the digital and analogue are entirely intertwined as a single entity, that it is also possible for the digital to be absent, and thirdly that the digital is only noticed by its absence. I propose that this imaginary is one of an entity which is at once fully permeating, ubiquitous, but also occluded. This may lead to a recast of essentially transhumanist fantasies of disembodiment, immateriality, and transcendence of the limitations of the biological human substrate. I will argue in this chapter that a postphenomenological approach allows for an opportunity to (re)gain much-needed theoretical purchase on the material, ephemeral, co-present, and embodied essences of the postdigital as lived experience. Building on previous work (Gourlay 2021a), I will present a heuristic-driven analysis following Adams and Thomson's (2016) approach to 'researching posthuman worlds'. In my analysis, I propose that campus-based digital surveillance technologies can be analysed as constituting elements of a broader culture of surveillance (Lyon 2018), linked to previous analysis of academic faculty individual bibliometric self-surveillance (Gourlay 2022). I go on to propose that postphenomenology, with its sensitivity towards the *moment*, is a methodology which may offer insights into the subtleties of the ephemeral, material, co-present embodied unfolding of higher education, in terms of not only lived epistemologies, but also the more-than-human ontologies of students and academic faculty, and the agency and liveliness of nonhuman actors. The chapter will conclude by building on recent work on postphenomeonology and digital technologies in education (e.g. Johnson 2020; Bonderup-Dohn 2022) with suggested implications of this analysis for theory and research into digital higher education, discussing the important differences between how posthumanism and postphenomenology theorize agency, and will also considering how this critique might contribute to the development of postphenomenology for future research into technologies in education.

THE 'POSTDIGITAL' UNIVERSITY

As I have argued elsewhere (Gourlay 2023), the notion of the 'postdigital' is one which requires close attention and, in my view, further theorization. One particular aspect of this idea is that digital technology has fully permeated all areas of life, to the point at which it can be claimed to be *inseparable* from the analogue. Put another way, the contention seems to be that there is no longer any such thing as the analogue existing separately from digital technology. This raises a series of questions regarding what it is we are in fact referring to with the two terms. In this chapter, I will explore this question in terms of three elements of being and practice: materiality, ephemerality, and copresence. I will argue that these characterize the analogue in particular ways, and also that they are undermined or even elided by the digital. Before that, I will make a case for what I consider to be an underpinning ideology concerning the digital in higher education.

The trope that digital technology allows us to transcend physical, biological, temporal, and geographical limitations is a pervasive one in society as a whole, and also in higher education. This claim is not without foundation, as clearly communication technologies in particular provide messaging, interaction, and broadcasting at distance in a range of ways that pre-digital technologies do not, and as such offer a great deal of convenience and potential. It might be argued that the mass pivot to online engagement during the COVID-19 pandemic has led to what might be termed a *discourse of inevitability*, in which the physical campus and face-to-face teaching is called into question. This can be seen in the use of the 'flipped classroom' approach, in which lectures are pre-recorded online, even in settings where the students are not at a distance but are attending the campus for other classes, such as seminars. This ethos can also be observed in the case of online academic conferences and seminars, beyond the period of lockdowns. What was a measure introduced in response to an emergency has arguably become mainstream. There are clearly a range of arguments for online engagement in terms of convenience, flexibility, environmental concerns, and so on. However, the underlying ethos, I propose, is that there are effectively only negligible differences to be concerned about, and that due to the capacities of technologies such as *Teams* and *Zoom*, all aspects of the educational encounter can be replicated, or at least approximated, online instead of face-to-face.

It is worth reflecting on what lies behind this notion. During the pandemic, taking the United Kingdom as an example, universities were strongly incentivized to avoid liability for refunds of student fees, and therefore it was expedient for them to present fully online teaching as equivalent in terms of 'quality'. It seems probable that this narrative was also propagated to maintain faculty morale, and to seek to convince students and other stakeholders that their education would not be damaged. It might be speculated that UK universities, in a financially dysfunctional system, may now post-pandemic be incentivized to hold lectures online in order to reduce costs and resource demands on campus. However, I would contend that this move to online is also driven by a broader underlying notion about human embodiment and materiality more generally. The discourse of 'Ed Tech' has for decades exhibited a tendency towards utopianism and visions of transcendence of human boundaries, which I would suggest have their roots in an implicit *transhumanism*; a worldview which seeks to augment and extend human biological limitations via technologies. The prototypical fantasy of the transhumanists is arguably the literal cyborg, defined by Webster's dictionary as 'A human being whose body has been taken over in whole or in part by electromechanical devices' (Webster 2023). It is instructive to observe the results of a Google image search for the search term 'AI in Education', which demonstrates the pervasive use of images of humanoid robot figures suggesting the central importance in these imaginaries of the human body in this imaginary. The images tend towards a science fiction sensibility, with what appear to be highly advanced cyborgs or humanoid robots depicted, which are in many cases not in current existence. The 'blue head with chips' image is particularly prevalent; and was in fact chosen as an infographic for one of my own papers—which can be viewed open-access (Gourlay 2021b)—which was (ironically) devoted to making exactly the opposite point. The head in the 'cyborg' image is transparent, the colour blue is used which does not appear in nature on human skin, and there are images of lights or strands emanating from the brain of this imagined being.

It might be argued that these are merely journalistic or illustrative tropes, but it is striking how the imagined effect of digital technologies on the human body and mind is placed repeatedly at the centre of these images. Extending this analysis, in the case of online engagement, the human body of the student may be invisible if video is not used or is rendered as a video image of the upper body and head, in the style of a portrait photograph, (upper bodies only are also used as avatars in the Metaverse). This is of course a common-sense approach to allow interlocutors to see one-another's faces. However, as I and colleagues found (Gourlay et al. 2021; Littlejohn et al.

2021) in a study of the impact of the UK pandemic lockdowns on academic faculty, there is a prevalence of 'staging' and 'clearing up' in order to present a particular 'professional' version of oneself, and also to conceal objects and people which are associated with private and domestic space, in a process I have suggested is a form of *symbolic hygiene*, with the screen operating as a *digital mask* (Gourlay 2022). Here, one's embodied self is not enhanced like the cyborgs in the images above; but is instead either concealed, disciplined, or rendered into an image. The material setting may also be de-emphasized by being staged or hidden by a background, or the interlocutor may engage while on transport or in any other location with a wifi connection or phone signal. Clearly, the lack of physical copresence of people is the defining feature of this form of engagement, as they are separated geographically. All of this, I suggest, speaks to a belief that embodiment, materiality and ephemerality are aspects of human interaction which can and even should be downplayed, and ephemerality and co-presence – being together physically in the same place – may be dispensable. This, I contend, is of a piece with a transhumanist ideology which seeks to displace these aspects of the human embodied and social experience.

SURVEILLANCE CULTURES

The previous section discussed the effects of digitization on the nature of interaction, touching on themes of visibility, invisibility, and self-disclosure. I argued that the human subject is rendered less visible and less present in online interaction. In that regard, the technology is replacing aspects of embodied, ephemeral, and co-present interaction such as bodily presence, gesture, stance, and evolved human reactions to being together with others in the same space. If these are seen as optional, or in a more extreme version, problematic aspects of subjectivity to be cleaned up and elided, then clear echoes of the ethos of transhumanism, which regards the human body as deficient, can be detected. The affective responses of humans to the company of others, and to the experience of being in a co-present gathering are therefore relegated to practical obstacles to be removed, arguably resulting in a stripped-out, sterile, and transactional form of controlled and mediated educational encounter. This is also rendered devoid of the rituals, textures, historical depth, continuity, and aesthetic rewards of being in a university campus.

However, paradoxically, while the effect of digitally mediated education renders the human subject less visible or tangible, it simultaneously enrols them in various forms of what Lyon calls 'the culture of surveillance' (Lyon 2018). Both online and in the physical campus, a range of digital technologies are deployed to monitor students' presence, whereabouts, and activities. Lyon differentiates between surveillance and *the culture of surveillance*, the latter being a more complex relationship in which the surveilled parties consent to surveillance in order to garner some gain or advantage, such as in social media use. The institutional context of the university represents a borderline case, however. Students may have given legal consent to be subject to the technologies which monitor them, but it might be argued that this is not consent given freely if there is no opt-out which allows for them to participate as a student unsurveilled. This paper will go on to consider an example of digital surveillance in the university; but will first set out how a postphenomenological methodology will be used.

THEORY AND METHODOLOGY

'Phenomenology is about wonder, words, and world' (van Manen 2014: 13). Van Manen proposes a means of inquiring about ' . . . the quotidian practices of everyday life' (2014: 15). As elaborated elsewhere in this volume, postphenomenology allows us to examine the relations between humans and technologies, as applied to 'real-life' contexts of practice (Ihde 1990). For Ihde, classical phenomenology lacked ' . . . a materially sensitive post-phenomenology' (Ihde 2018: viii), plus a recognition that all science is *technoscience*, involving material implementation. As Aagaard et al. (2018) set out, in methodological terms, postphenomenology has been critiqued for an overreliance on autoethnography (e.g. Mol 2010), and a related overemphasis on the perspective of the individual (Hasse 2015). They point the way in their volume for greater use of the empirical methods of social science in postphenomenology.

This relatively new movement in the philosophy of technology provides a rich but hitherto somewhat under-utilized theoretical and analytical resources to studies of technologies in education, potentially linking to the nascent perspective of sociomaterial and posthuman studies of digital education, with their related focus on human–non-human–material–technological relations and emergent agencies (e.g. Fenwick and Edwards 2010, 2012; Gourlay 2018). Adams and Turville (2018) provide an historical overview of postphenomenological studies of educational technologies. They define postphenomenology as ' . . . phenomenology that attends to specific technologies and the existential and epistemological difference they may be making to the lifeworld' (Adams and Turville 2018: 4). Drawing on *phenomenography of practice* (Adams and van Manen 2008; van Manen 2014), proposing a *postphenomenology of practice*. They identify five heuristics. These are (1) using autoethnography to conduct phenomenologically sensitive explorations, (2) observations of human action in interaction with technologies, (3) exploring lived experience through interviews, (4) using the concept of *epoche-reduction* to restore a thing to its wholeness, and (5) studying breakdowns to reveal what is taken for granted.

Aagaard (2017) provides an introduction to postphenomenological research in educational technology, which he draws on to provide a greater sensitivity to materiality, in the context of 'sociocentric' education research in which the material has a tendency to be 'blackboxed' (Walz 2004). He reviews *descriptive phenomenology* (Giorgi 2009) and *hermeneutic phenomenology* (van Manen 1990) and provides an overview of how Ihde's (2009) subsequent *postphenomenology* differs from these previous orientations in three ways:

> First, it is not concerned with immaterial subjectivity, but with bodily relations to technologies. Embodiment thus replaces Husserlian consciousness and its 'disembodied view-from-nowhere' (Ihde 2008: 3). Second, it is committed to the notion of multistability, which ontologically replaces the essences of Husserlian phenomenology and refers to a technology's various partially determined trajectories in different contexts (Ihde 2009). A lighter, for instance, is usually used to light things like candles or cigarettes, but can also be used to open bottles. There is thus no 'essential' use of the lighter. New technologies like laptops, tablets, and smartphones are even designed to incorporate such multistability (Ihde 2012). Finally, Ihde (2009: 63) holds that hermeneutic phenomenology remains under the influence of a linguistic turn in twentieth-century continental philosophy in which 'the silent privilege of the linguistic continued to hold sway'. Ihde replaces this postmodern world-as-text metaphor with a perceptual-bodily referentiality. The anchor in postphenomenology, in other words, is embodiment. This bodily focus owes a great debt to Maurice Merleau-Ponty. (Aagaard 2017: 525–526)

He reminds us that Ihde's postphenomenology emphasizes how technologies *mediate* our being in the world and that technologies are not neutral but are in fact *transformative* of our experiences (Verbeek 2005). Aagaard highlights how Verbeek (2011) linked postphenomenology to Actor-Network Theory (AN):

> Against the modern image of the autonomous subject, postphenomenology emphasizes an amodern, heteronomous subject whose comportment is always closely interwoven with the material environment in which it plays out (Verbeek 2011). This technological mediation of intentionality, however, does not imply an elimination of human freedom. Technologies do not determine what we do: They set up spaces of possibilities that enable and constrain certain actions and perceptions, but the room for maneuver within these spaces is what constitutes human freedom. Human freedom is not an absence of technological influences, but the practice of coping with such influences. 'Rather than clinging to a view of human freedom as absolute autonomy and sovereignty from technology, it

> seems wise to reinterpret freedom as a person's ability to relate to what determines and influences him or her' Verbeek 2011: 156). (Aagaard 2017: 527)

Aagaard likens postphenomenological analysis to craft, which explores ' . . . what things do' (Verbeek 2005), and as such it might be aligned with new materialism. In particular, he draws out how postphenomenology might resolve what Kalthoff and Roehl (2011) have characterized as a *semiotic indifference* generated by ANT's 'pitfall of symmetry', in which radical material-semiotic approaches lead to the notion of an undifferentiated 'semiological soup' (Scharff 2006: 136, in Aagaard 2017: 527). Humans are reinstated at the core of the theory, in a posthumanist sense which is sensitive to the distributed human-nonhuman nature of agency. He concludes that while ANT allows for an analysis of complex networks of relations viewed from a third-person perspective, postphenomenology allows for an exploration of relations between humans and technologies 'from within' (Rosenberger and Verbeek 2015).

As a research method, Aagaard suggests two possible approaches: (1) in-depth exploration of the typical use of a particular technology and (2) critical comparison of multiple versions of a technology. He draws on Adams and Thompson's (2011) *interviewing objects* approach, which bases itself on postphenomenology, ANT, media ecology, and hermeneutic phenomenology to investigate technologies and their material agencies. The second approach is illustrated with Roserberger's (2014) *variational cross-examination*, which looks at the multiple stabilities of a technology, looking at features such as contextual embedment, material make-up, and relation to embodied users. Using the example of a public bench, he explores how benches may be used to sit on or sleep on, but how some might be designed to prevent homeless people using them for sleep. This type of contrasting, he points out, has been used in educational research by Mangen's analyses of digital versus analogue books (Mangen and Balsvik 2016; Mangen and Kuiken 2014).

Adams and Thompson develop further their *interviewing objects* approach and associated heuristics in their book *Researching Posthuman Worlds* (2016), which I have used in a previous book-length analysis of posthumanism in the digital university (Gourlay 2018). In this chapter, I will use *interviewing objects* to explore technologies, which enable the surveillance cultures in higher education, looking at a particular instance of campus-based surveillance technologies. They base the approach on the etymology of the word 'interview', which has its roots in the old French verb *s'entrevoir*, a word which has two elements, '*entre*' meaning between or mutual, and '*voir*', to see. For them, 'To inter-view an object or thing is therefore to catch insightful glimpses of it in action, as it performs and mediates the gestures and understandings of its human employer, and as it associates with others'

(Adams and Thompson 2016: 17–18). They use eight heuristics: (1) gathering anecdotes, (2) following the actors, (3) listening for the invitational quality of things, (4) studying breakdowns, accidents, and anomalies, (5) discerning the spectrum of human–technology–world relations, (6) applying the laws of media, (7) unravelling translations, and (8) tracing responses and passages. These are based on various theoretical perspectives such as Actor-Network Theory and Idhe's (1990) postphenomenology of technics; and lend themselves to different types of analysis depending on the technology and setting under investigation.

For the purposes of this chapter, I will use the heuristic of 'tracing responses and passages', which they base on the work of social anthropologist Tim Ingold. Specifically, they make use of his concept of *co-respondence* (Ingold 2012a) to examine how human actors interact with nonhuman actors around them, and the nature of passages, as they put it ' . . . as entities thread their way through the ways of others' (Adams and Thompson 2016: 82). They refer to Ingold's *meshwork* defined in contrast to network, ' . . . the lines of the meshwork . . . are the trails along which life is lived. And it is in the entanglement of lines, not in the connecting of points, that the mesh is constituted.' (Ingold 2006: 26). They also refer to Ingold's concept of *improvising passages* (Ingold 2012b), where passages form lines in the meshwork. When more permanent practices gather at a certain point, contingent *knots* can be made. Ingold does not frame these concepts as phenomenology, but it is used by Adams and Thompson in a way which provides a means by which to investigate technologies within a loose family of ideas which can be used in postphenomenological research. In a recent work, Ingold (2023) addresses the question of what it would mean for an anthropologist to adopt a phenomenological stance, commenting on the nature of ethnographic writing, which he also portrays as part of a meshwork:

> What is the purpose of writing, after all, if not to allow readers to join in your deliberations; to open a path that they too can follow? This, at the end of the day, distinguishes truly phenomenological writing from the more conventionally ethnographic. Ethnography is writing about; it draws on what the people in your study have told you for what it tells about them. Here, lessons learned in the field become data for analysis; what readers encounter is the final outcome. Phenomenology calls for a different way of writing. It is a writing *with*, which welcomes others into the space of our deliberations as the hand, holding the pen, scours the mind for every next word. This is to join our own lines with the writing of the world, whether with the paths of human inhabitants as they find their ways around, or the tracks of animals, or the meandering vegetation. And just as our minds mingle with the world in writing, so the minds of readers mingle in turn with ours. All these lines, of living, thinking, writing, and reading, are braided in a meshwork which ravels and unravels as it goes along. In this way,

> life – as scholarship – does not progress towards a conclusion but carries on. And phenomenological inquiry, as I understand it, is dedicated thus, to the continuity of life. (Ingold 2023: 724–725)

Although Ingold is referring to a phenomenological sensibility towards fieldwork, this can equally be applied to using Adams and Thompson's heuristic to consider the nature of human engagement with technologies without direct observation. In the next section, I will consider one example of digital technology used in higher education – infra-red sensors – looking at it in terms of *tracing responses and passages*.

SMART CAMPUS SURVEILLANCE TECHNOLOGIES: OCCUPEYE

In the contemporary physical campus in socio-economically developed contexts, forms of surveillance technologies are commonly used to monitor the movement of people via CCTV cameras. Entrances to buildings are frequently controlled and recorded using electronic turnstile and access card technology. These are relatively well-established in the public sphere, and so may be regarded as neutral, to-be-expected technologies focused on safety and ensuring only those with appropriate credentials may enter buildings. Arguably, students and faculty may not find the presence of these technologies to be particularly noticeable or salient. However, there are other forms of digital surveillance technologies which may be more contentious in the contemporary 'smart campus', among them workplace sensors which use passive infrared (PIR) technology to monitor room and desk utilization, such as OccupEye, which has been deployed in some university libraries and staff offices. The manufacturers describe the product as follows:

> OccupEye, an FM:SystemsCompany, is an advanced sensor-based workspace monitoring solution, designed to accurately and efficiently analyse all types of room and desk utilization to a level of detail never before possible. OccupEye allows Facilities and Real Estate managers from both public and private sectors to make significant financial savings, while also enhancing staff experience. Through our state-of-the-art web-based dashboard, users enjoy complete control over data reporting and analysis. Various levels of detail can be applied to each sensor, allowing for quick and easy performance comparison in line with your project requirements. Our sensors are lightweight, non-intrusive and portable, allowing for flexible deployment in monitoring both room or individual workspace time after time in different buildings and offices; this makes OccupEye an excellent choice for any organisation. OccupEye 2023)

OccupEye has proved to be highly controversial when installed in workplaces and was removed from the offices of the UK newspaper *The Daily Telegraph* in 2016, after objections from the National Union of Journalists (The Guardian 2016). However, despite the controversy surrounding its use at this and other workplaces, OccupEye is used in some university libraries, including at University College London (Hammond 2018; Wu 2020). The occupancy of the seats in the library is visible to relevant UCL staff via a digital dashboard and is also linked to an app (UCL Assistant) allowing library users to see which seats are free. The rationale for the use of OccupEye is that it allows library staff to monitor usage patterns, and if used via the app, allows students to see where seats are available in the library in advance of going there. The infrared sensors detect the heat of the body, as opposed to other objects such as bags, to ascertain if the seat is occupied or not.

This technology can be analysed in terms of *responses* and *passages*, following Ingold, also with the notion of the *meshwork* in mind. For example, a student may be walking down the street near the campus planning to go to the library. She is forming a line in the meshwork by walking. She takes out her smartphone and checks the UCL Assistant app. At that moment, the student, the device, and the software could be said to form a *knot*, in Ingold's terms. The student is able to see a visual representation of the library seating plan, showing which desks do not have a human body sitting at them. Meanwhile, simultaneously, hundreds of students sit at desks in the library. They breathe, they hunch over, they shift in their seats, they touch their devices, they hold books. They look up, look around, see a friend, smile, or chat in whispers. They have all separately arrived at the library, some using public transport, then walking, forming lines in the *meshwork*, criss-crossing across London to converge on this large quiet room with rows of desks. Occasionally someone gets up to use the bathroom or to get a drink. The OccupEye sensor under their desk will register the desk as 'absent' meaning unoccupied, after 30 minutes.

The student in the street nearby looks at her phone and sees the library is almost full, but after some searching through various screens, she finds an empty desk. She immediately starts walking more quickly, while continuing to monitor the app, glancing occasionally at the screen to ensure the desk has not been occupied by another student. In her mind, she imagines other students also looking at the free desk on the app and converging on it. She reaches the entrance of the library, and reaches into her bag for her student card, which she uses to open the electronic turnstile, then puts the card around her neck with a lanyard to wear. She walks quickly to the lifts, then goes up to the floor where the free desk is located. She comes out of the lift, and walks intently over to the desk, which is still free. She sighs with relief and immediately sits down, placing her body on the seat in the knowledge that her biological form will activate the sensor and 'claim' the desk. In a room used by library staff, on the dashboard used by staff members responsible for monitoring occupancy, the desk shows as occupied. Across the university, thousands of students have access to the app, and if they look at it, they will now see that the desk is now taken. This, in Ingold's terms, might be termed another *knot* in the meshwork, plus a new mesh of lines which spread out to the app users. The student's body, the desk, the sensor, the software, and the app entangle at that moment, and stay in this provisional and temporary *knot* until it is untied by the student leaving the library.

As discussed above, Rosenberger (2014) proposes as a form of postphenomenological research *variational cross-examination*, which looks at the multiple stabilities of a technology. A slightly different analysis might look at responses and passages if the technology were not present. In the alternative scenario, the student is walking down the street, forming a line by walking. She does not take out her smartphone as there is no way of knowing whether there are any free desks in the library. She enters the library at a leisurely pace, walking around various floors and areas looking for a desk. She sees a friend and waves, then goes over to have a whispered chat. They arrange to go for coffee later. She continues to walk around and spots a book which interests her in the stacks. She takes it out, browses it and makes a note of its title to come back to later. She looks out of the window, noticing the light on the trees outside. She hears the sounds of another student packing their books and laptop, turns around, and sees a student is leaving his desk. She hangs back, he sees her, and they exchange a smile, he leaves and she sits down.

DISCUSSION AND CONCLUSIONS

As suggested at the beginning of this chapter, transhumanist fantasies of disembodiment underpin much thinking related to the use of digital technologies in education. In the case of the use of OccupEye, the body of the student is simultaneously foregrounded, elided, surveilled, monitored, and documented. When the body of the student enters into the knot of the meshwork, she becomes a highly complex entity. The technology seeks to circumvent any need for embodied co-presence for students seeking a desk, in simple terms, they do not have to walk around in the library looking directly at the desks and students. The 'eye' of OccupEye acts as a form of prosthesis, which actually senses heat rather than 'sees' the bodily presence. The technology is justified in terms of convenience, and the avoidance of potentially 'wasted time' looking for a desk, and as such it operates against a neoliberal logic that the student is simultaneously a customer whose needs must be met efficiently; and is also a person in a hurry who needs to avoid the redundancy of walking around several areas of the library looking for a desk. In terms of Lyon's culture of surveillance, it can be argued that the students in some sense derives a benefit to being subject to this surveillance, in exchange for the convenience it offers. However, as suggested above, the extent to which this can be regarded as consent is open to question, as the student in the library cannot opt out of being sensed and documented by OccupEye.

In terms of Ingold's concept of *passages*, as the imagined counter vignette above suggests, the wavy, longer, and slower line of the student walking around seeking a desk leads to several other *knots* in the meshwork being formed, knots which consist of copresent embodied humans meeting in an ephemeral manner face-to-face, seeing each other, talking, and forming relationships unmediated at that moment by digital technology. Arguably, this form of passage had more in common with Ingold's *wayfaring*, while the efficient, direct, and hurried line laid out by the OccupEye working with the app is closer to his *navigating* (Ingold 2007). The nature of OccupEye contains internal contradictions; it is profoundly concerned with the biological human in that it detects the heat of the body, but also renders the body into digitally documented data, obviating the need for evolved action of 'looking for' a place. The enactment of what I would argue is a fundamentally transhumanist ideology seeks to transcend the limitations of the biological human body, and in doing so, radically shifts the nature of the meshwork in and around the library, stripping out passages and pathways which are what I have termed *fugitive* (Gourlay 2023); existing in a mode in the meshwork which is embodied, copresent, and ephemeral, and 'under the radar' of the even more pervasive lines cast out by digital surveillance technologies. Undeniably, the gains offered by the technology are real, in that the 'risk' to students of spending a longer time searching for a desk, or even falling to secure one, is minimized. However, the losses of avoiding a wayfaring passage are subtle, less obvious, but I would argue crucially important. Serendipity, hesitation, waiting, changed plans, noticing a book by chance, the sense of uncertainty and incompleteness are all compromised or lost. And what is left is arguably an impoverished and stripped-out version of the potentially rich, messy, unpredictable, and fundamentally embodied campus.

ACKNOWLEDGEMENTS

This research is funded by a Leverhulme Trust Major Research Fellowship MRF-2020-135.

REFERENCES

Aagaard, Jesper. 2017. "Introducing postphenomenological research: a brief and selective sketch of phenomenological research methods." *International Journal of Qualitative Studies in Education* 30(6): 519–533.

Aagaard, Jesper, Jan Kyrre Berg Friis, Jessica Sorenson, Oliver Taftdrup, and Cathrine Hasse. 2018. "An introduction to postphenomenological methodologies." In *Postphenomenological Methodologies; New Ways in Mediating Techno-Human Relationships*, edited by Jesper Aagaard, Jan Kyrre Berg Friis, Jessica Sorenson, Oliver Taftdrup, and Cathrine Hasse, xi–xxv. Lanham: Lexington Books.

Adams, Catherine, and Terrie Lynn Thompson. 2016. *Researching a Posthuman World: Interviews with Digital Objects*. London: Palgrave Macmillan.

Adams, Catherine, and Joni Turville. 2018. "Doing postphenomenology in education." In *Postphenomenological Methodologies; New Ways in Mediating Techno-Human Relationships*, edited by Jesper Aagaard, Jan Kyrre Berg Friis, Jessica Sorenson, Oliver Taftdrup, and Cathrine Hasse, 3–26. London: Lexington Books.

Adams, Catherine, and Max van Manen. 2008. "Phenomenology." In *The SAGE Encyclopaedia of Qualitative Research Methods,* edited by Lisa M. Given, 614–619. Thousand Oaks, CA: SAGE Publications.

Bonderup-Dohn, Nina. 2022. "Investigating the background: Taking a Merleau-Pontian phenomenological approach to Networked Learning." In *Proceedings of the 13th International Conference on Networked Learning*, edited by Jimmy Jaldemark, Marcia Håkansson Lindqvist, Peter Mozelius, Lena Maria Ôberg, Maarten De Laat, Nina Bonderup Dohn, and Thomas Ryberg. Sundsvall University, Sweden.

Fenwick, Tara, and Richard Edwards. 2010. *Actor-Network Theory in Education*. London: Routledge.

———. 2012. *Researching Education through Actor-Network Theory*. Oxford: Wiley-Blackwell.

Giorgi, Amedeo. 2009. *The Descriptive Phenomenological Method in Psychology: A Modified Husserlian Approach*. Pittsburgh, PA: Duquesne University Press.

Gourlay, Lesley. 2021a. *Posthumanism and the Digital University: Texts, Bodies and Materialities*. London: Bloomsbury Academic.

———. 2021b. "There is no virtual learning: The materiality of digital education." *Journal of New Approaches in Educational Research* 10(1): 57–66.

———. 2022a. "Digital masks: Screens, selves and symbolic hygiene in online higher education." *Learning, Media and Technology* 47(3): 398–406.

———. 2022b. "Surveillance and datafication in higher education: Documentation of the human." *Postdigital Science and Education*. https://doi.org/10.1007/s42438-022-00352-x.

———. 2023 (In press). "Postdigital / more-than-digital: Ephemerality, seclusion, and copresence in the university." In *Postdigital Research: Genealogies, Challenges and Future Perspectives*, edited by Petar Jandrić, Alison MacKenzie, and Jeremy Knox. London: Springer.

Gourlay, Lesley, Donna M. Lanclos, and Martin Oliver. 2015. "Sociomaterial texts, spaces and devices: Questioning 'digital dualism' in library and study practices." *Higher Education Quarterly* 69(3): 263–278.

Gourlay, Lesley., Allison Littlejohn, Martin Oliver, and John Potter. 2021. "Lockdown literacies and semiotic assemblages: Academic boundary work in the Covid-19 crisis." *Learning, Media and Technology* 46(4): 377–389.

Hammond, Chris. 2018. "The magic behind UCL library seating: How UCL, and out API, seem to know everything." *Medium.* https://medium.com/ucl-api/the-magic-behind-ucl-library-seating-4fb1f494c789.

Hasse, Cathrine. 2015. *An Anthropology of Learning: On Nested Frictions in Cultural Ecologies.* Dordrecht: Springer.

Ihde, Don. 1990. *Technology and the Lifeworld: From Garden to Earth.* Bloomington: Indiana University Press.

———. 2008. "Introduction: Postphenomenological research." *Human Studies* 31(1): 1–9.

———. 2009. *Postphenomenology and Technoscience: The Peking University Lectures.* Albany, NY: State University of New York Press.

———. 2012. "Can continental philosophy deal with new technologies?" *Journal of Speculative Philosophy* 26: 321–322.

———. 2018. "Foreword." In *Postphenomenological Methodologies; New Ways in Mediating Techno-Human Relationships*, edited by Jesper Aagaard, Jan Kyrre Berg Friis, Jessica Sorenson, Oliver Tadtdrup, and Cathrine Hasse, vii–x. London: Lexington Books.

Ingold, Tim. 2006. "Up, across and along. Place and Location." *Studies in Environmental Aesthetic and Semiotics* 5: 21–36.

———. 2007. *Lines: A Brief History.* London: Routledge.

———. 2012a. "Toward an ecology of materials." *Annual Review of Sociology* 41: 427–442.

———. 2012b. "Looking for lines in nature." *Earthlines* 3: 48–51.

———. 2023. "Postscript: An anthropologist lands in phenomenology." In *The Oxford Handbook of Phenomenologies and Organisational Studies*, edited by Françoise-Xavier de Vaujany, Jeremy Aroles, and Mar Perezt, 719–724. Oxford: Oxford University Press.

Jandrić, Petar, Alison MacKenzie, and Jeremy Knox. 2022. "Postdigital research: Genealogies, challenges and future perspectives." *Postdigital Science and Education* 6: 409–415. https://doi.org/10.1007/s42438-022-00306-3.

Johnson, Mike. 2020. "Phenomenology and networked learning: Mobilage glimpsed from the inside through an online focus group." In *Proceedings of the 12th International Conference on Networked Learning,* edited by Marten De Laat, Thomas Ryberg, Nina Bonderup Dohn, Stig Børsen Hansen, and Jens Jørgen Hansen, 157–163. Aalborg University, Denmark.

Kalthoff, Herbert, and Tobias Roehl. 2011. "Interobjectivity and interactivity: Material objects and discourse in class." *Human Studies* 34: 451–469.

Littlejohn, Allison, Lesley Gourlay, Eileen Kennedy, Kit Logan, Tim Neumann, Martin Oliver, Jon Potter, and Jennifer A. Rode. 2021. "Moving teaching online:

Cultural barriers experienced by university teachers during Covid-19." *Journal of Interactive Media in Education* 21(1): 7.
Lyon, David. 2018. *The Culture of Surveillance: Watching as a Way of Life*. Cambridge: Polity Press.
Mangen, Anne, and Lillian Balsvik. 2016. "Pen or keyboard in beginning writing instruction? Some perspectives from embodied cognition." *Trends in Neuroscience and Education* 5(3): 99–106.
Mangen, Anne, and Don Kuiken. 2014. "Lost in an iPad: Narrative engagement on paper and tablet." *Scientific Study of Literature* 4: 150–177.
Mol, Annemarie. 2010. "Actor-network theory: Sensitive terms and enduring tensions." *Koelner Zeitschrift fuer Soziologie und Sozialspychologie* 50 (1): 253–269.
Negroponte, Nicolas 1998. "Beyond digital." *Wired*. https://www.wired.com/1998/12/negroponte-55/.
Rosenberger, Robert. 2014. "Multistability and the agency of mundane artifacts: From speed bumps to subway benches." *Human Studies* 37: 369–392.
Rosenberger, Robert, and Peter-Paul Verbeek. 2015. "A field guide to postphenomenology." In *Postphenomenological Investigations: Essays on Human-Technology Relations*, edited by Robert Rosenberger and Peter-Paul Verbeek, 9–41. Lanham: Lexington Books
Scharff, Robert C. 2006. "Ihde's albatross: Sticking to a 'phenomenology' of technoscientific experience." In *Postphenomenology: A Critical Companion to Ihde*, edited by Evan Selinger, 131–144. New York, NY: SUNY Press.
Quinn, Ben, and Jasper Jackson. 2016. "Daily telegraph to withdraw devices monitoring time at desk after criticism." *The Guardian*. January, 10. https://www.theguardian.com/media/2016/jan/11/daily-telegraph-to-withdraw-devices-monitoring-time-at-desk-after-criticism.
van Manen, Max. 2014. *Phenomenology of Practice: Meaning-Giving Methods in Phenomenological Research and Writing*. London: Routledge.
———. 2009. *Researching Lived Experience: Human Science for an Action Sensitive Pedagogy*. London: Routledge.
Verbeek, Peter-Paul. 2001. "Don Ihde: The technological lifeworld." In *American Philosophy of Technology: The Empirical Turn*, edited by Hans Achterhuis, 119–146. Bloomington: Indiana University Press.
———. 2005. *What Things Do: Philosophical Reflections on Technology, Agency, and Design*. Penn State: Penn State University Press.
———. 2011. *Moralizing Technology: Understanding and Designing the Morality of Things*. Chicago, IL: University of Chicago Press.
———. 2015. "A Field Guide to Postphenomenology." In *Postphenomenological Investigations: Essays on Human-Technological Relations*, edited by Robert Rosenberger, and Peter-Paul Verbeek, 9–41. London: Lexington Books.
———. 2016. "Toward a theory of technological mediation; A program for postphenomenological research." In *Technoscience and Phenomenology: The Manhattan Papers*, edited by Jan Kyrre Berg O. Friis and Robert P. Crease, 189–204. Lanham: Lexington Books.

Walz, Scott B. 2004. "Giving artifacts a voice? Bringing into account technology in educational analysis." *Educational Theory* 54, 157–172.

Wu, Jiaxu. 2020. "Using sensor data to explore library occupancy in smart campus." *UCL Connected Environments Blog*. https://connected-environments.org/guest-post/using-sensor-data-to-explore-library-occupancy-in-smart-campus/.

Chapter 7

Leveling Up

A Postphenomenological Perspective on Gamification

Stacey O. Irwin

As Johan Huizinga asserts, "Any game can at any time wholly run away with the players" (Huizinga 1944, 8). Humans play games. We don't "do" or "try" or "take" a game. We play a game. "This inexplicably makes the idea of play and the pursuit of (a) game linguistically and *culturally* linked" (Huizinga 1944, 37). We play guessing games and games of chance and games of skill and wit. We play sports games and board games—sacred games and Olympic games. Even poetry can be a game (Ibid., 40). The kinds of games that humans play are endless. (As an aside, animals play games too.) "The function of play in the higher forms which concern us here can largely be derived from the two basic aspects under which we meet it: as a contest for something or a representation of something" (Huizinga 1944, 13). The game absorbs the players in existential ways. Through time and across countries, anthropologists have tracked art, illustrations, artifacts, and letters that testify to the fact that humans *love* to play games. Dutch Historian Johan Huizinga's *Homo Ludens: The Study of the Play Element in Culture* (1944) delves deep into the history of the play, sharing that play is a social construction and a cultural factor of life (4). The book title translates to *Man the Player* and explores much about play and its closely associated concept, the game. Huizinga notes that players become absorbed in the game. Apostolos Spanos, in *Games of History: Games and Gaming as Historical Sources* (2021), shared that throughout different societies and times, humans actually need to play. Gameplay is a cultural and social phenomenon. Game studies theorists share that games and play are difficult to define and maybe should not be defined at all. The place that playing games has carved within human culture and society is prominently emphasized in this chapter. First, there was the game, and then came gamification. Now players enjoy technological gamification.

This chapter delves into the idea of leveling up, which suggests improvement, advancement, or moving to the next level of play. Studying the postphenomenology framework to understand technological gamification in educational contexts contributes to and advances the understanding of this widespread technology use in education. Don Ihde's embodiment, hermeneutic, alterity, and background relational schemas and a postphenomenological use case seek to identify multiple technological stabilities.

INTRODUCTION

Gamification is different from a game because gamification uses game details and design in previously non-game contexts (Domínguez et al. 2013). The etymology of gamification starts with the root word game, noted in early Old English, Old High German, and Frisian language as fun, merriment, play, sport, amusement, and joy. The added suffix *ifi-* or the longer connector *ification* is designed to connect back to the original root word to mean make, set up, put, or to become. Gamification is defined as *adding playful elements to educational practices* and typically involves mechanisms like points, feedback, levels, and sometimes badges or "unlocked" extras that capture the attention and motivation of the players. Andrew Stott and Carman Neustaedter (2013) note that a successful game-learning experience involves the freedom to fail, quick feedback, progression, and storytelling. The experience is also often combined with elements like collaboration and quests. Zamzami et al. (2020) note that gamification increasingly uses mobile apps and other digital media to foster the experience and positively affects learning by increasing enthusiasm for learning the content, providing feedback, receiving feedback, and promoting goal setting. Studies also note the adverse effects of gamification, such as jealousy and increased utility skills (Shurui et al. 2020).

For instance, when a learner plays an educational video game, this agency between the player, technology, and the world creates the autonomy for the player to choose experiences, characters, adventures, and tools to develop both expected and unexpected opportunities. These opportunities are woven into stories, giving the player quick feedback that progresses from one plot point to the next. Sometimes, the technology fosters a sandbox-like experience that allows the learner to engage in creative play and build, which does not have a predetermined ending. This creative crafting and building foster emergent play in an open-world concept and give quick gameplay feedback. The player—educational video game—world experience is co-shaped or co-constituted through the technology as the learner moves through the game.

Not everyone is motivated or learns the same way, and gamification can enhance and encourage learner motivation, autonomy, and individual skill building. In the United States, in the 1950s, the landmark *Brown v Board of Education* ruling required compulsory public education for all children, regardless of race, focusing on equal learning opportunities. This new educational shift toward equality for all required an expansion of pedagogical styles and the need to teach each student, regardless of race, gender, or ability. Up until this point, educational experiences were segregated. The idea of public education for all was based mainly on the logistics of integrating bodies in classrooms. The diversity of students and their divergent learning preferences increased in focus through the decades since the 1950s. Gamifying education has become one of the strategies to meet various students' educational needs.

Making learning "fun" has also been an educational focus for many years. Educational theorists felt that if the learning could be fun, more students would engage at a higher level of understanding. From the integration of educational television in the 1960s, through to cassette tape learning in the 1970s, video games in the 1980s and 1990s, and now learning through podcasts, TikTok, AR/VR, and shooter games, education has heralded in the age of technology (Obrega 2022). Much of the research has been afforded to positive or neutral outcomes toward technology use in the classroom (Aargard 2018, Adams & Turville 2018). Wearable game technologies have added a more integrated extension relation that "hyper" utilizes the body in barely studied ways (Rapp 2020). While some researchers have linked gamification with commodification through social media, education has made interesting gamified digital learning strides in ideas around mediation through digital epistemologies and pedagogy (Friesen 2011; Funk et al. 2020). Still, some say in many "educational technology circles, the mantra, 'it's just a tool,' is still used to describe the belief that technology is neutral, and that the teacher or student alone is responsible for constructing educational meaning" (Adams & Turville 2018, 20). In *Learning by Design: Good Video Games as Learning Machines* (2005), James Paul Gee explains that video games are the promise for today and for imagining the future. He notes a focus on engaged participation, customization, character identity, empowerment, and ability to problem solve, among other traits, with the strategy of accomplishing a goal. Later in that same piece, he shares "that there is a kinship between how the human mind works and how video games work, since video games are, indeed, perceptual simulations that the player must see as preparation for action or fail" (12). Central to Gee's ideas is the video game design. Commercial video games have had an educational focus for many years. Lemonade Stand and Oregon Trail in the 1970s, Sim City in the 1980s, and Loom and Where in the World is Carmen Sandiego in the 1990s were commercially available educational video games. Gee identified engaged participation, customization, character identity, empowerment, and problem-solving ability as essential components of these video games. Educational technology companies and publishers also develop games for online and downloadable play, have aligned gaming content with government-designed core competencies and learning objectives, and found ways for students and teachers to build games together. STEM and eSports and the more digitally savvy post-pandemic landscape with expanded connectivity also push increased interest in gamified education. Educational institutions have embraced specific technologies, online delivery modes, and cloud storage to focus on increasingly gamified educational environments, materials, and experiences.

Understanding the gamified experience requires a philosophy and methodology that looks at each educational technology experience separately and then wholistically. Postphenomenology fits nicely as a framework because this methodology studies the relations between users and technologies. For instance, postphenomenology asks, "How do technologies shape our choices, our actions, and our experiences in the world?" (Rosenberger 2015, 1). Additionally, postphenomenology highlights the mediating role which transforms the human–technology–world experience (Rosenberger 2012). "Applied phenomenology by way of postphenomenology provides a strong framework for the undertaking" of studying experiences, perceptions, interpretations, and trajectories (Irwin 2016, 32). Turnville and Adams, in their chapter "Doing Postphenomenology in Education" (2015), explain that this methodology attends to unique differences among specific technologies like "teaching practice, knowledge apprehension, and pedagogical meaning" (21) or encouraging prosocial behavior (Rosenberger 2013; Vindenes et al. 2021). "Postphenomenology may contribute to existing approaches to media by invoking new questions that push forward our understanding of the effects of contemporary media on humans and our surrounding world" (Van Den Eede et al. 2017, xix). In short, this framework provides tools that explore each component of the player—educational video game—world interplay. A technological use case can also help characterize how postphenomenology works to identify and illustrate elements of the gamification experience.

TURNING TO POSTPHENOMENOLOGY

My work in postphenomenology has centered on use cases of specific digital media technologies and artifacts and instruments like editing software, data mining, earbuds, and gifs (Irwin 2005, 2016, 2017, 2021). For this chapter, I use the sandbox educational video game called Minecraft as the use case. This game is procedurally generated, meaning the players create the world based on embedded digital tools and materials fostered by powerful algorithms that are both deterministic and random in their generation. This popular game was initially designed as a consumer game, but educational versions are also popular and have been researched and studied for their pedagogical use and learning experience (Tromba 2013; Gallagher 2015; Kenkel 2015). A 2022 article by Alessandro Bile shared that Minecraft's idea of cube logic makes all elements in the Minecraft world cube or block-like, so everything fits together. Complex structures can be made by delivering multiple blocks into space. Players move around the interface at will, creating objects and buildings, freely choosing the materials, and sometimes using scientific ideas from chemistry to make them. The Minecraft Education website shares that educators can "Connect students' passion with purpose through game-based learning" (Getting Started 6/11/23). The company also notes that the educational version of Minecraft can initiate meaningful learning, foster digital thinking, and build empathy through Minecraft play. The site includes educator training and guides, lesson plans, projects, and tutorials. Rick Marlatt, in "Literary Analysis Using Minecraft: An Asian American Youth Crafts Her Literacy Identity" (2018), explains that Minecraft is a great fit for teachers looking to enrich their pedagogy. "I encourage educators to consider how critical, technology-based investigations of literature can bolster students' agency as constructors of meaning" (Ibid., 65).

A postphenomenological framework integrating a use case focused on the human–technology–world experience can be used to study the relationship and connection between each component in this co-shaped experience. The human–technology–world schema centers on the human as the learner, the digital gamified educational entity as the technology, and the world as the lifeworld we live in. The analysis includes an exploration of the four existentials of lived time, lived place, lived body, and lived relation (Ihde 1990, 2002; Irwin 2005, 2016; Wellner 2016). In postphenomenology, technological co-shaping or co-constitution is called technological mediation. For example, technology mediates an educational video game players' knowledge of the world and changes the way they think of the world, or worlds, especially in an open world type digital environment with online connectivity. Technological mediation can magnify, reduce, reveal, conceal, enable, constrain, and/or involve and alienate the player, the technology, and the world. The two-sidedness is revealed in technological mediation as an indispensable methodological nexus to "investigate, anticipate, assess, and attempt to influence the technological shaping of our society" (Kiran 2015, 124). One of the foundational concepts of postphenomenology is the human–technology–world framework. How does the educational version of Minecraft co-constitute and mediate the experience of learning? This assessment of the relation between these elements provides a bodily-perceptual framework for how each works together toward a specific intentionality and multi-sidedness perception of experience.

A RELATIONAL SCHEMA

Don Ihde's (1990) relational model is often used in postphenomenological work to study embodiment, hermeneutic, alterity, background, and newer models that push the relational boundaries for the *Learner—Educational Video Game—World* experience and explore the technological mediation in the process (Rosenberger & Verbeek 2015) in the overall schema of educational video games.

From an embodied perspective, when playing Minecraft, the schema may look like *(Learner—Educational Video Game) → World*. This grouping focuses on transforming the "user's actional and perceptual engagement" in the world as shared by the directional arrow (2015, 14). The Minecraft game experience is taken into the user's body through various overt devices like a desktop or laptop computer, a tablet, or a phone. The game can be connected to the web or used in a closed system on a device not connected to the internet. Additional technologies may include a headset and microphone for multiplayer communication if that is part of the game structure. Minecraft uses a wireless game controller connected through Bluetooth, a trackpad, a keyboard, or a mouse. Some versions use VR technologies, so a VR headset is required. The learner experiences the Minecraft world through a fusion of learner and game, co-shaped or reshaped through bodily awareness toward the world. The more experienced the learner becomes at playing the game, the more knowledgeable and embodied the experience becomes. Embodied practice and habit take over through embodied awareness. The learner looks through the technology to engage in the world. There is a sense of magnification and reduction of embodiment as the learner engages in Minecraft play. As shared by Ihde, tradeoffs in this embodied experience include a lack of awareness of the material world during play, increased use of specific body parts like wrists, fingers, and eyes, and volume for hearing game audio. This schema illustrates an immersion in the educational experience toward the world outside of the experience.

From the hermeneutic perspective, the schema may look like *Learner → (Educational Video Game—World)*. The learner is reading the combined or co-shaped game and world together. This experience binds the game and world by interpreting the hardware and Minecraft software. A keyboard, mouse, gaming controller, or touch interface can execute functions like picking up a block, attacking and destroying, opening code builders, and toggling the interface, among many other options. On a keyboard, for instance, the space bar, shift key, letter F keys, and the escape key carry different functionality in the game. The learner can also change settings and worlds. The learner interprets the game as they read the experience and respond with specific key functions. The learner reads the display interface. A newer learner is slower to decode, interpret, read, and respond, but a more experienced learner responds quickly out of habit and embodied perception. The learner responds based on visual and auditory cues toward perceptual attunement while reading the combination of the game plus world action. This schema illustrates that the learner focuses on the game, device, and world melded together.

From the Alterity perspective, the schema may look like *Learner → Educational Video game—(– World)*. The focus is on the Technological Other in the game that somehow mirrors or connects with the learner's humanity (Irwin 2005). Learners choose one of two default human-like avatar skins to travel through the game, building and exploring in the sandbox environment with their tools and abilities. The algorithms used in Minecraft provide feedback through the characters the learner interacts with. For instance, auditory feedback often provides cause-and-effect sounds in this interactive environment and terrain. This techno-presence provides a body for the learner to interact with and helps learners collaborate for game experiences and solve problems. In the game, learners experience alterity relations with other learners, whom they might only have met through the game. Base characters in Minecraft have base default player skins for gendered characters called Steve and Alex. These characters can be altered by winning specific items and quests. This schema illustrates the learner looking toward the game, absorbed in the experience, without any focus or notice of the world, as noted by the double negative.

From the background perspective, the schema may look like *Learner → (Educational Video game) ←World*. The bracketed educational video game environment moves to the background of the experience, and the plethora of choices, opportunities, and quests at the edges, are also in the background, waiting as an option to be utilized. While the many components of the Minecraft game are technically absent, they are ready for use. The Learner knows they are available. Ihde calls this a present absence (Rosenberger & Verbeek 2015, 19). The distinct and blocky Minecraft environment shapes the learner experience as a backdrop for quests and experiences. The learner does not need to directly interact with the millions of optional opportunities and elements of the game. They are backgrounded until required and then perceptually called forward for play—the algorithm and data collection on the gameplay work in the background to heighten the experience. Gameplay is the product of the relations shared here. This schema illustrates the learner in a world where gamified educational experiences are available and ready for use.

From the Cyborg/hybrid intentionality perspective, the schema may look like this *(Learner/Educational Video game) → World.* This shift capitalizes on the use of Minecraft virtual reality worlds accessed by VR helmets and goggles that unlock the metaverse for play. In hybrid relations, "the physical boundaries between humans and technologies are blurred, and technologies merge with our bodies" (Ibid., 20). This relation is more intimate than the previous schema, shifting from intentionality toward integration to merge the learner and technology as one entity in the world. As the metaverse becomes a reality, the immersive world of *Learner ←→ Educational Video Game/ World* expands. The game and world meld together in a back-and-forth experience with the learner. Then in Augmented Relations, the immersion continues but is augmented to the point where the learner and game become one toward the world of the game within the natural world *(Learner—Educational Video game) →World*, the algorithms in the game create a feedback loop between the learner and the technological world, which is an immersion in a digital world called the Minecraft metaverse. The Minecraft educational VR game loses the perception of the learner and there is just the game and the world, which encapsulates this schema of *(Educational Video game)—World.* Each of these schemas alters the relational focus of the experience to bring forward the interplay and perception between the learner, the technology of engagement, and the world at large.

Across these relational schemas, cuts the multistability of the experience. Specifically, "How should we understand the way that technology at once in part determines our choices and actions, and yet at the same time itself remains open to our manipulation and interpretation" (Ibid., 25)? How might the *learner–educational video game–world* be meaningful to different learners in different ways? This next section brainstorms different potential stabilities for the *learner–educational video game–world.*

Thinking about ways learner–educational video game–world relations are formed and co-constituted is central to this next postphenomenological turn. One of the most used concepts in postphenomenology is multistability. In the simplest terms, stabilities are the constant or consistent components or characteristics that make up an entity. Working through the relational schema helps open possibilities for multiple stabilities or variations. Stability is a crucial dimension of a technology. Sometimes, one stability is at the forefront, and the "original" key element is the most discussed, advertized, or used. We need to ask what kinds of stabilities, as in perspectives or key elements, might be at play in the educational version of the Minecraft exhibit.

Altering the schema to *Learner–Minecraft–World* helps to identify and brainstorm stabilities. The educational Minecraft version called Minecraft-EDU will be the focus of this use case. The educational version shares increased game tutorials, materials for educators to create their own collaborative experiences, classroom management tools to manage student learning, the ability to archive gameplay, additional data and statistics on student experiences, and increased infrastructure compatibility between players and educators. The educational version and editions are locked down, so everyone is working with the same experience and structure, whether online or in a classroom environment.

Use Case Exploration

This use case explores three stabilities: technological infrastructure, givenness, and affordance. These jumping-off points highlight several technologically mediated experiences that feature or focus on the non-neutrality of an educational video game like Minecraft.

Technological Infrastructure

One stability we can identify for the Learner–Minecraft–World relational experience is some kind of infrastructure for the co-constituted experience to work. The infrastructure of the Minecraft educational video game is required, but many different networks and technologies can be employed for Minecraft play. As shared earlier, the educational game design, the media component, and the large-scale technological (infra)structure are needed to fully function at an expected level. Infrastructure may include the need for a computer with a specific kind of operating system, a particular CPU core, and a specific minimal RAM. Or an educational version can be played using an app on a tablet, desktop, or a mobile phone. Use requires a download from an online source and upgrades of operating systems for new versions. Heather Wiltse's 2017 study on mediating (infrastructures) illustrates the importance of digging into how technology cultivates relational meaning-making. Will the learner use a keyboard on a laptop or desktop, or use a phone or tablet version of the keyboard? The VR version uses goggles for an immersive experience. Along with infrastructure, lies a series of unseen technologies. Some experiences with the game are in closed networks, in individual buildings, homes, complexes, and schools, and others, like Massive Multiplayer, Multiplayer, and socially networked games, rely on connection to the web Internet connectivity is a standard stability for Minecraft learners, whether the game is used at home or in an educational setting. It may not be required, but it often is used. The gamified experience is open to other players when Minecraft is connected to the internet. While this increases the number of players available for gameplay, along with their perspectives and experiences, learners need to have increased media literacy about their expanded opportunities for play and to understand the infrastructure more fully to set up appropriate privacy settings. When players invite specific people to play, the circle of connection is decreased, and strangers or those looking to harm are reduced. Technology infrastructures exist beneath the surface, at levels typical users cannot easily access. At some level, educational institutions try to limit connections with out-of-network users, but the popularity of online education and increasingly faster network speeds propagates increased popularity for learners. Understanding the relationship between the technologies themselves and their multiple uses, with the addition of increased use of internet connectivity, electricity, AI, algorithms, and surveillance, creates what Heather Wiltse (2017) calls fluid assemblages. Wiltse explains, "fluid in the sense that they are continually changing, and assemblages in the sense that they are variations and possible multi-instabilities" (Ibid., 20) of Minecraft infrastructure, agencies, and intentionalities. Central to this idea of constant change is "an attunement to the constant becoming of the changing things that are (manifestations of) fluid assemblages" (Ibid., 20). The "always in the making" infrastructure and participation in ongoing design and configuration create complexities that are not easily identified. This facilitates constant troubleshooting, updates, and upgrades. The Minecraft software and program defines specific roles and pigeon-holed uses for a more expansive entry and even more user connectivity. Even single-player games are guided by algorithms based on information from many players. The educational version of Minecraft involves "things that are called into particular forms of being and relations through use, and this has a specific effect on learners (Ibid., 20)." Funding and fostering the use of educational video games like Minecraft without studying and understanding the complicated (infra) structure seems irresponsible at best and problematic on many levels. Wiltse shares the importance of following the relations and relevant questions: "It is critically important to follow the relations that are actually at play. Because in this context, the relevant question is not whether it involves the real or unreal, but rather *when, where,* and *how* it involves both" (Ibid., 21). While educators are working on gamifying concepts to engage learners, criminals are thinking of ways to use the vast infrastructure to connect and groom them for other purposes. For the educational version of Minecraft, all students need to be using the same version and edition to work together. This fosters stability under the giant banner of technological (infra) structure. Understanding the technological (infra) structures of Minecraft is an important task that will take time and much consideration in the increasingly expanding metaverse of the Learner–Minecraft–World experience.

Technological Givenness

The idea of technological givenness is more about the assumptions technology is granted in education and less about a specific philosophical movement of givenness. In the Learner–Minecraft–World experience, the givenness stability is pushed into the background more often as an intentionality but is still a key element. Specifically, the idea that faster and better technology is needed is baked into human–technology–world relations. The technological design has opacity and transparency (Unless it breaks down). An unquestioning accommodation occurs when an institution needs "initial technology" or "upgraded technology." Education may not specifically have the funds to purchase the technology it may "need." Still, the question is usually framed as "which" technology, and not "should" or "shouldn't" we spend money on this or that technology. Many times, parent groups will ask for more technology in the lab and the classroom. It seems like no one ever says, "Are you sure?" I name this idea technological givenness. It is a bit like the idea that technology is viewed by others, by educators, by parents, by learning agencies, as a neutral pass-through without negative effects. The addition of technology is almost always seen as a "help." One of the critical areas of the Learner–Minecraft–World experience is the quantification of reward structures. Because technology itself has become a given, the structural ways it is used in education have also become a given. Rewards and achievements honor the idea of keeping score. Many video games, whether educational or not, use rewards, perks, points, and incentives. This often unlocks the next level of play and provides dashboards, badges, and leaderboards for increased motivation to move ahead. All the while, algorithms gather statistics about the learner and tabulate and integrate them for continued engagement while feeding data information back into the game. And while rewards may motivate students, they also create a less collaborative structure overall, cultivating a patterned attitude toward work = reward. The compulsion feedback loop is another behavior in the video game reward structure. This loop creates rewards for engagement and is designed to release a neurochemical dopamine spike upon completion, leading to addictive training. Rewards mold expectations in specific ways. Not everyone is motivated by the same reward. And sometimes, the work that needs to be done to get the award is not easily obtained the same way by everyone. Finally, game strategies teach learners in a way that links to propaganda and specifically targeted training. Rewards promote linear thinking, leading to entitlement and the lack of intrinsic reward for completion. Technological givenness is a stability that names how the taken-for-granted can mold learning.

Technological Affordance

Technological affordance is the way we comply with the experience we are having with its use. In the Learner–Minecraft–*World* experience, the learner is afforded an encounter that makes learning fun. The technological game is valuable to the learner because they have an experience that affords learning in a specific way. Not only does the learner receive some sort of reward either during or at the end of Minecraft, but their entire experience also requires technological materiality to achieve the specific experience. An affordance is more than the social meaning of an educational game experience. It involves the materiality of the technology and the abilities, competence, techniques, and perspective of those interacting with it." Extending this into the social sphere; the role, function, and meaning the technology takes on in a society largely depends on the perspective and role it is perceived as having within the society" (Kiran 2015, 132). Kiran goes on to say that affordances deal with both actuality and potentiality. What does Minecraft enable learners to do in both full and potential ways? Minecraft can be played in various versions, such as the Creative Mode or the Survival Mode. The game is designed to allow learners to collect and purchase valuable items to explore and play in learning. The premise is to build and explore. To this end, learners solve problems and expand their imagination and creativity. Affordances negotiate the potentiality of the technology. In essence, what might be possible? Kirin notes, "Our awareness of ourselves is intimately connected to our engagement in our surroundings, but equally important is that our awareness of our surroundings is intimately connected to *our* engagement in them" (Ibid., 134). This idea connects with ethical considerations in computer gaming, specifically with a game like Minecraft. The kind of awareness learners have that is linked to experiencing the game and how the Minecraft infrastructure, the design, and the learner interface and connect are all part of the learning experience. In his book *Ethics of Computer Gaming: A Groundwork* (2022), Samuel Ulbricht states, "Computer games are fun. Thus, it is reasonable to think that this activity may play a weighty role in the habituation of virtues and vices" (Ibid., 62). Thinking about the ethics of video game use can become complicated, as Ulbricht shares in his book. Some experiences are fiction, some are more based on realistic tasks and experiences, and several different ethical constructs can be used to consider computer gameplay. It is essential to contemplate what happens when learners comply with the reality governing the technology they use, like the Minecraft game. How might the Creative Mode change the nature of affordances, compared with the Survival Mode? How might having all the resources available to the learner differ from collecting structures while surviving in their world while their health and hunger tallies deplete throughout gameplay?

This brainstorm opens the door to a more extensive, more thorough study of multiple stabilities of the Minecraft educational game and ideas for additional stabilities in the larger spectrum of gamified digital learning. "For the postphenomenologist, brainstorming a technology's multiple stabilities highlights technology's very context-dependent and materially situated relationality"

(Rosenberger & Verbeek 2015, 28). These stabilities highlight the complexities, multiple possibilities, and assumptions baked into the gamification experience.

CONCLUSION

Postphenomenology is a "practice" (Adams & Turville 2018, 11). This practice combines philosophical analysis with empirical investigation to invoke pragmatism and phenomenology for the most fully developed relational analysis because "[p]ostphenomenologists study the relationships that develop between users and technologies (Rosenberger & Verbeek 2015, 1)." When using the postphenomenology framework, the analysis intertwines in a way that negates neither the "pragmatism "nor the "phenomenology." Use cases are practical and pragmatic in terms of identifying a mediated experience, but because the described relations are lived, they are experiential and phenomenological. One of the main reasons I highlighted the orienting quote "Any game can at any time wholly run away with the players" is to set before us at the start of this postphenomenological perspective, the idea that the players become so engrossed and intertwined in the relationship of the game, the game pieces, the game worldbuilding, and the overall game experience that they are transported to another world (Johan Huizinga 1944, 8). The embodiment, hermeneutic, alterity, background, and cyborg/hybrid schemas highlight Learner–Minecraft–World in-game relational experiences that can magnify and shrink. The technological infrastructure, givenness, and affordance stabilities examine multiple ways of considering the intertwining of gameplay and technology, facilitating absorption and transparency. Contemporary technology is pushing into algorithmic and artificial intelligence territory we only know from science fiction novels and movies. If we don't start paying attention, we might become game to the point of complete transparency. Leveling up our understanding of gamification through a postphenomenological perspective reveals important understandings about technological mediation and human–technology–world relations and experience and its role in education.

REFERENCES

Aagaard, Jesper, Jan Kyrre Berg Friis, Jessica Sorenson, Oliver Tafdrup, and Cathrine Hasse. 2018. *Postphenomenological Methodologies: New Ways in Mediating Techno-Human Relationships*. Lanham: Lexington Books.

Bile, Alessandro. 2022. "Development of Intellectual and Scientific Abilities through Game Programming in Minecraft." *Education and Information Technologies* 27 (5): 7241–7256. https://doi.org/10.1007/s10639-022-10894-z.

Domínguez, Adrián, Joseba Saenz-de-Navarrete, Luis de-Marcos, Luis Fernández-Sanz, Carmen Pagés, and José-Javier Martínez-Herráiz. ("Gamification—Department of Education, Australian Government") 2013. "Gamifying Learning Experiences: Practical Implications and Outcomes." *Computers & Education* 63: 380–392. New York: Springer.

Fried, Samantha J. 2021. "Philosophize In It! Politicize With It! Postphenomenology and Earth Remote Sensing as Sites of Political/Scientific Intervention." In *Postphenomenology and Imagining: How to Read Technology,* edited by Samantha J. Fried and Robert Rosenberger. Lanham: Lexington Books.

Friesen, Norm. 2011. *The Place of the Classroom and the Space of the Screen, Relational Pedagogy, and Internet Technology. New Literacies and Digital Epistemologies.* Volume 50. New York: Peter Lang.

Funk, Michael, Bernard Dieber, Horst Pichler, and Mark Coeckelbergh. 2020. "Gamification and Trust in HRI." In *Culturally Sustainable Social Robotics. Proceedings of Robophilosophy 2020*, edited by Marco Nørskov, Johanna Seibt, and Oliver Santiago Quick, 632–642. Clifton, VA: IOS Press.

Gallagher, Colin. 2015. "What is Minecraft?" In *Minecraft in the Classroom: Ideas, Inspiration, and Student Projects for Teachers,* edited by Colin Gallagher, 3–12. San Francisco, CA: Peachpit.

Gee, James Paul. 2005. "Learning by Design: Good Video Games as Learning Machines." *E-Learning and Digital Media* 2(1): 5–16. https://doi.org/10.2304/elea.2005.2.1.5.

"Get Started: For Educators," Minecraft Education website, accessed June 11, 2023. https://education.minecraft.net/en-us/get-started/educators.

Huizinga, Johan. 1944. *Homo Ludens: A Study of the Play-Element in Culture.* Oxfordshire: Routledge.

Ihde, Don. 2002. *Bodies in Technology.* Minneapolis, MN: University of Minnesota Press.

———. 1990. *Technology and the Life World.* Bloomington, IN: Indiana University Press.

Irwin, Stacey. 2021. "The Gif Use Case." In *Postphenomenology and Imagining: How to Read Technology*, edited by Samantha J. Fried and Robert Rosenberger, 103–122. Lanham: Lexington Books.

———. 2016. *Digital Media: Human-Technology Relations.* Lanham: Lexington Books.

———. 2005. "Technological Other/Quasi Other: Reflection on Lived Experience." *Human Studies* 28 (4): 453–467. https://doi.org/10.1007/s10746-005-9002-5.

Kenkel, D. 2015. "Minecraft or Mindcraft: Co-designing and Co-learning in a Virtual World." *Adminfo* 28(2): 9–11.

Kiran, Asle. 2015. "Four Dimensions of Technological Mediation." In *Postphenomenological Investigations: Essays on Human-Technology Relations*, edited by Robert Rosenberger, and Peter-Paul Verbeek, 123–140. Lanham: Lexington Books.

Melatt, Rick. 2023. "Literary Analysis Using Minecraft: An Asian American Youth Crafts her . . ." Accessed July 30, 2023. https://ila.onlinelibrary.wiley.com/doi/abs/10.1002/jaal.747.

Obreja, Dragos. 2022. "Postphenomenology, Kill Cams and Shooters: Exploring the Code of Replay Sequences." *Games and Culture* 18 (3): 267–282. https://journals.sagepub.com/doi/10.1177/15554120221090972.

Rapp, Amon. 2021. "Wearable Technologies as Extensions: A Postphenomenological Framework and Its Design Implications." *Human-Computer Interaction* 38 (2): 79–117. https://doi.org/10.1080/07370024.2021.1927039.

Rosenberger, Robert. 2013. "Mediating Mars: Perceptual Experience and Scientific Imaging Technologies." *Foundations of Science* 18 (1): 75–91.

———. 2012. "Embodied Technology and the Dangers of Using a Cell Phone While Driving." *Phenomenology and Cognitive Science* 11: 79–94.

———. 2009. "The Sudden Experience of the Computer." *Ai & Society* 24: 173–180.

Rosenberger, Robert, and Peter-Paul Verbeek. 2015. *Postphenomenological Investigations: Essays on Human-Technology Relations*. Lanham: Lexington Books.

Shurui Bai, Khe Foon Hew, and Biyun Huang, 2021. "Does Gamification Improve Student Learning Outcome? Evidence from a Meta-analysis and Synthesis of Qualitative Data in Educational Contexts." *Educational Research Review* 30: 100322. https://doi.org/10.1016/j.edurev.2020.100322.

Spanos, Apostolos. 2021. *Games of History: Games and Gaming as Historical Sources*. Abingdon, Oxon: Routledge Taylor & Francis.

Stott, Andrew, and Carman Neustaedter. 2013. "Analysis of Gamification in Education." *Surrey, BC, Canada* 8 (1): 36.

Tromba, Peter. 2013. "Building Engagement and Knowledge One Block at a Time with Minecraft." *Learning & Leading with Technology* 40 (8): 20–23.

Turville, Joni and Catherine Adams. 2018. "Doing Postphenomenology in Education." Essay. In *Postphenomenological Methodologies—New Ways in Mediating Techno-Human Relationships,* edited by Jesper Aagaard, Jan Kyrre Berg Friis, Jessica Sorenson, Oliver Taftdrup, and Cathrine Hasse, 3–25. Lanham: Lexington Books.

Ulbricht, Samuel. 2022. *Ethics of Computer Gaming: A groundwork*. Basingstoke: Palgrave Macmillan.

Van den Eede, Yoni, Stacey O'Neal Irwin, and Galit P. Wellner. 2017. *Postphenomenology and Media: Essays on Human-Media-World Relations*. Lanham: Lexington Books.

Vindenes, Joakim, and Barbara Wasson. 2021. "A Postphenomenological Framework for Studying User Experience of Immersive Virtual Reality." *Frontiers in Virtual Reality* 2: 40.

Wellner, Galit. 2016. *A Postphenomenological Inquiry of Cell Phones Genealogies, Meanings, and Becoming*. Lanham: Lexington.

Wiltse, Heather. 2017. "Mediating (Infra) Structures." In *Postphenomenology and Media: Essays on Human–Media–World Relations,* edited by Yoni Van Den Eede, Stacey O'Neal Irwin, and Galit Wellner, 1. Lanham, MD: Rowman & Littlefield.

Zainuddin, Zamzami, Samuel Kai Chu, Muhammad Shujahat, and Corinne Jacqueline Perera. 2020. "The Impact of Gamification on Learning and Instruction: A Systematic Review of Empirical Evidence." *Educational Research Review* 30: 100326. https://doi.org/10.1016/j.edurev.2020.100326.

Chapter 8

Skype, Zoom, and the Zoombies

Reflections on Artistic Play, Malfunction, and the Traits of the Trade-Offs

Annie Kurz

This chapter seeks to analyze the technology of videoconferencing in educational settings employing the parallel examples of the applications Skype and Zoom to extract technological ambiguities to be considered from both the lecturer's and the student's perspective. American philosopher of technology, Don Ihde often uses the terms technological *trade-offs*, *malfunctions,* or *side-effects* to discuss unavoidable technological ambiguities (e.g., 1990, 99, 2008, iv, 39). According to Ihde, the ambiguity of technologies cannot be understood in absolute or essentialist terms, but within an inter-relational framework, analyzed through situated use cases. With Ihde's integrative approach in mind, the guiding questions in this chapter seek to elaborate on how to phenomenologically detangle the technology of videoconferencing from the negative contexts of the COVID-19 pandemic. How to comprehend the educational implications of videoconferencing? Particularly, how to conceptualize the potential trade-offs (frequently but not necessarily tied to technological malfunctions), without being unduly influenced by the extreme circumstances during "online education" amid the lockdowns? To work toward a nuanced and a playful perspective, I rely on our recorded findings during an art and research project developed 2011 in Germany, at Reutlingen University, in partnership with the RMIT University in Melbourne, Australia. The interdisciplinary initiative known as SkypeLab, set out to analyze the medium of Skype through experimental means without setting a strict research or artistic agenda. This chapter intentionally adopts a playful, and, at times, less formal tone to capture the spirit of SkypeLab and the students' experiences. The approach thus seeks to embrace artistic modalities and experimental speculation as avenues for evaluating videoconferencing technologies within educational contexts.

The empirical framework of this exploration therefore entails a comparative juxtaposing of two distinct experiences engaging with videoconferencing technologies:

(1) Our research findings from SkypeLab[1] (Eichinger and McCormick 2013, 2016),[2] wherein Skype was voluntarily explored as a mediating agent within the creative milieu of art and design education. (2) Collective as well as subjective encounters, often under duress with Zoom (or similar applications) during the COVID-19 pandemic. Being both a design lecturer and a PhD student in 2020, alongside the role of a parent supporting school children in adapting to online learning, allows for a multi-layered perspective to frame the discussion. The different sides of the "educational screen" shed light on tensions or the dialectics that may remain obscured when solely considering the lecturer's perspective.

The title of this chapter, along with its inquiries, draws upon Galit Wellner's article titled *The Zoom-bie Student and the Lecturer* (Wellner 2021). Expanding on her examination of the lecturer's challenges posed by the passive *Zoom-bie student* confined within a black Zoom box, this text endeavors to incorporate additional phenomenological insights uncovered by the art and design students of SkypeLab.

Videoconferencing came to the rescue in 2020, suddenly bringing much of Western social life onto the Zoom screen (Lorenz, Griffith, and Isaac 2020). This situation catapulted the question concerning this technology into the forefront of pedagogical discussions, injecting new aspects into already versatile debates (e.g., Gillies 2008; Turville and Adams 2018; König, Jäger-Biela, and Glutsch 2020; Salhab 2022). Based on Don Ihde's *relational ontology* and his core postphenomenological schema: *human–technology–world*, the adequate questions into (digital) technologies are not whether these technologies bring changes but what the guaranteed changes—the *non-neutrality*—mean for our relationship to ourselves, others, and the world (see Rosenberger and Verbeek 2015). Peter-Paul Verbeek's mediation theory, also considering Bruno Latour's actor-network theory, builds onto Ihde's postphenomenology, a layer of normative evaluations exposing concrete *co-constitutive* and manipulative agencies within human–technology–world relations (Verbeek 2005, 2011). Through this theoretical framework of Don Ihde's postphenomenology, this inquiry delves into videoconferencing as a *non-neutral* tool, a medium, and an environment.

Virtual reality (VR)[3] and videoconferencing are the closest digital variations of "being there" in RL (real life). Sound (audio), akin to the telephone, is complemented by live visual imagery (video) during online conferences. Such meetings may involve two or multiple participants spanning various geographical locations. In 2020 this technology enabled individuals and institutions in Western nations to sustain educational activities and academic gatherings via Zoom or comparable apps,[4] notwithstanding the constraints of lockdown measures. Benefiting from its most conspicuous advantage, individuals can attain a form of *bodily extension* through these applications, transcending the physical constraints of ordinary vision to convene in "cyberspace" despite geographical separation. In 2020/21, like many, I seized the opportunity to participate in additional online seminars as a PhD candidate. Ordinarily, the two-hour journey to Frankfurt would have rendered such spontaneous attendance impossible. To cope with the augmented cognitive demands, I used the online possibility to take screenshots of relevant slides facilitating later in-depth analysis. The ease and accuracy with which students can create recordings for future reference emerged as another significant advantage of online teaching and learning. Conversely, educators have the option to record teaching sessions, or even pre-record them, introducing a layer of flexibility and freeing both instructors and learners from temporal constraints. The widespread availability of high-speed internet, also referred to as large bandwidth, has rendered videoconferencing technology a ubiquitous medium for educational institutions.

The argued for guiding thesis suggests, that while the advantages of videoconferencing are apparent, its trade-offs can sometimes elude detection—such as the physical and mental strains, background distractions, or the loss of world (loss of real-life buzz). However, these aspects may not only be discerned through playful artistic exploration of the medium, as undertaken by SkypeLab, but the very glitches can serve as catalysts for creative endeavors how to "deal" with the technology. Consequently, addressing videoconferencing within education may involve employing creative, art-informed methodologies, which are more likely to be discovered within the context of technological malfunction itself, rather than through a nostalgia for the traditional classroom.

In dialogue with the collective findings of over 100 SkypeLab artists and research participants documented during the active years (2012–2019), this exploration delves into some of the specifics of videoconferencing, commencing with its historical evolution.

EARLY VARIANT: VIDEOPHONE 2500

Many assume the technical possibility to have a life image accompany a voice call, to be a recent innovation, but foundations of virtual communication trace back nearly six decades. First experiments even date back prior to World War II (Times 1925). The so-called *Picture Telephone* was introduced publicly at the 1964 World Fair in Queens, New York (The year Marshall McLuhan's *Understanding Media: The Extensions of Man* was published). Forecasts already then were enthusiastic about the future of the Bell Labs device (Egido 1988). In 1992, AT&T offered a version of the regular home phone with a small screen enabling video-calls between parties owning a *VideoPhone 2500* (Early, Kuzma, and Dorsey 1993). In his book *Bodies in Technology* (2002), Don Ihde draws parallels to contemporary concepts of virtual reality, highlighting the enigmatic decade it took to reach a wider recognition of the Philips variant of the audio-visual telephone. Ihde describes how the Philips Corporation struggled with the puzzle why the audio-visual telephone was not a success. Their market researchers discovered that potential customers often disliked the idea of a

> visual phone catching them unawares—for example on the toilet, or not made up, or unshaved. Philips proposed to counter this by building into the technology a selection of avatars, that is, self-selected flattering images. When you ring and pick up, the audiovisual phone projects me sitting comfortably in my study, all made up with proper hair and a nice turtleneck. So here we are, back to virtuality. (Ihde 2002, 83)

The AT&T version, the *VideoPhone 2500* was rather costly and similarly failed to integrate into typical households (Liao 2020). It would require high speed internet and the smart phone with a front-facing camera to prime the masses for the concept of videoconferencing. In 2003 Skype, the peer-to-peer voice over Internet Protocol (VoIP) telephony service was launched.

In addition to real-time presence in online environments, video conferences afford individuals the capability to disseminate documents or display screens to their audience. Augmenting these interactions, a sidebar equipped with chat functionality presents a selection of preset emojis, facilitating the expression of emotions. Many intuitively understand that online social interactions are not the same as face-to-face encounters, but it is difficult to put the finger on the "trouble" with this technology. What factors contribute to the heightened sense of exhaustion during long and frequent online meetings leading to the sometimes humorously referred to phenomenon of the "Zoombie"? Galit Wellner uses the neologism, of the *Zoom-bie student* (2021) combining the words Zoom and zombie to refer to lifeless students, hidden behind digital conversation boxes of Zoom not showing signs of life. Zombies are not foreign to philosophy (see Amy Kind (2011) discussing David Chalmer's *zombie argument*). Famously adopted by popular culture, the zombie is a creature between life and death. An empty corpse animated not through consciousness but through a brainless compulsion often programmed for destruction and not seldom induced by a zombie virus.

As education had to be reinvented in many countries in 2020 the term "zoom fatigue" caught on quickly (Bailenson 2021), yet it is difficult to pinpoint what this technology does to suck out the lives of users. Digital technologies have infiltrated human life and communication into almost all facets. Education is no exception. A phenomenological inquiry into the experience of "playing," teaching, and learning mediated by Zoom, Skype or similar technical variations, provides nuanced insights into the Janus-Face of the technology. In other words, into our love-hate relationship to videoconferencing. The postphenomenological toolbox (Ihde 1990) allows for a nuanced depiction of this ambiguous relationship. Ambiguities with Ihde are understood under the umbrella term "*multistability*"[5] that cannot be anticipated by the designer (Ihde 2008, 19–30, 2012). Shifts or switches of stabilities bring along *trade-offs* (that may or may not be due to malfunctions) and cannot fully be anticipated in a similar manner.

Don Ihde identifies four types of relations to phenomenologically differentiate the ways humans may relate to the world mediated by technologies. Ihde's seminal diagram frames his "general intentionality relations" as: *human–technology–world* (Ihde 1990, 89). In chapter 5 of his book *Technology and the Lifeworld,* he also provides a schema for the "auditory embodiment relation" with the early variant, the telephone: (I-telephone)-you (1990, 78). Within the context of videotelephony I proposed to similarly begin with the embodiment relation: (I-Zoom)-you. In educational settings the subject can be specified, and "you" becomes a more general "world" due to the possibility of multiple parties joining the video-call. However, Zoom is more complex than the audio telephone also providing a visual and world-like environment. This exposes that Zoom acts as a "mediating (infra) structure": a *tool*, a *medium* plus the *world* (environment) itself (compare Wiltse 2017). A further expanded postphenomenological formula describes the general embodiment relation *(I -technology) → world* (Ihde 1990, 89; Rosenberger and Verbeek 2015, 14). It shows that during embodiment the technology merges with the body to some extent (expressed through the parentheses) enclosing in this example, the semi-entity (Student/Lecturer-Zoom). Intentionality (indicated by an arrow) directs *toward* the world in this case toward the *ZoomWorld* (Student/Lecturer- Zoom) ZoomWorld. The role of the technology deviates from Ihde's classic embodiment relation[6] through its double position. The technology is here simultaneously experienced as *tool* and an *environment/ world*, flickering between the two stabilities. The rest of the RL world remains a factor, demanding reduced attention (Student/Lecturer→ ZoomWorld)—(world). RL (world) steps into the background. Zoom itself becomes an increasingly *transparent* background (requiring less attention) when embodied and fully functioning. During any type of malfunction this relation *breaks,* now requiring full attention.[7] While there is more to be discovered within the embodiment relation to Zoom, the merging into a "ZoomWorld" indicates a *hermeneutic relationship* (involving interpretations or translations of the data in the (Zoom)world plus the self within). The impossibility of a dissection into singled out relations shows what is the case for all complex technologies, specifically for digital technologies; multiple relations are at play concurrently, composing technological mediation (compare Rosenberger and Verbeek 2015, 13–22). Heather Wiltse (2017) consequently breaks down digital mediating structures into useful and often referred to lenses for analysis: *Technology / Media / Environment.*[8] All three lenses describe different *non-neutral / mediating* structures, in her words, *(infra)structures.*

When schools and universities squeeze their teaching agendas into the "box" of Zoom, the expectation is to achieve an experience as close as possible to setups of traditional pedagogy.

During the academic year of 2020/21, while instructing two design seminars at Reutlingen University in Germany, like many, I faced the sudden imperative task to make them fit for "the virtual." Simultaneously, I attended online seminars for my PhD studies, benefiting from the flexibility of this format by accessing additional sessions that would have been logistically challenging to attend in person. This simultaneous experience made me aware of both pedagogical sides of the Zoom box—the side of the student as well as the lecturer.

Another viewpoint briefly considered herein pertains to the sometimes-challenging role of parents in facilitating technological infrastructure and proficiency necessary for online learning while overseeing the homeschooling of their children. In the context of my daughter's third-grade education, optional sporadic online class gatherings were provided to supplement the independent tasks outlined in to-do lists distributed to parents via email in early 2020.

The SkypeLab exploration and research on videotelephony present a notable juxtaposition. In 2012, participants engaged in voluntary and playful experimentation with the prevalent application of the time, Skype, to delve into the medium of videoconferencing.

The interdisciplinary art and research venture was preoccupied with the medium Skype as a tool, a medium and a new creative environment for artists and designers. The choice in this paper for a comparative investigation is to be able to take the technology out of the negative context of the pandemic, while allowing for a postphenomenological dip into the rich pool of experiences collected and documented during the active years of SkypeLab. The art students gathered a list of questions into the *human-Skype relation*, which propelled my own subsequent research.[9] Our investigations at SkypeLab only later, with Ihde, I came to label *experimental phenomenology* and *naïve realism.*

STAYING (OR PLAYING) WITH THE TROUBLES OF SKYPE AND ZOOM

The international art and research project SkypeLab was conceptualized by Maggie McCormick and Henning Eichinger in 2011 to experimentally investigate the medium Skype. The primary objective was to leverage Skype's technology as designed, fostering virtual meetings among participants, and thoroughly exploring the platform's technical capabilities. Subsequently, participants delved into potential uses beyond the application's intended or conventional scenarios, probing its boundaries and potential (stabilities). To kickstart open artistic experimentations and to ease the crossing of cultural boundaries, students were encouraged to begin their Skype sessions with a simple and quick drawing exercise portraying each other exploring whatever they encountered via the screen.

> (. . .) The framework favours purposeful positioning of dialectics across long time frames and distances—art/technology, hand/digital, here/there, day/night, summer/winter, you/me. Within such a laboratory artists connected for the first time across screens in drawing sessions that bridged the physical and digital worlds that over time grew to become transdisciplinary (. . .)." (Eichinger and McCormick 2019 (skypelab.org)

Eichinger, McCormick and their students were interested in the sociocultural implications of this technology and what they meant technically for artists, designers and for the creative industry. McCormick and Eichinger were ahead of numerous art and research programs preoccupied with similar questions (exploding in numbers in recent years, especially post pandemic). SkypeLab had two initial goals: (1) to playfully explore the role of the medium Skype as a technological tool and a meeting space as well as its role within art and design, in an increasingly technological world. (2) Bringing together art students from different educational institutions and cultures thus exploring intercultural communication. In a nutshell: SkypeLab began in 2012 with coordinating online meetings between art and design students in Germany, Reutlingen University and Australia, RMIT University, Melbourne. Eichinger and McCormick's home universities. Later SkypeLab also connected students from East China Normal University, Shanghai, Federal University of Rio de Janeiro, Brazil as well as the University of Atlántico, Barranquilla, Colombia. (Speculations for a possible follow up project are brewing in Rio de Janeiro using the *SkypeLab method*, hoping to connect with students in Africa[10]). The initial tasks given to the students were simple; to attend regular Skype meetings and to draw each other using a specific technique called blind contour drawing (BCD),[11] After a certain period, new

Skype pairs were made and occasionally group meetings were scheduled. BCD is a fast one-line technique that does not allow looking at the hand while drawing. It trains the *eye-hand* coordination without privileging premature self-judgments. This technique helps students to ease the hand and to pay attention to details (persons/objects/pixels on the screen). This artistic warm-up acts as an "icebreaker" for conversation. BCDs are usually funny and a great start for an online meeting inducing an affective state toward each other even before "knowing" anything else of the person "faced" with. Students had no limitations in material choices or how to proceed artistically from the starting point of the drawings. The initial project name *Skypetrait,*[12] emphasizes this connection to drawing and the portrait (Eichinger and McCormick 2013). After each meeting, a short protocol had to be filled out answering a few basic questions about the experience of using Skype to meet and draw an unknown person. Within the scope of almost ten active years, meetings happened until 2018. Numerous art exhibitions and two major catalogues followed. The monumental installation that meant to close SkypeLab in 2019 titled "1000 Pixels" was shown at the Baden-Württemberg embassy in Berlin, Germany.[13] Due to revitalized interests more installations followed. It would go beyond the scope of this chapter to describe granular details of the artistic output. Instead, I turn to useful insights for the task at hand, iterating my way into depth of SkypeLab as well as the more recent "Zoom environments" along the points I am making about the technology itself.

Upon reflecting on SkypeLab, it's notable that Skype held a similar prominence in the 2010s to what Zoom does in the 2020s. The shift from using "skyping" to "zooming" as the common verb for virtual calls raises questions regarding the timing and reasons behind this evolution that would require further research to fully be clarified. One reason Zoom was adopted by academia so quickly is its simple integration and easy-to-use interface in a comparably bug free state providing stable call environments.[14] Zoom moreover aims to connect conceptually to other social platforms through technical features such as the face "touch up" improving online appearance, smoothing out the skin tone of the presenter like an Instagram filter.[15] Changing the phone-like character of Skype into a *social space* achieves Zoom through providing links to meetings without having to gather contacts. This change from "calling" contacts to letting people into a *real-time (cyber)space* helps to cut associations to the "old" telephone creating an environment. This environment however appears inherently fragile and susceptible to disruption. In this section, I aim to distinguish between malfunction as an anticipated concern during videoconferencing sessions, and malfunction as an opportunity for playful and artistic experimentation—a phenomenon worthy of further exploration based on the findings of SkypeLab students. Malfunctions can stem from either technological glitches or user error. Whether addressed by IT experts or non-specialists, both variations must be considered. Through the lens of postphenomenology, malfunctions, such as sound issues on Zoom, can be viewed as disruptions within the human-technology relationship. The user (when malfunctioning) might not properly identify where or how to enable the microphone OR the issue could be a straightforward technical failure. The interface location for enabling the microphone might not be perceived as intuitive or "logical" by the user. This suggests a third type of malfunction—*bad design*. Phenomenologically therefore the user and malfunction cannot be detangled. It is self-evident (and well documented within the field of design and human-computer interaction (HCI)) that technologies are conceived as "better," when enabling better human-technology relations not only better functioning features. Sometimes it is an obviously locatable malfunction, other times it remains a mystery—forever unclear why the sound system in a specific case wasn't working. A teacher today is not only trained pedagogically how to educate and care for students but also how to "care for" a growing variety of (digital) technologies avoiding malfunctions.

Not fully prepared for an abrupt online shift in 2020, the sense of being overwhelmed in situations like a pandemic is not surprising and lies partially beyond the technical realm. Instead, it must be understood within the broader context of the disruptive nature of the global crisis. Distinguishing the impact of the technology from the broader stressors of 2020/21 poses a significant challenge for educational institutions as they navigate the future of online and hybrid teaching. Despite this, an increasing number of institutions worldwide are already offering online seminars or even full degrees. Many students embrace, and in numerous cases, even prefer online study options primarily for the flexibility they afford. Therefore, an integrative approach, a possible "eutopia" (Botin 2017) must be worked out for video pedagogy. To be able to rid of utopia as well as doomsday dystopian perspectives a *dialectical exercise* helps to uncover what Ihde on many occasions calls *ambiguities*, *trade-offs* and *multistabilities*. From a postphenomenological standpoint it is unusual to begin with malfunction[16] and the troubles with a technology. However, it is one of the main reasons educators may dislike or even reject having new technological gadgets introduced. Additionally, to their raison d'être (teaching), repair and *maintenance* of a growing number of technological gadgetries needs to be learned and re-learned and regularly performed. This adds to the workload that is often ignored or put aside as a generational problem. Not properly functioning tools (or relations) are often considered a secondary issue, only a matter of time until the users get more skilled and technologies even better. I argue that *non-neutrality* and *multistability* are already to be found within the sheer growing number of (pedagogical) gadgets in need of maintenance and attention. Malfunction of a tool can be thought of with Heidegger's famous *broken hammer* (Ihde 1990,

31–34), in postphenomenological terms also through the question of transparency. Transparency within postphenomenology describes situations when the technology itself is not perceived as such and users' focus turns away from the technology to the task at hand (in this case in collaboration with others online). The ideal scenario is for the medium and the tool of Zoom to seamlessly blend into the background, becoming an invisible and "magical" environment in which interaction happens (Liberati 2017). There is no need for a major technological breakdown to disrupt the "magical" state destroying the transparency of Zoom. A small sound disruption or a stuck slide are enough to remind of the technology as an object *present-at-hand* in need of attention, analysis, and "repair." Minor disruptions like a sound or visual glitch can shatter the seamless "magical" experience of Zoom, highlighting the requirement for full attention. A *(gestalt)switch* occurs from a semi-transparent *room* to a broken *tool*. In this case the background/environment technology becomes a tool in the center of everyone's attention within the meeting. In a physical classroom a technical breakdown (e.g., not functioning projector) is just as likely to disrupt the teaching and learning environment, through improvisation however, simple low-tech alternatives are possible (whiteboard, printouts). The process of repair thus might require only the attention of a few, ideally outside of the class in progress. In other words, malfunction is as ubiquitous as technology itself and a factor of *contingency* is constantly calculated—the more technology, the more malfunction possible. One never knows when and therefore constantly anticipates the teaching session to be disrupted (online with little alternatives to improvise). Present on both sides of the educational screen, the banality of the anticipated malfunction should not be underestimated as an online stressor. While this is certainly so for many technologies, in the case of videoconferencing, the issue is amplified through the technology occupying multiple roles: tool, media, environment (Wiltse 2017, 3–26). All participant's focus usually will gravitate toward even a small malfunction (sound issues, visual distortions, (not) turned off microphone…). This could be put aside as marginal, however considering a larger virtual environment with say, eighty students, it is obvious that the attention of such a big group gathered around one small mistake the lecturer, or a fellow student made, is a distortion or amplification of relevancy. Minor technological or human imperfections might go unnoticed in a large lecturing hall. Fellow students can continue work or thought without feeling distracted and interrupted by small technical troubles. In physical classrooms students normally start conversations while the lecturer or his or her assistants fix the problem. Within a "technosphere" (Wiltse 2017) all attend to the problem by being interrupted, annoyed and the feeling of wasting of time creeps in (indicating reasons why "hybrid" meetings are even more complicated). If a session pause is called in while repairs are performed, no student exchange happens online but all go their ways, maybe to grab a cup of tea. The experience reveals a "loss of world." As the online environment breaks down (minor malfunction is enough, as I hope to have shown) the world in which students casually would socialize is lost, which Wellner calls the "buzz" (Wellner 2021, 4). Technological malfunctions provide socializing opportunities offline but usually not online. From the SkypeLab research we know that this is reversible. "Malfunction" can be turned into an opportunity for experimentation and play and possibly opening the way to new kinds of online socializing opportunities not found in RL[17].

In 2020 installing Zoom (or a similar application) became an imperative. Whatever malfunctions came along, had to be dealt with. In corporate environments the shift from office to home office happened slightly smoother than the change from school to home schooling. Higher education institutions also did better than schools for the younger. Especially within work environments that are considered office and desk-bound involving tech-savvy protagonists, the shift to online was less hazardous. This reveals that education should not be considered a desk-bound practice directly translatable to online. Some of the contemporary app variations for online teaching are Webex, BigBlueButton, Microsoft Teams, BlueJeans, and Google Hangouts, to name a few.[18] Lecturers typically prefer not to alter an adequately functioning technological system once established due to a new set of malfunctions that potentially need dealing with. The tech savvy or the IT specialists might smile at such worries, but I like to be persistent around *anticipation* and the *opportunity of malfunction* as this showed up to be one central finding of SkypeLab students and artists[19]. Creative ways to work with malfunction of all sorts developed. Slow internet—more problematic in the 2010s than it is now in the 2020s—was the "favorite malfunction" of the SkypeLab students. Stagnating internet connection usually leads to interesting sound effects, "beautifully" pixelated images, reminiscent of painter Francis Bacon's "distorted faces." Many of the students started collecting screenshots documenting the contorted images (see figure 8.1). Chinese artist Funding Sun created sound recordings (named *Middle Layer Fantasy*, 2015) capturing strange background noises that appeared to be produced by technological glitches (Eichinger, McCormick, and Ma 2016, 50–51). Later collaborative monumental artworks re-used the aesthetics of the discovered visual glitches. Pixels found within the earlier produced artworks were printed on 30 × 30 cm acrylic tiles. Three installations followed in Germany, titled: "1000 Pixels," "468 Pixels," and "Echoes")[20] developed by Henning Eichinger, artist, and architect Javiera Advis and I. The pixel installations became metaphors for an all-encompassing digital lifeworld and our favorite glitches within. Pixels turned "inside out" becoming like sculpting material (see the SkypeLab.org website). This is fun if you are an artist set out exclusively to explore pixels, other glitches and

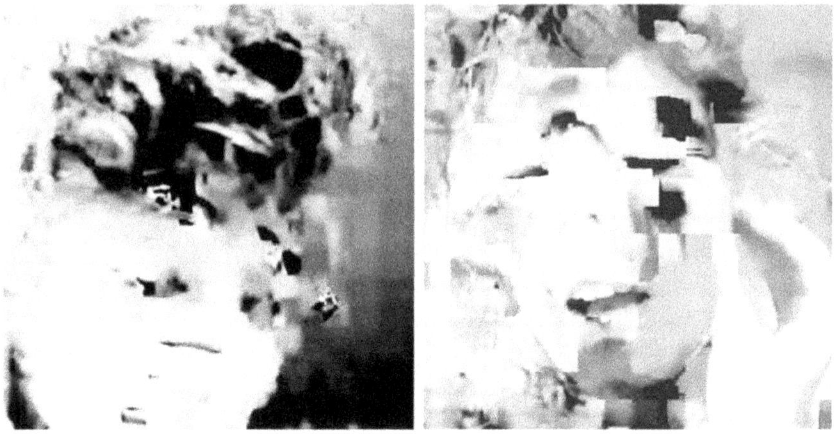

Figure 8.1 Screenshots Taken by German SkypeLab Student Uta Krauss, 2012, Documenting Visual "Skype Pixelations." © SkypeLab 2023. Used by permission.

the technology producing them. With a concrete teaching or learning agenda, random pixelation or other malfunctions usually are annoying, experienced as a disturbance, a waste of valuable time. The SkypeLab discoveries however provoke speculations[21] on the possibilities of what *could* be "achieved" with or despite malfunction in a pedagogical, not necessarily artistic but art-informed setting. Could malfunction scenarios during videoconferencing be used for "play," creating new ways of social connection and communication? Not replacing but re-designing casual social "buzz"? Answers might be found within existing *online practices* rather than in physical classrooms. Another widely acknowledged and more complex malfunction tied to video conferencing can be identified as *trolling*. A New York Times article describes the phenomenon as *Zoom-Bombing* (Lorenz 2020). A "troll" finds a link, plus password, possibly on social media to enter a meeting uninvited with the goal to disturb through sharing inappropriate content or noise.[22] SkypeLab artists have not experienced trolling (as Skype during the SkypeLab years worked without such links) but encountered another *digital pathology*—intruding *spam* messages sent to individuals via Skype. To close notes on malfunction my final point is, that the contingency of malfunction is intractable and must be anticipated when it comes to video calls adding a layer to the "fatigue" phenomenon. Consider Steve Jobs in 2010 presenting the new iPhone and amongst others introducing the novel built-in application *Facetime* at the Worldwide Developers Conference. Jobs has trouble making a connection to the internet failing several times. Finally, forced to ask the audience within the theater to turn off their Wi-Fi connection, so that he would be able to get necessary access.[23]

BACKGROUND AMPLIFICATIONS

Within a laboratory style framework SkypeLab used artistic free play and experimentation to better grasp the medium Skype. Unsurprisingly the primary visual focal point during an online session is the face and the "gaze" (discussed in last section). The "background" providing the *stage* for the face came up a bit later as an object to study. The Australian artist Giordano Biondi (Eichinger, McCormick, and Ma 2016, 24–25) developed artwork patterns using only *bits* of the background. Galit Wellner (2021) describes how Zoom offers virtual background options. Those who use the technology understand what she means with "docile body," as each movement distorts the body-image partially blending it into the background. For postphenomenologists the background is an intriguing notion built around Ihde's background relation (e.g., Ihde 1990, 108–112; Rosenberger and Verbeek 2015, 18–19). In this section I use the background relation formula as proposed by Galit Wellner I: → *(technology –) world* (2021, 5). Often named references of *background technologies* are the heater, AC, or the fridge. The parenthesis shows that the technology is not in the center of attention. The background technology is not "in-use" as in hermeneutic, embodiment and alterity relation but withdraws into the environment, operating "silently" creating context. Background technologies occasionally get noticed consciously but typically do not require engagement (unless malfunctioning). Ihde's background relation and the "world" as a whole (Michelfelder 2015, 241) might hold some of the puzzles postphenomenology yet must solve. A closer reading of what the Zoom / Skype background does, allows for insights into the relationships that *co-constitute* the video-meeting experience. The Zoom background has different variations; the chosen backdrop image can be the "naturally" showing background behind the user within the video-frame—not necessarily observed as technologies but rather as the environment around a person. Descriptions of the background within an online call, however, make clear that what the observer usually encounters *are* technologies. Even a white wall or minimalist curtain fits the common definition of a technical object.[24] Here Ihde's background technologies are literally showcased in the *actual background* behind the person in the video-frame—bookshelves or other furniture, clothing, or a window (possibly leading to unflattering light). The third option is the blurring of the "real" background. Blurring acts like a filter, slightly distorting whatever is behind an individual making it impossible to identify the background objects, reintroducing the issue with the "docile body." This variation suggests that less distracting backgrounds are advisable—a quick solution for messy rooms or offices (this variation however often achieves the opposite, attracting even more attention while the observer (more or less consciously) might try to figure out what the distorted objects could be). SkypeLab artists noted this amplification of the background, which demands attention on both sides of the camera. For each SkypeLab session the "designing" of the background became a commonly described ritual comparable to setting the *stage* for a performance or having someone visit at home (coming backstage). In RL classrooms this is far from people's concerns, rarely one cares what exists behind us in class. Online the background is *amplified*, its "innocence" is lost, potentially becoming a burden. In RL a window or coats hanging behind the students, looking messy are no issues, not the students' responsibility. Bad lighting in RL does not affect us in our understanding of ourselves or others. Nor does it change hermeneutic perspectives on someone's sensitivity, professionalism, competence, or identity. In RL people must oversee body-appearance (clothing, hair . . .) but one thing everyone may relax about is whatever happens to be behind us as we move through the day. My daughter had a few online meetings during her third grade in 2020 (using Microsoft Teams). As a parent I had to make sure that all needed technology was ready, sound and light were set. I forgot about the background for their first meeting. My daughter sat at her desk as she normally would, her shelf with a few books and other kids' things behind her. Her piggy bank (a white ceramic pig with golden dots) on the shelf. The pig looked like standing exactly on top of her head. This was a hilarious *distraction* not of the worst kind. Parents were not allowed in the background or in the room because of privacy concerns but I could hear the laughter from next door, obviously the piggy was not the only interesting background discovery. The teacher clearly had difficulties bringing concentration into their Zoom-room. For the next meeting my daughter made sure that piggy was sitting somewhere else. With Ihde and the question of relational ontology a *co-constitutive* relationship unfolds between us and the things in the Zoom background. *Environment-objects* come forward becoming an amplified responsibility within a theater-like surrounding—made fit the box. This can add a significant level of anxiety to the Zoombie's world. As Zoom habits set in, the background might regress becoming "silent" again, but constant re-evaluations need to be performed.

What SkypeLab artist Giordano Biondi emphasizes is the importance of the *background fragments* our brains try to make sense of as well. The natural privilege of the face however prevails throughout an online session.

While discovering this dialectic (person/background), one of the SkypeLab groups (Germany–Brazil) devoted most of their attention to the objects found in the background while developing an inventory, later presented as large-scale infographics designed by German artist Susanna Krupp. The artists discovered that several "hermeneutic umbilical cords" link identities to the technical objects involved in video-meetings. To *frame* ourselves properly, the laptop position must be optimized (usually too low)—external cameras are easier to adjust for a more flattering angle. Those sophisticated and ambitious in optimizing their (home) office technological gadgetry, soon arrive at a "plane-cockpit-like" set up always ready for "take-off."[25] Responsibilities further extend as the online environment demands control over proper lighting and the acceptable sound environment without undesired noise (clamant street, others in the room . . .). Care for the "background" takes time and effort. I argue that it is one of the reasons some students prefer turning cameras off, becoming Wellner's "Zoom-bie students" visually protected by the black "box."

THINKING INSIDE THE BOX

SkypeLab students were required to arrange meetings within different timezones (e.g., + 10 – Germany – Australia). Time zones have a restricting box-like character. But even within the same time zone, different life schedules with dictating factors require online meetings to be carefully worked around. RL meetings on the other hand synchronize *time-space boxes*—bodies together now act in a shared *mind and body* zone making social interaction more probable.

It can be useful to begin an online teaching session a few minutes before the official start to allow a gradual entry to cyberspace, simulating the RL shared zone scenario. Insecurities of the lecturer (or SkypeLab artist) how to handle such basic tasks of when to start a meeting or what to do about awkward silence when additional entry time is granted, shows that an etiquette "how to behave" does not automatically translate into the online but must be re-invented. As a lecturer I typically keep an eye on the attendee number growing. Numbers going down again irritate, leading to wonder if something is wrong. Despite efforts, audio conversations seldom happen during the entry process. SkypeLab artists were able to counter this uncomfortable initial hesitation to "interact" through the drawings they did upon arrival in the virtual room. The tendency to multi-task is welcomed online by the chat box opting (unless turned off by the host). Some type "hello" instead of an audio greeting. A few of the name-boxes eventually turn into faces. Depending on the group size, it can be helpful to suggest a "virtual reality face-to-face" greeting moment. This is not a *double oxymoron*[26] but the attempt of a lecturer to cope with the wall of black name-rectangles. Teaching can be a challenge considering the lack of facial and bodily clues, normally useful indicators of (a) presence, (b) state of interest or the lack of it, and (c) level of mutual understanding of the discussed topic. All in need of active clarifications online. Direct questions trying to understand a) b) and c), again might result in uncomfortable moments of silence but if allowed to linger long enough, some might feel ready or pressured by the silence to comment. From the perspective of a student, the larger the attendee number, the less responsible one feels for aiding the conversation. Even within small groups, some students prefer hiding behind their lifeless boxes failing to "constitute well-functioning *embodiment relations*" (Wellner 2021, 3–4). One of the pedagogical challenges is, how to deal with these faceless and often voiceless boxes. Those with cameras turned on are overwhelmingly more active, participating with their own ideas. Considering the amplified background however (outlined in previous section), the demand for "all cameras on" does not provide an obvious solution. Some dislike to be on video indicating that the self-image can turn into a burden, too. The face—the center of the video frame needs to be "protected" or made to "fit" the box. In my daughter's third grade on the other hand everyone wants to "show up" online. Discomfort or the fear of the video-box seems to grow as individuals get more mature and self-aware.

To better grasp how to "deal" with the black boxes, next I drill deeper into the box-like character of the video frame and the screen that the self-image must be squeezed into. In an account of virtuality, Ihde distinguishes between the *here-body* and the *image-body* as a disembodied *over-there body*. "The bistability of these two perspectives may be expanded and made complex in many social and cultural activities" (Ihde 2002, 6). In a slightly different context (virtual reality) Ihde refers to the here-body as the physical body that finds itself in a fixed position mostly motionless when looking at the screen—similarly trapped within the frame when video conferencing. The experience of the here-body when on an online call can be understood as a *restricted body* in a technical cage or box. Being online doesn't allow looking around to see different perspectives—you only have *one view*. A tiring experience as we feel obligated to produce an acceptable image-body—restricting speed and size of our movements. A five-hour workday videoconferencing is therefore closer to the experience of a truck driver sitting in a fixed position performing repeated small body gestures than to a person moving freely in a classroom. A lecturer in a RL classroom usually can move freely according to his or her methods and available equipment. While students in a RL classroom are also expected to be sitting still at their desk, they can always at least shift positions without having to keep an eye on the camera angle. In RL group collaborations, when attention is no longer on the lecturer, students gain even more possibilities for bodily movements. This trade-off can only be met through taking frequent breaks during video conferencing sessions or through allowing to turn off cameras enabling more movement without the worry of strange camera angles. The cage of the screen-box however remains constant as we still need to sustain proximity to the screen to see the content presented.[27] Companies such as Meta (formerly Facebook) are working on a VR conferencing sphere, where movement is said to be re-naturalized through VR glasses eliminating restricting screens. "Can VR replace RL?" Ihde asks "Only if theater can replace actual life" (2002, 11). VR technologies aim to fully absorb the body into the *technological cocoon*—only "face sucker" (Ihde 2002, 10) goggles remain to strangle the here-body. "The ultimate goal of virtual embodiment is to become the perfect simulacrum of full, multisensory bodily action" (Ihde 2002, 7). Ihde dismantles virtual utopias like the one proposed by Meta by reminding of the *non-symmetrical* position of the here-body. Ihde argues along Merleau-Ponty's notion of embodiment as the central epistemological and ontological magnet of consciousness. The videoconferencing body therefore finds itself surrounded by a technological *stage-cage* that does not serve, nor fully understand the experience of the here-body, leaving *amplifications* and *reductions* to cope with.

SkypeLab artists discovered that laptop and tablet cameras make the head and the face online appear disproportionally large. Depending on the kind and angle of the camera this distortion can be adjusted slightly. According to Ihde technological *amplifications* are accompanied by *reductions* on other fronts. With videoconferencing the "amputation" of the lower body is easy to detect. Other visual distortions (lost 3D perspective) appear next as well as the failings to transcend full sensory experience of touch and smell. This brings the here-body into confusion of a *double position* having to attend to both the online as well as the offline bodies (here-body / image-body). SkypeLab artists noticed the disproportionately exaggerated virtual faces late, not until meeting the first time in person.[28] Technological amplification became obvious as in RL everyone appeared smaller than indicated by the Skype video. Reductions were noted within SkypeLab protocols as the "missing" lower body parts and the sensory void, connected to smell, touch, and lacking full spatial perceptions. To compensate or experiment with the "losses," SkypeLab artists started walking with their Skype devices widening frames to help their bodies out of the fixed position. Outside the box, they found body parts in movement in an environment distorted through fluctuation represented with delays via Skype.[29] They refer to the guesswork their brains had to perform to understand what was happening on the screen when someone was moving at a "normal pace" through a foreign city or room. The artists started recording perceived amplifications/reductions through collecting screenshots and video *fragments* of what they saw.[30]

THE CAMERA-GAZE

SkypeLab artists also noted confusions associated with the "gaze"—Where to look when on camera or when speaking? Videoconferencing "shows" the eyes of the other, while it is impossible to identify directions of the "look." Looking into the camera feels unnatural, as the others and the environment cannot be observed while staring at the light. Looking at the other person's face feels more like a face-to-face encounter, but frustratingly the eyes are never allowed to meet. Upon several parties entering the conference, attention splits and jumps between the equally partialized rectangles with the speaker or content full screen. Switching to gallery view allows scrolling through the boxes to see who is or isn't present. Another disorienting experience is the self-image gazing back, like in a mirror being too close. The virtual self "*exaggerates itself*" in this setup through its constant presence and amplified demand for aesthetic evaluations. The distorting/amplifying character of the camera might come as a surprise or shock as usually people find discrepancies between expectation vs. actuality of the way they look. Leading to the discovery of an ideal "state" that wants to be obtained, despite fatigue settling in. Especially those with reading glasses suffer from this "straitjacket position"—wrong movements might produce reflections on glasses hiding the eyes through light patterns. A regular checkup allows to make sure that the image produced is not out of control. Similarly, pro-active check-ups are performed to ensure continuous WiFi connection, that personal sound is off when not speaking or properly working when speaking. This ties back to the amplified responsibilities of the background offering another layer of empathy with the *Zoom-bie student*. The impulse of having to check the self-image is reduced in RL to bathroom routines plus times similar to being in mirrored elevators. Otherwise, we are "free from" the self-image. What I call the "gaze" or here "camera-gaze" is well established in *existential phenomenology* as well as *feminist theory* (Ihde 2002, 24–27). Jean-Paul Sartre puts his finger on the "gaze," what he also describes as the "look."[31] The look of the other complicates (co-constitutes) the object-ness of the self's *being-for-itself* (e.g., Sartre 1992, 340). Digital technologies can also fill the role, a *quasi-other* producing *alterity relation*—experienced when entering an empty online room. The pure *camera- gaze* is judgmental enough to make us change positions, adjust light, glasses. . . . Another scenario when the camera-gaze makes an appearance is when no online session is in progress, yet we check the "light" before putting on pajamas to ensure "nothing" is watching. This anxiety provoked by the camera was amplified during SkypeLab years due to frequent virtual meetings.

CONCLUDING "CHAT"

Picking up on current technological developments in education and based on Ihde's notion of *experimental phenomenology* (Ihde 2012), I conclude

Table 8.1 Chat with ChatGPT on February 28, 2023, about the "Dialectics of Videoconferencing"

AK:	*Please Name Advantages and Trade-Offs of the Videoconferencing Technology*
ChatGPT:	Advantages of videoconferencing: Cost-effective: Videoconferencing can be a cost-effective solution for businesses that need to communicate with colleagues, customers, or partners in different locations. It eliminates the need for travel, which can be expensive and time-consuming. Increased Productivity: Videoconferencing allows employees to communicate and collaborate in real-time, regardless of their physical location. This can lead to increased productivity as it saves time spent on travel and allows for quicker decision making. Enhanced Communication: Videoconferencing provides the ability to see and hear other participants in real-time, which can help to improve communication and collaboration. This is especially important when dealing with complex issues or sensitive topics where nonverbal communication is critical. Flexibility: Videoconferencing allows participants to join from anywhere with an internet connection, which increases flexibility and reduces scheduling conflicts. Accessibility: Videoconferencing allows for participation of individuals with disabilities or mobility issues who may not be able to travel to a physical meeting location. Trade-offs of videoconferencing: Technical Difficulties: Technical difficulties such as poor internet connection, software compatibility issues, and hardware failures can cause disruptions and delays during a video conference. Distractions: Participants may face distractions such as background noise, notifications on their devices, and interruptions from others in their physical environment. Lack of Personal Interaction: Videoconferencing may not provide the same level of personal interaction as face-to-face meetings, which can affect the quality of communication and collaboration. Security and Privacy Concerns: Videoconferencing may pose security and privacy risks, especially if sensitive information is being discussed. Companies need to ensure that appropriate security measures are in place to protect confidential information. Fatigue: Participants may experience fatigue and burnout due to prolonged periods of staring at a screen and the lack of physical movement. This can affect productivity and engagement.

with the practice-oriented playfulness that I propose to be complementary to postphenomenology. A chat with OpenAI's ChatGPT, assists me in summarizing my topic of the dialectics of videoconferencing. After many hours to write this chapter, it is intriguing to see that the AI *did* produce points not yet mentioned (see table 8.1), whether this is due to my ignorance or the lack of space, I shall leave open. The suspicion that AI "hallucinations" might be at play, indicates an interesting new type of malfunction on the educational horizon—AI is the next stage technology, that is going to cause a manifold of possible troubles (not only within educational contexts) demanding philosophical attention as well as artistic explorations building possible pathways toward a "response-ability" handling (Haraway 2015).

NOTES

1. Specific roles I held at SkypeLab: among the first participating artists while an M.A. design student 2012/13. Later, administrative assistant and thereafter research coordinator 2014–2017 as well as contributing graphic designer, and co-curator.

2. Eichinger and McCormick received several educational awards for SkypeLab, including financial backing by the Baden-Württemberg Stiftung within the framework of the Baden-Württemberg-STIPENDIUM for university students—BWS plus, a program of the Baden-Württemberg Stiftung. See website for details: https://www.skypelab.org.

3. VR (virtual reality): Don Ihde grants reference to Jaron Lanier to have popularized the term.

4. In the years building up to 2020, a noticeable mass shift away from the prior popular application Skype could be observed as alternatives showed up on the market. Further research is needed to understand this shift.

5. *Multistability* can be thought of as: use or perceptual variations. Within the context of digital technologies, the term *multi-instability* is also discussed (e.g., Redström and Wiltse 2015).

6. The classic example for an embodiment technology: reading glasses; people look through glasses and not at them. What they see is the world mediated through technology (a sharper image of the world than without glasses).

7. Compare Martin Heidegger's hammer analysis discussed by Ihde (1990, 31–34).

8. Heather Wiltse (2017) deals specifically with social media (such as Facebook) in this text. In the case of videoconferencing, it is likewise important to differentiate between technology as a *tool*, which Wiltse (2017, 11) associates with *action* and *perception*. Videoconferencing technologies as *media* allow for *communication* to happen, while the *environment* holds the functions for *dwelling*. Here people can "be" either *active* or *passive* and the technology falls into the background.

9. SkypeLab students were encouraged by the project initiators and by each other to keep an experimental approach throughout the years. With a "Socratic" attitude they kept asking and collecting over *"100 questions"* preoccupied with an increasingly technology-oriented world (presented at Sympolab Synopsis, November 21–22, 2018, at RMIT Europe, Barcelona).

10. The idea of trying to connect to students in Africa was mentioned during the ARTE2 UFRJ interview in 2022: https://www.youtube.com/watch?v=brbyp-XVtqA. Accessed August 7, 2023

11. Blind Contour Drawing (BCD) goes back to the work of artists Kimon Nicolaïdes and Betty Edwards.

12. Eichinger and McCormick do not disclose their choice to add the word "Lab" within SkypeLab after the beta project Skypetrait. But an increasing "laboratory" style openness away from the focus of the initial portrait marked the developments during that time.

13. See for example: https://www.skypelab.org or https://www.tex.reutlingen-university.de/aktuelles/news/news-td/1000-pixel-skypelab-berlin/. Accessed August 7, 2023

14. Compare interview of Eric Yuan, CEO of Zoom at Stanford Graduate School of Business on October 13, 2022: https://www.youtube.com/watch?v=_V0Ke4bi-ypQ. Accessed June 5, 2023

15. Social media platform (owned by Meta) known for the heavy use of predefined photo-filters enhancing shared photographs. The Zoom aesthetic "touch up" can happen unknowingly on both sides. For instance, in cases when played around in settings.

16. Wellner (2021, 3) points out that "only a few postphenomenologists" have dealt with mal-functioning technologies. Markus Bohlmann reminds in his editor's notes of the following important paper: Tafdrup, Oliver Alexander, Bjarke Lindsø Andersen, and Cathrine Hasse. 2019. "Learning to Interpret Technological Breakdowns: A Path to Technological Literacy." In: *Translational Systems Sciences*, 139–52. https://doi.org/10.1007/978-981-13-7725-9_8.

17. Online groups could for example agree on making a quick (blind contour) drawing every time disruptive malfunctions occur. This exercise has the potential to engage students in a fun way when later sharing their artworks. It can break the "ice" in communication as it is well described by SkypeLab participants.

18. Why the previously widely used app Skype was no longer a popular choice provokes speculations but requires more research to be better understood. Compare article by Chris Stokel-Walker 2020 "How Skype Lost Its Crown to Zoom." *WIRED UK*, May 12, 2020. https://www.wired.co.uk/article/skype-coronavirus-pandemic. Accessed August 4, 2023

19. Some of the SkypeLab students continued online meetings and contributed even after finishing their degrees.

20. Printed pixels for the monumental SkypeLab installations were not produced through making of new artwork but rather were "found" within the large quantity of artworks previously produced by all participants.

21. The role of such speculations yet must be defined within postphenomenology. Some prominent voices call for more speculative approaches, others are critical of them.

22. Coping mechanisms develop to counter new trolling tactics, so do prophylactic measures on platforms. Users now are more cautious when sharing a meeting link.

23. See Steve Jobs' demo fail (CNET News): https://www.youtube.com/watch?v=znxQOPFg2mo

24. Peter-Paul Verbeek (2005) uses the following definition of technology: "science-based" technological devices of the sort that began to emerge in the last century.

The term "technical" is broader and describes all sorts of techniques and their associated objects.

25. With large headphones and several screens, the advanced video conferencing setup provokes associations to sophisticated computer gaming arrangements. In computer gaming the comparison to the airplane cockpit is likewise used to describe the "take-off" into a different place, disconnected from RL time—a "flight" into a different world.

26. The first oxymoronic layer within the concepts: "virtual reality—face-to-face"—in common understanding either something is virtual OR real, not both. The second oxymoronic layer within the words: "virtual—face-to-face"—a meeting in common language is either virtual OR face-to-face, not both.

27. On the other hand, this trade-off also might be contrasted by online meetings that people only attend or listen to but do not need to participate in, during which they can move around freely using wireless headsets (contrary to being trapped around a table in a chair during RL meetings).

28. After several months of Skype meetings, SkypeLab groups had the chance to meet in China, Brazil, Germany, and Australia.

29. For this experiment SkypeLab participants usually used mobile devices (tablets or smartphones) to walk around providing views into the surroundings (their apartments or outside showing sights of their hometowns). During this time face and audio contact was either lost completely or re-established regularly.

30. Can screenshots be considered photographs? One of the questions provoking SkypeLab discussions.

31. In Sartre's phenomenology, the "gaze" (1992) is not conceptualized within this context of *technological mediation.*

REFERENCES

Bailenson, Jeremy N. 2021. "Nonverbal Overload: A Theoretical Argument for the Causes of Zoom Fatigue." *Technology, Mind, and Behavior* 2 (1). https://doi.org/10.1037/tmb0000030.

Botin, Lars. 2017. "Sublime Embodiment of the Media." In *Postphenomenology and Media*, edited by Van Den, Stacey O'Neal Irwin, and Galit Wellner, 167–183. Lanham: Lexington Books.

Early, Scott H., Andrew Kuzma, and Eric Dorsey. 1993. "The VideoPhone 2500-Video Telephony on the Public Switched Telephone Network." *AT&T Technical Journal* 72 (1): 22–32. https://doi.org/10.1002/j.1538-7305.1993.tb00519.x.

Egido, Carmen. 1988. "Video Conferencing as a Technology to Support Group Work: A Review of Its Failures." *Proceedings of the 1988 ACM Conference on Computer-Supported Cooperative Work—CSCW '88.* https://doi.org/10.1145/62266.62268.

Eichinger, Henning, and Maggie McCormick. 2013. *Skypetrait: Transcontinental Faces 2012/13.* Stadt Reutlingen, Reutlingen.

Eichinger, Henning, Maggie McCormick, and Yonglei Ma. 2016. *SkypeLab, Transcontinental Faces and Spaces 2015/16.* Berlin: KERBER.

Gillies, Donald. 2008. "Student Perspectives on Videoconferencing in Teacher Education at a Distance." *Distance Education* 29 (1): 107–18. https://doi.org/10.1080/01587910802004878.

Haraway, Donna. 2015. "Anthropocene, Capitalocene, Chthulucene." In *Art in the Anthropocene: Encounters among Aesthetics, Politics, Environments and Epistemologies*, edited by Heather M. Davis, and Etienne Turpin, 255–270. London: Open Humanities Press.

Ihde, Don. 1990. *Technology and the Lifeworld: From Garden to Earth*. Bloomington, Ind.: Indiana University Press.

———. 2002. *Bodies in Technology*. Minneapolis: University of Minnesota Press.

———. 2008. *Ironic Technics*. Copenhagen: Automatic Press/Vip.

———. 2012. *Experimental Phenomenology: Multistabilities*. Albany: State Univ. Of New York Press.

Kind, Amy. 2011. "Chalmer's Zombie Argument." In *Just the Arguments*, edited by Michael Bruce and Steven Barbone, 327–329. Malden, MA: Wiley-Blackwell.

König, Johannes, Daniela J. Jäger-Biela, and Nina Glutsch. 2020. "Adapting to Online Teaching during COVID-19 School Closure: Teacher Education and Teacher Competence Effects among Early Career Teachers in Germany." *European Journal of Teacher Education* 43 (4): 1–15. https://doi.org/10.1080/02619768.2020.1809650.

Liao, Shannon. 2020. "AT&T Launched the VideoPhone 2500 in 1992. It Sold for $1,499 | CNN Business." *CNN*. June 12, 2020. https://edition.cnn.com/2020/06/12/tech/att-video-phone-vault/index.html.

Liberati, Nicola. 2017. "Magic, Augmentations, and Digital Powers" In *Postphenomenology and Media*, edited by Den, Van, Stacey O'Neal Irwin, and Galit Wellner, 63–77. Lanham: Lexington Books.

Lorenz, Taylor. 2020. "'Zoombombing': When Video Conferences Go Wrong." *The New York Times*, March 20, 2020. https://www.nytimes.com/2020/03/20/style/zoombombing-zoom-trolling.html.

Lorenz, Taylor, Erin Griffith, and Mike Isaac. 2020. "We Live in Zoom Now." *The New York Times*, March 17, 2020. https://www.nytimes.com/2020/03/17/style/zoom-parties-coronavirus-memes.html?action=click&module=RelatedLinks&pgtype=Article.

McLuhan, Marshall. 1964. *Understanding Media the Extensions of Man*. New York: McGraw-Hill.

Michelfelder, Diane. 2015. "Postphenomenology with an Eye to the Future." In *Postphenomenological Investigations: Essays on Human-Technology Relations,* edited by Robert Rosenberger and Peter-Paul Verbeek, 237–246. Lanham: Lexington Books.

Rosenberger, Robert, and Peter-Paul Verbeek. 2015. "A Field Guide to Postphenomenology" *In Postphenomenological Investigations*, edited by Robert Rosenberger and Peter-Paul Verbeek, 9–41. Lanham: Lexington Books.

Salhab, Reham. 2022. "ZOOM or NOT to ZOOM: STUDENTS' ATTITUDES towards USING ZOOM at PTUK." *Journal of Southwest Jiaotong University* 57 (1): 488–501. https://doi.org/10.35741/issn.0258-2724.57.1.45.

Sartre, Jean-Paul. 1992. *Being and Nothingness. The Principal Text of Modern Existentialism.* New York: Washington Square Press.

SkypeLab Website. n.d. "SkypeLab". Accessed February 1, 2023. https://www.skypelab.org.

Times, New York. 1925. "Pictures of President Coolidge's Inauguration Sent by Telegraph throughout the Country, Seven Minutes for Each Photograph." *The New York Times,* March 5, 1925. https://timesmachine.nytimes.com/timesmachine/1925/03/05/98817014.html?pageNumber=3.

Turville, Joni, and Catherine Adams. 2018. "Postphenomenology in Education". In *Postphenomenological Methodologies*, edited by Aagaard, Jesper, Jan Kyrre, Jessica Sorenson, Oliver Tafdrup, and Cathrine Hasse, 3–25. Lanham: Lexington Books.

Verbeek, Peter-Paul. 2005. *What Things Do: Philosophical Reflections on Technology, Agency, and Design.* University Park, Penn.: Pennsylvania State University Press.

———. 2011. *Moralizing Technology.* Chicago: University of Chicago Press.

Wellner, Galit. 2021. "The Zoom-Bie Student and the Lecturer." *Techné: Research in Philosophy and Technology* 25 (1): 153–61. https://doi.org/10.5840/techne2021121132.

Wiltse, Heather. 2017. "Mediating (Infra)structures: Technology, Media, Environment." In *Postphenomenology and Media*, edited by Van Den, Stacey O'Neal Irwin, and Galit Wellner, 3–25. Lanham: Lexington Books.

Chapter 9

Let's Study Together

A Postphenomenological Investigation of "Study with Me" Content in South Korea

Sou Hee Yang

In South Korea, as young people have become "digital natives" who readily use technology in many aspects of their lives (Cha and Park 2020, 84), their interest in educational technology has also increased. In particular, "study with me" content (SWM) is widely popular. SWM is content in social media that simulates the experience of learning side by side. Such content is used in a wide variety of educational settings, for example, by primary school pupils doing their homework or office workers doing further training. In particular, YouTube is one of the most widely used platforms for uploading and watching SWM content.

The effects of this new human–technology relation on learning culture have not yet been investigated. In this paper, SWM content is critically analyzed using the tools of postphenomenology.

WHAT IS "STUDY WITH ME"?

SWM content refers to videos and online real-life streaming filmed by students themselves. The video content is distributed with the gesture of an invitation to learn side by side. SWM content shows a person, the creator of an SWM and host of an SWM-session, immersed in study, taking notes or working on a computer for long periods of time but without "social components" such as conversations or instructions (Lee et al. 2021, 1). Nevertheless, there is a silent invitation to learn side by side, in that the creator of a SWM directly involves the viewer with the video. In this way, "study with me" content follows other "with me" YouTube content, for example, "get ready with me," "draw with me," and "clean with me."

The following three salient characteristics of SWM content have been identified by Lee et al.: "(1) the creator studies on camera during their actual study session, (2) there is no direct instruction delivered from the video, and (3) there is a distinction between viewers and creators" (2021, 4).

In the run-up to this paper, I would like to make a fundamental distinction between two types of SWM based on their phenomenologically analyzable experience:

1. Interactive SWM. In these "study with me" sessions, all participants use conference software such as Zoom. Although there is also a creator of the SWM, the social experience is more that of a learning group (see YouTube link 1 in the appendix).
2. Unidirectional SWM. These study-with-me sessions are characterized by the fact that the creator of the content does not see the participants and they also know that the creator does not see them. The content is often recorded, that is, time- and location-delayed, but can also be live. The technologically mediated social experience of this content is more comparable to learning in a public library, where a stranger also learns in a concentrated and undisturbed manner — even while being observed (see YouTube link 2 in the appendix).

In the following, I will focus on the second type. It is the more common type, and the YouTube video platform is the typical technological solution for creating and viewing SWM content (Cha and Park 2020, 84). From the perspective of postphenomenological analysis, the SWM relation established with YouTube is particularly interesting. It turns the YouTube video platform into a self-discipline technique.

To inform potential viewers of a SWM session's characteristics, content titles offer appropriate descriptions, for instance, "calming wave sounds in the background" or "autonomous sensory meridian response (ASMR)," which indicates tingling or goosebumps spreading down the skull, neck or spine in response to certain audio or visual stimuli (see YouTube Link 3 in the appendix). Other descriptions such as "let's pull an all-nighter together," "let's concentrate for just two hours," and "you should study harder than me, an elementary-school student" aim to attract certain viewers (see YouTube Link 4 in the appendix). Some content contains the host's status or study purpose, such as "studying for civil service exam" (see YouTube Link 5 in the appendix). Although interactive SWM sessions often attract those studying for a common purpose, unidirectional SWMs are generally not expected to attract viewers with the same study status or purpose.

On YouTube, SWMs depict a variety of learning environments. Some show the desks (see YouTube link 6 in the appendix), and others show the face of the presenter (see YouTube link 7 in the appendix). Still others show the scenery behind the desk outside the window where the host is actually studying, for example, a panoramic view of the ocean or skyscrapers (see YouTube link 8 in the appendix). With such scenery, viewers can feel the passage of time as the sun sets, for example, thus attracting viewers who prefer a continuous visual experience with SWM content. The sounds used by creators also vary, with some using soft music or nature sounds as "white noise" (see YouTube link 9 in the appendix), while others' content is mostly silent, with only the occasional sound of a page being turned (see YouTube link 10 in the appendix).

From a post-phenomenological perspective, two things are obvious at first glance. I present these here as working theses:

1. SWM content on YouTube builds a relation of the Other, but it is designed very differently from relations of the Other in alternative motivational software, such as the ToDo apps treated by Bergen and Verbeek (Bergen and Verbeek 2021). For example, SWM content (a) represents an actual person or at least hints at the presence of an actual other learner and (b) SWM content does not address the learner directly. The learning of the other person thus disappears technologically in the background and can be sketched cursorily by the following human–technology relation:
   I (–Technology → Other/World)
2. There is a clear cultural component to this technology. SWM content is used in Asia, and it is especially popular in South Korea. Even though study programmes in other countries also include long and difficult learning phases, macroperception seems to have a decisive influence in this human–technology relationship. Macroperception, a term coined by Don Ihde, means: "the ways in which cultures embed technologies" perceived by the user in human–technology relations (Ihde 1990, 124).

In order to analyse these theses in more detail, I will proceed in three steps. First, I will describe macroperception in the use of technology by presenting the collective educational culture in South Korea. Then I outline two cases to illustrate the microperceptive phenomena (see Ihde 1990, 27) of learning with SWM content, an interview ("Lee") and a story sketched from current research and my own experience with SWM ("Kim"). I conclude with a post-phenomenological analysis of "study with me" on YouTube.

COLLECTIVE EDUCATIONAL CULTURE IN SOUTH KOREA

South Korea has a collective educational culture. From preschool to college, studying with others is deeply rooted in the educational culture of South Korea; students share hardships and develop camaraderie while also competing against each other. While group study (i.e., classes) is a common form of education globally, South Korea's collective educational culture is unique in that students often study together after school as well. In fact, learners often share educational experiences throughout their lifetime, for example, in "co-studying spaces" (Choi et al. 2022, 326), like *dock-seo-sil* and *hak-won*.

First, as a historical symbol of collective study culture, dock-seo-sil refers to a rental reading room with a desk. Created in the 1960s by those who studied together and shared information about national examinations, dock-seo-sil expanded exponentially in the 1980s (Kang 1990, 84). Kang reported that 80% of 150 high school students in Seoul had used dock-seo-sil in 1990, when such rental study facilities were rare in the world (ibid., 83). Unlike libraries, dock-seo-sils are not designed for reading, but only for studying. They often offer various services to promote concentration and comfort, such as designated desks and personal lockers. Over time, dock-seo-sils have evolved to offer different environments: traditional quiet, dark rooms where a desk lamp is the only source of light; open spaces similar to a library reading room where students sit with others; and a relaxed reading room where students can talk and drink with others while they study. From those in public education to those studying for professional exams, many South Koreans use dock-seo-sil on a daily basis. Hak-won, another unique characteristic of South Korea's collective educational culture, has emerged from private education. Many Korean students begin additional private education in kindergarten, and nine out of ten primary school students receive one or more forms of private education (Oh 2022). A very common form is hak-won, where students receive academic instruction in a school-like environment. Typically, an elementary school student goes directly to hak-won for several hours after school (Kim 2016, 54). Satisfaction with hak-won varies according to the quality of education and the characteristics of the students. Here is a second-year middle school student remembering hak-won: "I am motivated to study harder when students of similar level indulge in studying together" (ibid., 119). Another student states,

> [s]tudying with similar level students made me feel competitive. This would be a kind of competition in good faith. Studying alone never made me feel that way before. Studying together with other students in a hak-won helped me to find passion for academic achievement and to work harder. (Ibid. 184)

Because South Korean students study and compete with others in hak-won or dock-seo-sil during and after school, they internalize the collective style of education, which motivates them to seek study companions when applying for jobs and even when studying for professional development (Choi 2022; Shin 2020). However, with the development of technology and SWM content, an increasing number of South Korean learners have begun to replace real-life collective learning with SWM content (Shin 2020). This macroperception of South Korean educational culture must be taken into account when analysing SWM content as a technology, especially in the normalization of technologies for self-discipline. The following descriptions of the microperception of "Lee" and "Kim" are strongly influenced by the macroperceptual interpretation.

INTERVIEW WITH AN SWM VIEWER ("LEE")

I conducted a thirty-minute in-person interview in Korean with a South Korean SWM viewer, here pseudonymously referred to as "Lee." The interview was recorded and transcribed (Interview Transcript). Lee uses SWM to study for his foreign language proficiency exam. Lee prefers using SWM content with white noise and the host's desk (Ibid., 3–4). However, he does not watch content that shows the host's face, as it "distracts" him (Ibid.). Lee has started using SWM content because he is "used to study with others since I was young. I guess it became a habit" (Ibid., 2).

He reports two advantages of using SWM content to study. First, SWM content gives him the feeling of studying together with other people (Ibid.). It also gives him the feeling of "being supervised while studying" (Ibid.). He describes the motivation to study with SWM content as "peer pressure" (Ibid., 12). And he directly says that SWM content makes him feel like being in dock-seo-sil (Ibid., 1). Lee can't concentrate well using SWM content with timers, as it irritates him to see how much time has gone by (Ibid., 13). He states that the only downside of transitioning from collective studying to SWM is that he cannot ask others questions (Ibid., 2–3). While Lee would generally choose SWM content over studying with other people in general, studying with others allows him to ask questions and share information, which becomes an even greater advantage as the exam date approaches (Ibid., 11). Lee states he never left a comment for SWM content on YouTube because he "does not like the exposure" (Ibid.).

Lee describes using SWM content on YouTube as "the most convenient, quickest, and most accessible" way to feel like he is studying with others, as "I can use SWM at my convenience" (ibid., 8). He explains, "YouTube does not distract me from studying . . . because I do not see other features [of YouTube, such as suggested videos] when SWM content is in full-screen mode" (ibid., 10). In addition, Lee states that while his laptop distracts him from studying before using SWM content, he can avoid using it for non-educational activities when SWM content is displayed in full-screen mode (ibid., 7). While he can concentrate fully on learning when SWM content is on, there is some temptation for distraction before the content starts (ibid, 9–10). He explains: "The search bar is a big problem. There is a lot of temptation [to play non-educational content]" (ibid, 10). Although Lee believes that the use of SWM content is conducive to academic success, he does not see YouTube as an educational tool but as an 'entertainment' platform because he spends most of his time on YouTube watching non-educational content (ibid, 9–10). Nevertheless, Lee explains the effectiveness of studying with SWM content for him: "It's surprising that I can study effectively by [watching] others study. . . If SWM content is developed further, people may stop studying together in person" (ibid, 14).

KIM'S STORY

To further illustrate the microperceptual relation between SWM content and South Korean learners, this section offers the hypothetical account of Kim, who is transitioning from studying with her friends to using SWM content, which I created based on information from news articles, research, and my own experience (Shin 2020; Lee et al. 2021; Kim 2016; Kim 2022).

As a middle school student, Kim usually studies with two friends after school, except on Tuesdays and Thursdays when her friends have hak-wons. Kim and her friends are "ka-gong-jok," they are following the recent trend of studying in a café. Before they can start studying, however, they usually have to check out two or three cafés to find one that is suitable for studying for hours.

Sometimes Kim's friends break her concentration by talking during her study sessions, and one of them habitually clears his throat, which disturbs Kim. Despite the search for a café and some other annoyances, Kim knows that she studies better with her friends. Kim also enjoys spending time with her friends, getting a coffee, taking a break, and then walking home. Kim's friends tend to do better in exams than Kim, so Kim can rely on them to ask questions. In addition, when Kim does not feel like studying, the sight of her friends studying motivates Kim.

However, since the COVID-19 pandemic began, Kim has not been able to meet with her friends. All the cafés were closed, and her school and hak-won classes have gone online. Kim had to study at home. On the phone, Kim complains about the difficulty of studying at home alone, and a friend recommends SWM content. At first, Kim finds it difficult to adapt to the SWM content on YouTube, but she soon finds certain content particularly helpful. As Kim cannot concentrate for hours, she likes content with 50-minute sessions and 10-minute breaks. By trying different SWM content, Kim has learnt that she does not like complete silence, loud music, or ambient noise, so she chooses content with low levels of white noise. Kim particularly likes content created by university students because watching it motivates her to study so that she can get good grades and get into a good university.

Kim usually watches content uploaded by her favourite SMW hosts. When Kim does not feel like studying, she clicks around YouTube looking for other helpful SWM content — sometimes for more than an hour. Sometimes the search leads her to watch other (entertaining) YouTube content, but once the SWM content is on, Kim usually concentrates well. If Kim is able to study until the end of the SWM content, she will leave comments on the video reporting that she concentrated well. Sometimes the hosts respond with compliments, motivating Kim to study harder the next day.

Overall, Kim finds that using SWM content is a good way to study. The most beneficial aspect is that the SWM video plays on her phone, so Kim is not distracted by other applications. When Kim was studying with her friends, texting on the phone was a constant distraction. Although Kim misses her friends, the ability to schedule study sessions at her own convenience is also helpful. Kim is no longer distracted by chatter from other café tables, the sound of throat clearing or friends scribbling.

Interim remark: COVID-19 as the catalyst for "study with me." According to Shin's article, a sharp increase in COVID-19 cases caused new memberships in a start-up SWM platform to jump by 405% (Shin 2020). Jae Bung Choi explains: "with the rapid spread of COVID-19, new and unfamiliar methods of online studying and teleworking have been utilized, and the public is experiencing increased efficiency for working and studying" (Ibid.). "Google Trends" reported that search interest associated with "study with me" quadrupled from January 12–18, 2020 (25) to January 15–21, 2023 (98). In 2020 SWM has become a widely used educational tool in South Korea (Shin 2020).

Worldwide, South Korea is considered one of the most tech-savvy countries, with "the most digitally connected population" (Evans 2016) and the highest proportion of cellphone users in the world at 95% (Silver 2019). Because South Korean SWM content was available on YouTube years before the pandemic, many students might have already discovered that familiar technology, such as cell phones, tablets, and laptops, provided an efficient way to simulate collective study.

Additionally, YouTube serving as both an educational tool and a distraction yields various implications for technologies used in such a way.

DISCUSSION FROM A POSTPHENOMENOLOGICAL PERSPECTIVE

Against the background of the microperceptions reported in the stories of "Lee" and "Kim" and the cultural dependencies that South Korea's education system and the COVID-19 pandemic have brought with them, I will now conclude by applying the tools of postphenomenology to analyse this technology.

I will proceed in three steps. First, I will analyse the human–technology relationship here in more detail and describe self-discipline as a form in which power manifests itself in the educational system. Then, in addition to analysing microperception, I will describe the multi-stabilities of SWM content against the background of Wellner's attention theorem (2019). Finally, I will analyse the cultural embedding of the technology and clarify to what extent this macroperception would also be possible in other cultures.

SWM content as a disciplinary technique

Michel Foucault describes in *Discipline and Punish* how "docility" has two sides, a mechanical side and an educational side (2012, 136). For example, according to Foucault, Prussian King Frederick II was not only obsessed with automata, early mechanical replica of the body, but also with docile mouldability in schools and the military. In these institutions disciplinary techniques were performed to form the body as a function of power. Interestingly, for Foucault these techniques sometimes require what he calls "enclosure," that is "a place heterogenous to all others and closed in upon itself" (Ibid., 141).

The working thesis at the beginning of this paper focusses on the relation of the other, as proposed by Bergen and Verbeek with regard to to-do list apps and especially the gamified habitica (2021). These apps confront the "I" with the "self" (Ibid., 334). The interesting thing about SWM content is that another person is shown in the content, but disappears in the background, as we have heard in the stories of "Lee" and "Kim," the microperception is essentially about others, who are otherwise active co-learners, disappearing without disturbing. The situation of side-by-side learning is the important thing here. In fact, as already assumed above, it is about creating a background in which you yourself are integrated as a docile body. The fade-outs are the decisive factor in this human–technology relation:

I (–Technology → Other/World)

On another level, we have seen that this background is essentially there to control attention by what it first adds (the technology as a real human other) and then takes away.

SWM Content, Multi-stabilities, and attention

Multi-stability, "the idea that any technology can be put to multiple purposes and can be meaningful in different ways to different users," is a key concept for understanding technology from a postphenomenological perspective (Rosenberger and Verbeek 2015, 25). According to the concept of multi-stability, "[n]o technology is "one thing,' nor is it incapable of belonging to multiple contexts" (Ihde 1999, 47). Multi-stability allows examination of "both the constraining character of technology and its open-endedness," implying that "the 'same' technology may exhibit a variety of quite different 'stabilities': variations of their materialities and uses that come to surface when we inquire into practices of their use in particular historical settings" (Sergeeva 2022, 5). Examination of multi-stabilities is conducted by a close analysis of "active perceptual engagements' with those technologies, that is, how they are enrolled by the bodily active humans" (Ibid.).

In a phenomenological analysis of the affordances of laptops, Aagaard explained that although laptops can be used for educational purposes, they "constitute a backdoor through which students may occasionally escape" (2018, 6), as students are habitually distracted when they open educationally irrelevant webpages. However, SWM content allows learners' laptops to become a tool for fostering an effective study environment by preventing themselves from using them for noneducational activities. During the interview, Lee explained that although his laptop often distracted him from studying by making him want to "browse stuff, social media, like Facebook" he can avoid using it for noneducational activities when SWM content is displayed in full-screen mode (Interview Transcript, 7). Lee reported that once the content was playing, it should not be disrupted. Lee explained: "even if I want to do something else [on the laptop] because the content is on, I cannot stop it. Of course, [technically] I can stop it, but I do not want to stop it" (Interview Transcript, 7). YouTube's multi-stability as a platform for SWM content provides another interesting insight from an educational perspective. As a media platform offering myriad content and providing content suggestions based on one's viewing preferences, YouTube offers various stabilities. Kim accessed various SWM content on YouTube, thus allowing her to experience various study settings and, as a result, understand which study environment best improved her concentration. However, many, like Lee, associate YouTube with entertainment.

With SWM content, attention is the only phenomenal factor that can create stability. Wellner (2019, 48) explains that "digital multitasking," during which attention is simultaneously "distributed among several objects," is an appropriate way of understanding how people attend to their environment with the developed technology (Ibid., at 53). As Koreans have become "digital natives"

(Cha and Park 2020, 84), many may have developed an ability to simultaneously allocate attention to various objects to utilize available technologies. Even before using SWM content, Korean learners were aware of the benefits of studying together. If these learners could focus on only one thing, that is, the material they are studying, a collective learning environment with stimulations created by the presence of other people, which may distract learners from the single object they are studying, would have disturbed their learning process. During the interview, Lee explained that when SWM content is playing, he can fully "concentrate" on studying. Such concentration does not require a learner's attention to one thing; rather, it is achieved all together simultaneously, as Lee focuses on the sound from the SWM content and his learning material.

SWM Content in the South Korean Context

SWM content gives viewers control of how they study by allowing them to decide which aspects of collective learning experience they want and do not want. This is also a technological way to relate to the specific educational culture in Korea. Those who want to share their studying experiences actively with others do so by hosting SWM content or participating in interactive SWMs. To a lesser extent, others can communicate by leaving comments on content. Like Kim, users can share comments about their study experience, including what subject they studied, why, for how long, and how effective the SWM content was.

If users want to keep their study experiences private, like Lee, they can choose to do so by viewing unidirectional SWM content without engaging in such activities as leaving comments. That is, users can feel as if they are studying with others—but in private. Whatever social activity one engages in, of course, interaction is give-and-take. In studying with others, the potentially annoying and distracting sounds, and movements they make—flipping papers or pages, coughing, or responding to material aloud—must be tolerated to gain the positive effect of collaboration. However, using SWM content puts undesirable aspects, including others' opinions, schedules, and preferences, to rest. On the other hand, even when SWM content users engage in social interaction, it necessarily differs from in-person interaction. Most prominently, Kim, for example, cannot socialize with friends before and after sessions and during breaks. Neither can Kim ask them questions. Something is missing.

CONCLUSION

South Korea's collective educational culture has attracted creators and viewers on YouTube to SWM content as a convenient way to "learn with others."

This paper's postphenomenological investigation of SWM content suggests that such content has transformed laptops and other technologies, traditionally seen as distractions, into background-generating tools, they form an enclosed space, a place to study. Moreover, digital multitasking allows Korean learners to replace real-life collective learning experiences with SWM content, as they consciously select the conducive aspects of such experiences while focusing on their learning materials. In South Korea, replacing traditional methods of studying with others with SWM content provides learners with a less socially active but more individually tailored "collective" learning environment. Some aspects, such as the opportunity to ask questions and exchange ideas after a learning unit, are similarly omitted.

APPENDIX

YouTube Link 1: https://www.youtube.com/watch?v=vYvuaDf4xnQ
YouTube Link 2: https://youtu.be/R7TXQF8bW_Y
YouTube Link 3: https://www.youtube.com/live/66dAJYv9RNk?feature=share
YouTube Link 4: https://www.youtube.com/watch?v=vYvuaDf4xnQ
YouTube Link 5: https://youtu.be/OVgFHgDEDsQ
YouTube Link 6: https://www.youtube.com/watch?v=R7TXQF8bW_Y&t=42s
YouTube Link 7: https://www.youtube.com/watch?v=HMyn6eMV0OA&t=1s
YouTube Link 8: https://www.youtube.com/watch?v=VH6RUpmh4hU
YouTube Link 9: https://www.youtube.com/watch?v=8QFqUkaKmPQ
YouTube Link 10: https://www.youtube.com/watch?v=23vzcXLtslI

REFERENCES

Aagaard, Jesper. 2018. "Magnetic and Multistable: Reinterpreting the Affordances of Educational Technology." *International Journal of Educational Technology in Higher Education* 15 (1). https://doi.org/10.1186/s41239-017-0088-4.

Bergen, Jan Peter, and Peter Paul Verbeek. 2021. "To-Do Is to Be: Foucault, Levinas, and Technologically Mediated Subjectivation." *Philosophy & Technology* 34 (2): 325–48. https://doi.org/10.1007/s13347-019-00390-7.

Cha, Seungbong, and Hyejin Park. 2020. "The Effect of Expected Consistency, Cognitive Attitude, and Emotional Attitude on Reuse Intention to Use YouTube Learning." *Journal of Korea Society of Digital Industry and Information Management* 16 (2): 83–93. https://doi.org/10.17662/KSDIM.2020.16.2.083.

Choi, Seo In. 2022. "주경야독'에 쉴 틈 없다 . . . MZ직장인 업무·이직 스터디 열풍." *The JoongAng*. March 6. https://www.joongang.co.kr/article/25053190.

Choi, Woojin, and Moon-Hwan Lee. 2022. "Let's Study Together: Designing 'Study-With-Me' System with the Concept of Social Translucence." *Archives of Design Research* 35 (4): 325–41. https://doi.org/10.15187/adr.2022.11.35.4.325.

Evans, Michelle. 2016. "Meet the World's Most Connected Population and Future Home for Digital Commerce Growth." *Forbes*. December 14. https://www.forbes.com/sites/michelleevans1/2016/12/14/meet-the-worlds-most-connected-population-and-future-home-for-digital-commerce-growth/?sh=6ffaaf1dcef1.

Foucault, Michel. 2012. *Discipline and Punish. The Birth of the Prison.* Translated by Alan Sheridan. New York: Vintage.

"Google Trends." *Google*. Accessed January 20, 2023. https://trends.google.com/trends/explore?date=today%205-y&q=study%20with%20me.

Ihde, Don. 1990. *Technology and the Lifeworld. From Garden to Earth*. Bloomington, Ind.: Indiana University Press.

———. 1999. "Technology and Prognostic Predicaments." *AI & Society* 13 (1–2): 44–51. https://doi.org/10.1007/bf01205256.

Kang, Hye Won. 1990. "하숙에서 비디오까지 다양해지는 사설독서실." 중등우리교육 7: 82–89.

Kim, Chung Yeon. 2022. "온라인에서 '따로 또 같이' 공부해볼까." *Hankyoreh*. September 12. https://www.hani.co.kr/arti/society/schooling/1058318.html.

Kim, Young Chun. 2016. *Shadow Education and the Curriculum and Culture of Schooling in South Korea: Hakwon Education as a Worldwide Curriculum Question.* London: Palgrave Macmillan.

Lee, Yoonjoo, John Joon Young Chung, Jean Y. Song, Minsuk Chang, and Juho Kim. 2021. "Personalizing Ambience and Illusionary Presence: How People Use 'Study with Me' Videos to Create Effective Studying Environments." *Proceedings of the 2021 CHI Conference on Human Factors in Computing Systems*, May 6. https://doi.org/10.1145/3411764.3445222.

Oh, Jiyeon. 2022. "친구와 놀고 싶지만, 친구가 학원에 있어요.' . . . 어린이 목소리①." 교육희망. May 3. https://news.eduhope.net/24155.

Rosenberger, Robert, and Peter-Paul Verbeek. 2015. *Postphenomenological Investigations: Essays on Human-Technology Relations*. Amsterdam, Netherlands: Adfo Books.

Sergeeva, Anastasia V. 2022. "A Postphenomenological Perspective on the Changing Nature of Work." *Computer Supported Cooperative Work (CSCW)*, September 6, 2022. https://doi.org/10.1007/s10606-022-09447-2.

Shin, Mukyung. 2020. "취준생-공시족 온라인서 모이자' 캠스터디 폭증." *Donga*. March 2. https://www.donga.com/news/Society/article/all/20200302/99972444/1.

Silver, Laura. 2020. "Smartphone Ownership Is Growing Rapidly Around the World, but Not Always Equally." *Pew Research Center's Global Attitudes Project*. August 26, 2020. https://www.pewresearch.org/global/2019/02/05/smartphone-ownership-is-growing-rapidly-around-the-world-but-not-always-equally/.

Wellner, Galit. 2019. "Chapter 4. Onlife Attention: Attention in the Digital Age." In *Hyperconnectivity and Digital Reality: Towards the Eutopia of Being Human*, edited by Kathrin Otrel-Cass, 47–65. Cham: Springer Nature. https://doi.org/10.1007/978-3-030-24143-8_4.

Part II

CONCEPTUAL CONVERGENCE

Chapter 10

Ambiguous Relations

A Postphenomenological Reflection on Technological Multistability in Education

Patrizia Breil

Much of the scholarly debate around technology in education focuses on how technology can enhance learning processes. Rather than focusing on how technologies can be used to optimize what we know and what we can do, this chapter discusses the ways in which relations to technologies are always necessarily characterized by underlying experiences of ambiguity. The same technology can be experienced differently in different contexts. For example, students may use a tablet in class to create content such as educational videos, but they may also use the tablet at home to watch a movie. Constantly navigating between different ways of engaging with technology—while knowing that there may be other unknown possibilities—characterizes the notoriously ambiguous relation to technology in general. At first, it might seem that this experience is related to the fact that the technologies themselves are becoming more and more complex in nature. Experiences of ambiguity would then actually be experiences of one's own helplessness and inability to understand what is going on. Using the postphenomenological notion of multistability, I will show in this chapter that there is more to it than that and that an experience of ambiguity can actually be a valuable part of pedagogical practice.

First, I will briefly sketch both the role of ambiguity in educational research and an attempt by educational theory to improve the reputation of ambiguity, particularly in situations of technology use in educational settings. A look at postdigital and posthuman theory shows that they offer effective ways of shifting the focus beyond an output orientation and a rigid juxtaposition of the separate entities of human and technology to the in-between of human–technology relations. These relations unfold as a dense web between earlier binarisms such as analog and digital, nature and culture, or human and technology.

Actor-network theory (ANT) and critical constructivist theories in general have been made useful in challenging traditional pedagogical beliefs on a broad societal scale. I want to show that a postphenomenological perspective can add yet another layer to the discussion, allowing us to consider how the various human–technology relations that constitute pedagogical practices are actually being experienced from an individual point of view. With this classical phenomenological focus on experience, postphenomenology offers a way to conceptually grasp the quality of human's everyday interactions with technology. The multistability of human–technology relations is the basis of an interaction that has no clear initiators but is created in between. After outlining the connection between multistability, ambiguity, and transparency in postphenomenology, I argue how these concepts can be made useful for educational theory. With a postphenomenological perspective on human-technology-world relations, it becomes clear that our situational actions are imbued with technological intentionalities that require different pedagogical actions.

TECHNOLOGIES OF USE IN EDUCATION

In general, using the words *technology* and *education* in the same sentence evokes a fairly predictable set of associations. Empirical educational research, in particular, is concerned with questions concerning the use of (digital) media in the classroom. The underlying research questions of most of these studies can be summarized as follows: How can the use of technology improve teaching? How can technological devices be integrated into educational processes in ways that maximize student achievement? The common assumption is that technologies can and should be used to improve learning and teaching in some way. There are several issues to consider. First, there must be equipment and a stable Internet connection, both in schools and in the homes of students and teachers. But infrastructure changes alone will not make much of a difference. Second, it is critical that people who use these technologies know what they are doing.

With the TPACK model, Punya Mishra and Matthew J. Koehler show that, in addition to the mere availability of technological devices, teachers need to have a quite differentiated set of skills in order to use technology competently in educational settings and to acquire technological pedagogical content knowledge (i.e., TPACK; Koehler and Mishra 2009); that is, they should know (a) how to use the devices, (b) how to use them for pedagogical purposes, and (c) how to use them to get students to learn specific content.[1] In turn, the students themselves should end up with certain competencies. What exactly they should know is still under discussion and has been given many names. In addition to political initiatives such as the European Union's Digital Education Action Plan (European Commission 2021), concepts such as digital sovereignty (Pohle and Thiel 2020) or media literacy, digital literacy, or information literacy (Wuyckens and Fastrez 2022) populate the research landscape. While a common definition is difficult, the general objective of this "neoliberal logics of efficiency and optimization" (Galea and Roussel 2022: 660) is clear: We must be better. We must be better at using technologies and we must be better at living in a technologically mediated world. And more often than not, technological devices serve as instruments to achieve this kind of optimization. In this logic, technical failures and creative blackouts in the use of technological devices are obstacles to be overcome.

In addition to the focus on using technology in the classroom to improve student achievement, there are approaches in educational theory that emphasize the value of indeterminacy. This indeterminacy is also often associated with complex technologies and is based on the assumption that most technologies today are simply too complicated to understand and that therefore an autonomous use of technology is hardly possible (Allert and Asmussen 2017: 34). On the one hand, the complexity of the technologies includes the material components of their operating principles, which are hidden in a black box for the perception of the ordinary consumer. On the other hand, their structural influences on society are not easy to grasp as well. This has led to the conclusion, that indeterminacy, which is ontologically fixed and not socially made (Allert, Richter, and Asmussen 2019: 224), cannot be overcome and that we will never know, for example, how our relationships will be affected by technologies (Allert, Richter, and Asmussen 2019: 230). Therefore, Heidrun Allert and Michael Asmussen propose a transactional concept of education (*Bildung*), where transactionality is to be understood as a form of engagement with this (technologically induced) indeterminacy (Allert and Asmussen 2017: 35f.). Instead of stepping back, their reflexive approach calls for a proactive step forward and encourages interacting and living with indeterminacy. While a more welcoming approach to indeterminacy is certainly crucial, a critical and more nuanced view of technological indeterminacy in educational contexts is still needed.

At first glance, declaring technologies as inexplicable sounds like an all-too-simple coping mechanism. If there is no way to fully understand how certain technologies work, why even bother in the first place? Surely this is not quite what is meant. Van Den Eede's (2011) distinction between transparency of use and transparency of context, which will be discussed later in this chapter, can be useful here. Van Den Eede encourages the use of technologies without constantly reflecting on their existence and function. While using a technology, the technology itself can become transparent. In addition to the transparency of use, he emphasizes the importance of occasionally taking a closer look at the transparency of context that goes along with our use of technology. What becomes transparent then are the situational circumstances required for and constituted by our use of technology. Put simply, if you want to communicate with your friend via Messenger, do it. You do not need to know how encryption works. But if you care about secure communication and want to make sure that no one can read your messages, then you have to put a lot of good, hard thought into it. Depending on how we relate to technological indeterminacy, different pedagogical intentionalities may emerge in the situation.

Postdigital and posthumanist theories offer approaches that do not focus on individual technologies alone. Instead, they emphasize the relation between humans, technology, and the environment, and have already been made useful for pedagogical reflection. Although the prefix post- means different things to postdigitality and posthumanism, the approaches share a common focus on the situational relations that defy binary categorization and on the relevance of (digital) technology for our everyday practices. This shift in perspective leads to new and different understandings of the subject, its teaching, and its learning, which I will briefly outline below.

Postdigital Pedagogy

With the notion of the postdigital, which dates back to Nicolas Negroponte's 1998 WIRED column (Negroponte 1998), researchers criticize an overly stubborn focus on the digital. Yet, according to Negroponte, the digital is nothing special anymore and the truly revolutionary things will happen elsewhere. Besides, it makes no sense to make a strict distinction between the digital and the analog. What is digital is already woven into the analog lifeworld, much like the mycelium of a fungus that cannot be seen with the naked eye (Jörissen 2017). This means that research that seeks to address the challenges of digitalization must necessarily also address the invisible. Since the postdigital is "hard to define; messy; unpredictable" (Jandrić et al. 2018: 895), postdigital science and education must find ways to deal with this kind of messiness. Here is an example: Even students' handwritten essays are infused with the digital: Whether the students did research on their laptops, chatted online about their ideas, made a digital copy, or were broadly inspired by their digital surroundings and media use. The final essay may seem analogue, but in the context of its creation, it is not. The goal of postdigitalist theory is not to overcome this situation of blurriness and uncertainty, but to accept and deal with it as a contemporary condition of (co)existence in general. From the perspective of educational research, the notion of heterogeneity is helpful in this regard. Instead of juggling with an unclear distinction between offline (not necessarily analog) and online (definitely digital) tasks, teachers can describe students' situations as heterogeneous in terms of digital and analog components that are inextricably intertwined. This is not a utopian vision, but simply a way of orientation in our "multiple, layered, intra-woven rhythms of world-making" (Macgilchrist 2021: 665).

Posthumanist Pedagogy

In critical posthumanist theory (cf. e.g., Braidotti 2013), agency is no longer limited to the human realm. Non-human actors as well as technologies influence a situation in which they find themselves and co-constitute the meaning that emerges in this situation. Thus, meaning-making in technologically equipped environments takes place in between and cannot be reduced to humans doing things with technologies. For technologies are not mere instruments, but actors themselves. In education, this requires a major change of perspective. Teachers must take into account that the outcome of technologically mediated learning processes is not only a matter of teacher and student competencies, but is also influenced by the agentive qualities of the devices used. The objects of posthumanist study in education are precisely these kinds of "material-discursive entanglements" through which meaning emerges (Sheridan 2020: 1278). Learning is thus seen as an "embodied, yet collective, process that entangles humans and non-humans" (Hasse 2018: 362). With her "'messy', speculative approach" (Gourlay 2020: 21), Lesley Gourlay demonstrates what a posthumanist perspective on learning processes can look like. By conducting interviews with specific technologies, she uncovers the inherently postdigital nature of posthuman epistemic practices.

Both postdigital and posthumanist theories are good starting points for thinking about pedagogical practice. Their pedagogical value lies in their focus on recognizing the necessarily complex and entangled pedagogical situation that requires a careful ethical approach that includes technological devices as equal and constitutive partners in meaningful situations. In this chapter, I will not focus on how the "messiness" of the pedagogical situation is distributed among different actors. Instead, I want to go back to the human, without questioning the anti-anthropocentric approach of posthumanism. How is messiness being experienced? How are my actions influenced by the fact that technological devices are co-constituents of meaningful (learning) situations?

When it comes to learning as an experience, phenomenological pedagogy has proven to be a promising approach (Meyer-Drawe 2019). While classical phenomenology is not too concerned with issues of technology, postphenomenology rises to the challenge. Combining phenomenology's first-person perspective with a focus on human–technology relations, postphenomenology paints a rich and nuanced picture of technologies (Cilesiz 2020: 157) and of what our relation to technology means for us and our technologically influenced and mediated experience. By doing phenomenology with specific technologies, as is the intention of Catherine Adams' and Joni Turville's "*post*phenomenology of practice" (Adams and Turville 2018: 4), this theoretical approach offers new insights into the "messiness" that is attributed to the postdigital condition. Following a brief recap of the main concepts of postphenomenology and their differences from constructivist approaches, the notion of multistability is highlighted as a promising way of thinking about ever-changing technologically induced situations.

POSTPHENOMENOLOGICAL RELATIONS

Postphenomenology is an interdisciplinary approach that is commonly described as combining the empirical focus of science and technology studies with a phenomenological focus on experience from a first-person perspective. While phenomenology is said to have a romanticized view of technology that focuses on alienation, postphenomenology offers a more nuanced approach to theoretically grasping human–technology–world relations. At the same time, the meaning of these relations for human existence continues to play an important role, in contrast to research in science and technology studies that largely neglects this experiential dimension (Rosenberger and Verbeek 2015: 10f.).

An important premise of postphenomenology is that all experience is mediated. Therefore, human–world relations are considered in terms of mediated human–technology–world relations. These relations are the main object of research in postphenomenology and can take on different forms. Ihde's categorization of embodiment, hermeneutic, alterity, and background relations (Ihde 1990: 72–112) describes different entanglements of human, technology, and world in which the technologies shape the human experience in different more or less noticeable ways. Peter-Paul Verbeek has added cyborg relations to this categorization, taking into account a more active role of technologies and thus defending postphenomenological theory against the charge of anthropocentrism. The mediated, hybrid, or composite intentionalities associated with cyborg relations show that a situation emerges against the backdrop of a highly intertwined set of humans and non-humans in which meaning-making cannot be attributed to either site alone. Human and technology are both shaped by their relation (Verbeek 2008). With this focus on relations, postphenomenology seems to come close to actor-network theory.

Postphenomenology and Actor-Network Theory

Postphenomenology is strictly non-essentialist in that it assumes that neither humans nor technologies have a rigid essence that would withstand all situational changes. Instead, starting from a relational ontology, different relations to technologies open up new stabilities of these inherently multistable technologies. This focus on co-construction places postphenomenology in close proximity to constructivist approaches such as actor-network theory (ANT). And yet, the postphenomenological focus on individually experienced technological mediation allows for an entirely new theoretical discussion.

Although concrete contexts are the starting point of phenomenological research, its considerations are not limited to questions of mere use. Focusing on the implications of technologies for specific dimensions of human existence, the more general question is then "how technologies help to shape knowledge, politics, aesthetic judgments, normative ideas, religious experiences, et cetera" (Rosenberger and Verbeek 2015: 31). To consider the non-neutrality of technologies is not only to question whether one's own intentionality in using or being in close proximity to the technology is really one's own alone. It also means examining, on a broader scale, how one's experience in general has been changed by particular technologies. A vivid example of this is given by Don Ihde, who shows that the perception of mountains as beautiful is actually the product of technologies that somehow help humans to climb or pass over them; before that, mountains were basically just scary (Ihde 1990: 11f.).

Postphenomenology may begin with the situated experiencing individual, but it does not stop there. However, a recurring criticism of postphenomenological research still is that it is not well equipped to deal with social issues. Robert Rosenberger, in particular, has found a way to overcome this criticism by combining the approaches of postphenomenology and ANT, with the latter focusing on collective actions in societies and the world at large (Rosenberger 2020). While postphenomenological theory is tied to the first-person perspective, actor-network theory analyzes assemblages from the third-person perspective. An important feature of the ANT approach is that it allows for technological artifacts to be agents as well, creating symmetry in human–technology relations. In contrast, postphenomenology explicitly does not describe technologies as actors, so as not to anthropomorphize them by attributing human-like capabilities to them (Rosenberger and Verbeek 2015: 20). Its postphenomenological focus is not on individuals—be they human, animal, or object—but on relations. Thus agency is something that happens in between.

Looking at technologies from an ANT perspective also means looking at situational assemblages rather than separate objects without contexts. From this perspective, the agentive properties of technologies often turn out to be substitutes for those of human actors who have delegated their tasks to certain technologies (Rosenberger 2014: 374). The notion of mediation in ANT seeks to capture the role of specific technologies in a broader societal context, considering that each situational assemblage is likely to be a response to another (Rosenberger 2014: 373). Thinking about mediation here transcends the realm of individual experience and focuses on interdependencies that may be invisible to the naked eye or simply lie in the past. Because of the postphenomenological focus on experience, on both the individual and the societal level, such connections necessarily slip out of the postphenomenological perspective.[2] It is precisely the contextualization of human–technology relations in larger social contexts that raises questions of a political nature. On a methodological level, the question arises to what extent postphenomenology lacks a critical dimension (Müller 2020: 181f.), or to what extent social critique falls within the remit of critical constructivism or postphenomenology at all (Van Den Eede 2020).

Postphenomenology and Critical Constructivism

There are several approaches that combine both postphenomenological and critical constructivist theories. In each of these approaches, three of which I briefly outline below, constructivist theories add a critical dimension to the detailed postphenomenological analysis of human–technology relations.

With variational cross-examination, Rosenberger develops a method that combines the strengths of both postphenomenology and ANT (Rosenberger 2014). Following Husserl's variational analyses, this multi-perspective approach reveals different stabilities of multistable technologies. The non-essentialist version of variation is not to look for the essence that remains the same in every variation, but to look for the differences that bring about a change in the corresponding human–technology relations. In short: It is not about the things themselves, but about the relations (Rosenberger 2021). With these relations come different ontological stabilities. By including ANT in this methodological approach, it is possible to analyze why a certain stability has proven to be prevalent with a certain technology (Rosenberger 2014: 373).

Tom Børsen's account on staging the dialogue between postphenomenology and critical constructivism consists in pointing out the different subject areas of the paradigms in question. Following a didactic experiment, Børsen differentiates the theoretical strands according to their respective research focus on the thematic triangle user–expert–artifact (Børsen 2020). While postphenomenology focuses on user-artifact and expert-artifact relations, critical constructivism is more interested in user-expert and expert-artifact relations, with a special focus on possible oppression through technologically mediated relations (Børsen 2020: 241).

Markus Bohlmann shows how these different emphases can be made useful for teaching purposes in schools (Bohlmann 2022). While critical constructivism and postphenomenology can be compared in many ways, he makes a rough distinction between two categories of technologies that illustrate the different subject areas. On the one hand, according to Bohlmann, there are distal technologies that are rather distant from the bodily subject of experience. These technologies are of particular interest to critical constructivism and include, for example, school-owned clouds. On the other hand, there are proximal technologies that are close to the body and should be analyzed from a postphenomenological point of view, such as videoconferencing tools like Zoom (Bohlmann 2022: 147).

Of course, the distinction between these two categories is not clear-cut, just as the distinction between the research areas of critical constructivism and postphenomenology is not clear-cut. In general, ANT and critical constructivist approaches have been proven more than useful in highlighting the power of technologies or rather our situational entanglement with technologies on a broad societal scale. With a critical perspective, technology use can be interrogated against the background of the normative situatedness of our cultural practices. In combination with postphenomenology, these (normative) dimensions of situated human–technology relations can be understood as something that has consequences for the individual. Accordingly, the above approaches demonstrate that the combination of a more experience-based postphenomenological approach and a critical constructivist perspective on the implications of technologies at the general, societal level can provide a rich and detailed analysis of technological mediation at multiple levels. However, this should not obscure the fact that postphenomenology also claims to offer insights at the macro-perceptual level. Observations at the micro-perceptual level may well constitute the main body of postphenomenological research, but this is based on the premise that every micro-perception is grounded in and influenced by a certain macro-perception that must also be taken into account (Ihde 1990: 42). In postphenomenology, there is no either or in terms of micro- or macro-perception; it is always both, and it always starts from experience.

Multistable Relations

One way to explore how technologies shape our experience and are co-constituents of everyday practices is through the postphenomenological premise of the multistability of technologies. There are different things you can do with a certain technology, different ways of relating to technology. The specific nature of the relation has corresponding ontological consequences for the technologies, which in turn have consequences for the experiencing human and her or his perception. At the heart of it all are human–technology–world relations that, in theory, are always on the verge of becoming something else, paving the way for experiences of ambiguity. The postphenomenological concept of multistability attempts to account for this kind of entanglement. A closer look at multistabilities reveals the many facets of this notion: the multistability of technologies on both synchronic and diachronic levels, the multistability of our relations to technologies, and the limiting factors for technological stabilities. What the concept primarily seeks to capture is that "technology can be put to multiple purposes and can be meaningful in different ways to different users" (Rosenberger and Verbeek 2015: 25). This description is aimed at at least two different aspects. First, technology can be used for specific purposes, it serves as an instrument with certain material properties that allow for a limited number of uses. Second, a specific use of this instrument can still have different meanings for different users, which accounts for a multistability of perception alongside the more obvious multistability of use.

Kyle Powys Whyte (2015) describes these two aspects of multistability at a methodological level. He distinguishes between imaginative and practical multistability; the first is the imaginative attempt, from a first-person perspective, to relate to a particular technology in different ways and to evoke different meanings. With a little imagination, the optical illusion of the Necker cube can be seen in different geometric stabilities (e.g., the front is on the bottom left or instead on the top right). However, practical multistability fails at the limits of imagination because the different stabilities require a "strong embodiment" to be grasped. The stabilities cannot be imagined, they must be embodied. Using an example from Ihde, different practices of archery are described as practical stabilities that require embodiment beyond a purely imaginative approach (Whyte 2015: 73f.). Postphenomenology must therefore not be reserved for philosophers alone, but is of potential interest to any kind of practitioner who has exclusive access to the practical multistabilities in her or his technology relations. With this in mind, Galit Wellner argues that the concept of multistability is itself multistable and has different meanings for, for example, designers and engineers (Wellner 2020: 120f.).

Furthermore, as Don Ihde shows, multistability can be studied not only on a synchronic but also on a diachronic level.[3] There may be different stabilities of a certain technology over time: The cell phone, which twenty years ago was primarily a telephone, has now become a camera and a computer. These stabilities also exist at the synchronic level; one moment I am using the cell phone to make a phone call, the next moment I am using it to navigate my way to the nearest supermarket.

Frequently asked questions about multistability are how a technology can be multiple and stable at the same time, and why the stabilities are limited in the way they are. How is it that the smartphone at our ear does not suddenly become a paperweight in the middle of a conversation, but later it does? How is it that the same smartphone can be an object to be thrown, but not a car? One answer to these questions lies in the materiality of the technology, which allows or does not allow certain stabilities (Rosenberger 2014: 377). In addition, there are other, more subtle, circumstances that limit certain stabilities and favor others. Rosenberger's variational cross-examination recognizes that the stabilities of a technology can also be influenced by external circumstances that contribute to certain stabilities being more stable than others (Rosenberger 2020).

A similar argument exists in affordance theory, which builds on Gibson's (2015) work on perception: How something is perceived cannot be attributed solely to the quality of the perceiver's sensory organs, nor can it be fully explained by the physical properties of the perceived. Perception is also shaped by possible or usual modes of interaction between the perceiver and the perceived. The way we perceive a chair is largely due to the fact that "sitting on it" hovers over it like a promising offer. Gibson calls these offers "affordances." Bringing together postphenomenology and affordance theory, de Boer argues for taking a closer look at the environment of the human–technology relation. How we relate to technologies, de Boer argues, is influenced by the normative structure of our shared lifeworld, which encourages certain interactions and prohibits others (de Boer 2021: 10). The combination of affordance theory and postphenomenological multistability has also proven helpful for educational theory: Through a postphenomenological reinterpretation of technological affordances in educational contexts, technologies come into view in their more agentive role, not only offering but even inviting certain uses (Aagaard 2018: 1). In addition to teachers and students, technologies have an intentional and co-constitutive influence on the meaning of a situation.

CONDITIONS AND EFFECTS OF AMBIGUITY

Ambiguity and Multistability

With relations as the focus of postphenomenology, it is clear that multistability cannot be attributed to technology or to humans and their perception alone. There is no longer a thing in itself, but only contexts. And these contexts are multistable and account for the "ambiguity of technology in general" (Ihde 1990: 70). Elsewhere, Ihde speaks of bodily perception as having a "multistable range of ambiguity" (Ihde 1990: 30) that allows for different relations in multiple contexts. Multistability seems to lie in between, while humans and technology each have an ambiguous structure that serves as a condition for generally multistable relations that can be experienced as ambiguous. Technological ambiguity is expressed in the fact that technologies in certain uses both are and are not what they are used as. Any stability is unstable. The magnification-reduction structure that postphenomenology describes in terms of embodiment relations can also be observed at the level of stabilities. In a standard embodiment relation with glasses, things become visible while other things become invisible. You may be able to see an advertisement some distance away that you did not see before, but at the same time, you may not be able to see out to the sides as easily because of the frame of your glasses. In the same way, a certain stability allows you to perceive a technology in a certain way, while other stabilities elude your field of perception. When we enter a stable relation with a particular technology, there are always several other stabilities that we have chosen, consciously or unconsciously, to ignore.[4]

The relational ontology of technology is (physically) expressed in different human–technology relations and different stabilities of the technologies at hand. Depending on the current stability, the perceived environment changes accordingly. If another stability comes to the fore, the whole situation can change. Every human–technology relation takes place in the face of multiple possibilities. This back and forth between stabilities is experienced as ambiguity; there is no fixed meaning to what we relate to. The meaning we ascribe to the technology is the product of a situational relation that is (for the moment) bound by a certain stability, without being generally stable. Experienced stability is a possible mode of being of the technological device. But what's essential is that there is no essence at all. The classical phenomenological search for the "thing itself" necessarily fails. There is not one. There are many.

When it comes to ambiguity as a condition of our being-in-the-world, or even as a mode of our being, phenomenology provides helpful theoretical approaches that can also be used in relation to technology. Merleau-Ponty, in particular, is considered a philosopher of ambiguity. In his phenomenology of the body, ambiguity is linked to the double sensation of the body, which can simultaneously touch and be touched (Merleau-Ponty 2010: 118). Like Ihde's ambiguous structure of the body (Ihde 1990: 30), Merleau-Ponty's lived body is the central part of the human condition that participates in the meaning-making in interaction with others. In his later works, Merleau-Ponty devotes more attention to a relational being-in-the-world, and develops a theory of the flesh that captures the ambiguity of this in-between. The flesh is not matter, but the principle of being itself, from which all being is chiasmatically woven (Merleau-Ponty 1994: 183f.). Flesh is the basis of all that is visible, but itself not visible. In relating to anything, one is always confronted with this simultaneity of visibility and invisibility. The core of ambiguous relations lies in their foundation in the flesh. Postphenomenological research has already made use of Merleau-Ponty's notion of the flesh and shown the extent to which this chiasmatic entanglement can be mediated technologically (de Boer and Verbeek 2022).

An existential-phenomenological approach to ambiguity is offered by Simone de Beauvoir. Again, ambiguity is grounded in the lived body. Beauvoir's phenomenology is non-essentialist; human beings do not have a fixed essence to fall back on. Instead, they are forced to constantly (re)invent themselves. The ambiguity of human existence lies precisely in the fact that these projects have no meaning except the meaning that people give to them themselves. In this process, others are also co-constitutively involved in meaning-making. Postphenomenology takes this interconnectedness one step further by including technologies and their intentionalities in this process of meaning-making. Nevertheless, Beauvoir shows how to deal with ambiguity in a moral way. Accepting ambiguity as a condition of existence and of relations to others (and other beings and things) is at the heart of her moral philosophy. With this acceptance, the relations need not be characterized by fear of instability, but can be entered into responsibly and openly (de Beauvoir 1997: 83f.).

These different kinds of ambiguity within human existence, which have not yet been sufficiently addressed in postphenomenological research, can also be found in human–technology relations. Moreover, it is precisely in the ambiguity of the relations that a great pedagogical value lies. Although it has been shown elsewhere that this lived ambiguity should be a central goal of education (Breil 2021: 260–271), postphenomenological research on this ambiguous relation to and through technologies remains to be done.

Ambiguity and Transparency

In addition to multistability, there is another way in which invisibility and ambiguity find their way into human–technology relations. In a sense, the technology itself becomes transparent when in use. This is obvious in the case of embodiment relations. The more accustomed you are to your glasses, the less likely you are to be aware that you are wearing them. They become transparent in two ways: you see through them, and you no longer notice them. This is also true of an implant, which can be registered as a kind of embodiment^{+} relation, where the boundaries between human and technology are blurred beyond reversibility, as in Verbeek's fusion relation. Your heart implant allows you to perceive the world, in a sense you experience the world through the implant. At the same time, after initial feelings of alienation, you may forget that the implant is there. While the embodied glasses can be easily removed, it would require invasive surgery to remove the implant from the body, making it a more extreme form of embodiment because it is not as easily reversible. In hermeneutic relations, technology becomes more transparent the better you know how to read it. One look at the thermometer is enough to know how to dress to go out. The thermometer is the means to perceive the heat outside and is hardly noticed itself. This changes when the temperature is no longer displayed in Fahrenheit but in Celcius or vice versa. The technology is not as transparent as before, and you become aware that you are not perceiving the weather outside, but numbers on a technological device.

If all experience is mediated, as postphenomenology suggests, it is all too easy to overlook the concrete nature of this mediation. Drawing on various theoretical horizons, Yoni van den Eede distinguishes two types of transparency: One is the transparency of use, as discussed with the glasses through which one can see; the other is the transparency of context, in which the context becomes transparent and disappears from the field of perception (Van Den Eede 2011: 154). From a postphenomenological point of view, the transparency of context is probably best explained in terms of the magnification-reduction structure. To bring something into focus means to put something out of focus at the same time. As shown above, the same is true for different stabilities. To enter a particular stable relation means to disregard several other possible stabilities. The challenge is to get a hold of several transparencies at the same time. While Van Den Eede believes that transparency of use is reasonable, at least in situations where the technology is actually used and needs to be used, he argues strongly for being aware of the transparent contexts in relations, especially in theoretical research. If structures are to change, we must also be aware of what we might be missing: "It's here. And it's not here. It's in between us" (Van Den Eede 2011: 157).

One goal of pedagogical practice is to look at these transparencies now and then. It may not be possible to unblack every box, but we can become aware that the boxes are just that: boxes, and not the stuff of nightmares that dystopian narratives see in them. Especially in educational contexts, the doubt that inhabits human–technology relations can be productively used to create space for creativity (Grünwald 2021: 257f.). Postphenomenology offers concepts that help to open up educational practice to ambiguous and productive relations with technologies. In particular, the concepts of multistability and transparency allow for a theoretical understanding of ambiguity as a space for creative engagement and change.

ENDURING AMBIGUITIES

With the increasing convergence of media, it has become normal that technologies can be used for different things and therefore have different stabilities. When using technologies in educational contexts, the various corresponding relations need to be considered as influences on pedagogical practice. For example, tablets can be used as televisions and as dictionaries, among other stabilities. When using tablets to search for information, students may be tempted to succumb to a different stability and relate to the tablet in a student-entertainment or a student-scribble manner (see Bohlmann, Wilmer in this volume). Not only resisting these underlying affordances, but becoming aware of the different stabilities that are constitutive of the situation can be seen as an important aspect of media literacy. As the number of stabilities increases, so does the complexity of competent media use. From the perspective of empirical educational research, the competent use of technology in educational settings is a matter of optimization and limiting ambiguity. Postphenomenology disrupts this linear thinking. Here, the intentional and competent relation to technology involves the knowledge of the inherently ambiguous structure of the relation, which defies any quantitative determination.

Incorporating technology into education is not just about working with the current stabilities in human–technology relations. Instead, the currently unrealized stabilities also play an important role in how we relate to technologies. The postphenomenological concepts of multistability and transparency have proven helpful in understanding why different people in different times relate to technologies differently, and why this can lead to experiences of ambiguity. The relation to technologies is not a one-way street. What can be experienced as "messy" and ambiguous turns out to be a set of possibilities that run through our daily lives with technologies. With such a non-foundational approach, pedagogical theory must learn to deal with specific technologies as well as with their multistable properties. The pedagogical potential of a transactional understanding of *Bildung* can be applied to dealing with technologies that are not indeterminate per se, but which are and are not what they are used for, and which can evoke an experience of ambiguity and lack of understanding. By taking a postphenomenological point of view on human–technology relations, we can distinguish between a lack of understanding caused by a lack of technical knowledge, which must be the subject of education, and a second, positive lack of understanding, which lies in the ambiguous relations and the multistable ontology of objects. Thus, the desire to understand technology is not only rooted in a humanistic, performance-oriented obsession with control, but rather in the need for an autonomous approach to technology that always remains open and calls for change.

NOTES

1. The same structure holds true for the notion of DPACK, which puts an emphasis on the specifics of digitization (Huwer et al. 2019).

2. With the concept of sedimentation, Rosenberger takes into account that human–technology relations can be strongly influenced by previous experience (Rosenberger and Verbeek 2015: 25). For example, interpreting MRI-images becomes easier the more times you have done it. However, this notion of sedimentation only accounts for individual sedimentation. Individuals can benefit from a kind of collective knowledge, but they are still bound by what they experience themselves.

3. Don Ihde's essay on the history and phenomenology of instrumentally mediated music would be an example of a postphenomenological take on multistability that researches on a diachronic level (Ihde 2007).

4. Rosenberger argues that the field of perception can be altered even more radically through use of technology. He uses the term *field composition* to describe phenomena in which entire physical environments in plain view fade into the background while other parts stand out: In a movie theater, we focus on the screen, not (at best) on the people in front of us (Rosenberger and Verbeek 2015: 24).

REFERENCES

Aagaard, Jesper. 2018. "Magnetic and Multistable: Reinterpreting the Affordances of Educational Technology." *International Journal of Educational Technology in Higher Education* 15 (4). https://doi.org/10.1186/s41239-017-0088-4.

Adams, Catherine, and Joni Turville. 2018. "Doing Postphenomenology in Education." In *Postphenomenological Methodologies. New Ways in Mediating Techno-Human Relationships,* edited by Jesper Aagaard, Jan Kyrre Berg Friis, Jessica Sorenson, Oliver Tafdrup, and Catherine Hasse, 3–25. Lanham: Lexington Books.

Allert, Heidrun, and Michael Asmussen. 2017. "Bildung als produktive Verwicklung." In *Digitalität und Selbst. Interdisziplinäre Perspektiven auf Subjektivierungs- und Bildungsprozesse*, edited by Heidrun Allert, Michael Asmussen, and Christoph Richter, 27–68. Bielefeld: Transcript.

Allert, Heidrun, Christoph Richter, and Michael Asmussen. 2019. "Produktive Dinge. Fragen, die wir nicht gestellt haben." In *Pädagogik als Disziplin und Profession. Historische Perspektiven auf die Zukunft,* edited by Manfred Böge and Marc Fabian Buck, 217–239. Berlin: Peter Lang.

Bohlmann, Markus. 2022. *Bildung—Philosophie—Digitalisierung. Eine Curriculumtheorie.* Berlin: J. B. Metzler.

Børsen, Tom. "Bridging Critical Constructivism and Postphenomenology at Techno-Anthropology." *Techné: Research in Philosophy and Technology* 24 (1/2): 218–246.

Braidotti, Rosi. 2013. *The Posthuman.* Cambridge: Polity.

Breil, Patrizia. 2021. *Körper in Phänomenologie und Bildungsphilosophie. Körperliche Entfremdung bei Merleau-Ponty, Waldenfels, Sartre und Beauvoir.* Opladen, Berlin, Toronto: Budrich.

Cilesiz, Sebnem. 2020. "Making Meaning Through Lived Technological Experiences." In *Phenomenological Inquiry in Education. Theories, Practices, Provocations and*

Directions, edited by Edwin Creely, Jane Soutcott, Kelly Carabott, and Damien Lyons, 148–160. London: Routledge.

de Beauvoir, Simone. 1997. "Für eine Moral der Doppelsinnigkeit." In *Soll man de Sade verbrennen? Drei Essays zur Moral des Existentialismus*, edited by Simone de Beauvoir, 77–192. Translated by Alfred Zeller. Reinbek bei Hamburg: Rowohlt.

de Boer, Bas. 2021. "Explaining Multistability: Postphenomenology and Affordances of Technologies." *AI & Society*, 38: 2267–2277. https://doi.org/10.1007/s00146-021-01272-3.

de Boer, Bas, and Peter-Paul Verbeek. 2022. "Living in the Flesh: Technologically Mediated Chiasmic Relationships (in Times of a Pandemic)." *Human Studies* 45: 189–208. https://doi.org/10.1007/s10746-022-09625-7.

European Commission. 2021. *Digital Education Action Plan 2021–2027. Resetting Education and Training for the Digital Age.* https://ec.europa.eu/education/education-in-the-eu/digital-education-action-plan_de.

Galea, Simone, and David Roussel. 2022. "Rethinking Teaching: Alternative Ontologies of Educational Praxis and Thought." *Discourse: Studies in the Cultural Politics of Education* 43 (5): 659–665.

Gibson, James J. 2014. *The Ecological Approach to Visual Perception.* New York: Psychology Press.

Gourlay, Lesley. 2020. *Posthumanism and the Digital University. Texts, Bodies and Materialities.* London, New York: Bloomsbury Academic.

Grünwald, Jan. G. 2021. "Embracing Doubt: Teaching in a Post-Digital Age." In *Post-Digital, Post-Internet Art and Education. The Future is All-Over*, edited by Kevin Tavin, Gila Kolb, and Juuso Tervo, 257–267. Cham: Palgrave Macmillan. https://doi.org/10.1007/978-3-030-73770-2_15.

Hasse, Catherine. 2018. "Posthuman Leaning: AI From Novice to Expert?" *AI & Society* 34: 355–364. https://doi.org/10.1007/s00146-018-0854-4.

Huwer, Johannes, Thomas Irion, Sebastian Kuntze, Steffen Schaal, and Christoph Thyssen. 2019. "Von TPaCK zu DPaCK—Digitalisierung im Unterricht erfordert mehr als technisches Wissen." *Der mathematische und naturwissenschaftliche Unterricht* 5: 358–364.

Ihde, Don. 2007. "Technologies—Music—Embodiments." *Janus Head* 10 (1): 7–24.

———. *Technology and the Lifeworld. From Garden to Earth.* Bloomington: Indiana University Press.

Jandrić, Petar, Jeremy Knox, Tina Besley, Thomas Ryberg, Juha Suoranta, and Sarah Hayes. 2018. "Postdigital Science and Education." *Educational Philosophy and Theory* 50 (10): 893–899. https://doi.org/10.1080/00131857.2018.1454000.

Jörissen, Benjamin. 2017. "Subjektivation und 'ästhetische Freiheit' in der post-digitalen Kultur." *kubi-online*. https://doi.org/10.25529/92552.421.

Koehler, Matthew, and Punya Mishra. 2009. "What is Technological Pedagogical Content Knowledge (TPACK)?" *Contemporary Issues in Technology and Teacher Education* 1: 60–70.

Macgilchrist, Felicitas. 2021. "Theories of Postdigital Heterogeneity: Implications for Research on Education and Datafication." *Postdigital Science and Education* 3: 660–667. https://doi.org/10.1007/s42438-021-00232-w.

Merleau-Ponty, Maurice. 2010. *Phänomenologie der Wahrnehmung*. Translated by Rudolf Boehm. Berlin: de Gruyter.

———. 1994. *Das Sichtbare und das Unsichtbare. Gefolgt von Arbeitsnotizen*. Translated by Regula Giuliani and Bernhard Waldenfels. München: Fink.

Meyer-Drawe, Käte. 2019. "Lernen als Erfahrung (2003)." In *Phänomenologische Erziehungswissenschaft von ihren Anfängen bis heute. Eine Anthologie,* edited by Malte Brinkmann, 423–434. Wiesbaden: Springer.

Müller, Oliver. 2020. "Postphänomenologie. Über eine technikphilosophische Methode." *Phänomenologische Forschungen* 2: 165–184.

Negroponte, Nicolas. 1998. "Beyond Digital." *Wired* 12. https://www.wired.com/1998/12/negroponte-55/.

Pohle, Julia, and Thorsten Thiel. 2020. "Digital Sovereignity." *Internet Policy Review* 9 (4). https://doi.org/10.14763/2020.4.1532.

Rosenberger, Robert. 2021. "Technological Multistability and the Trouble with the Things Themselves." In *The Oxford Handbook of Philosophy of Technology,* edited by Shannon Vallor, 374–392. Oxford: Oxford University Press. https://doi.org/10.1093/oxfordhb/9780190851187.013.42.

———. 2020. "On Variational Cross-Examination: A Method for Postphenomenological Multistability." *AI & Society* 38: 2229–2242. https://doi.org/10.1007/s00146-020-01050-7.

———. 2014. "Multistability and the Agency of Mundane Artifacts: from Speed Bumps to Subway Benches." *Human Studies* 37: 369–392. https://doi.org/10.1007/s10746-014-9317-1.

Rosenberger, Robert, and Peter-Paul Verbeek. 2015. "A Field Guide to Postphenomenology." In *Postphenomenological investigations. Essays on human-technology relations,* edited by Robert Rosenberger, and Peter-Paul Verbeek, 9–41. Lanham: Lexington Books.

Sheridan, Mary P., Amélie Lemieux, Ashley Do Nascimento, and Hans Christian Arnseth. 2020. "Intra-active Entanglements: What Posthuman and New Materialist Frameworks Can Offer the Learning Sciences." *British Journal of Educational Technology* 51 (4): 1277–1291.

Van Den Eede, Yoni. 2020. "The Purpose of Theory. Why Critical Constructivism Should, 'Talk' and Postphenomenology Should 'Do'." *Techné. Research in Philosophy and Technology* 24 (1/2): 114–137.

———. 2011. "In Between Us: On the Transparency and Opacity of Technological Mediation." *Foundations of Science* 16: 139–59. https://doi.org/10.1007/s10699-010-9190-y.

Verbeek, Peter-Paul. 2008. "Cyborg Intentionalities: Rethinking the Phenomenology of Human-Technology Relations." *Phenomenology and the Cognitive Sciences* 7: 387–395. https://doi.org/10.1007/s11097-008-9099-x.

Wellner, Galit. 2020. "The Multiplicity of Multistabilities: Turning Multistability into a Multistable Concept." In *Reimagining Philosophy and Technology, Reinventing Ihde*, edited by Glenn Miller, and Ashley Shew, 105–122. Cham: Springer. https://doi.org/10.1007/978-3-030-35967-6_7.

Whyte, Kyle Powys. 2015. "What is Multistability? A Theory of the Keystone Concept of Postphenomenological Research." In *Technoscience and Postphenomenology. The Manhattan Papers* edited by Jan Kyrre Berg O. Friis and Robert P. Crease, 69–81. Lanham: Lexington Books.

Wuyckens, Géraldine, and Pierre Fastrez. 2022. "Untangling Media Literacy, Information Literacy, and Digital Literacy: A Systematic Meta-review of Core Concepts in Media Education." *Journal of Media Literacy Education* 14 (1): 168–182. https://doi.org/10.23860/JMLE-2022-14-1-12.

Chapter 11

Freedom through Restriction?

Merleau-Ponty, Dewey, and Foucault on Habit-Formation

Jesper Aagaard

For the past ten years, I have been researching the relationship between digital devices, attention, and distraction as these phenomena coalesce and intertwine in the classroom. In that period, I have become increasingly intrigued with the idea that contemporary schools should be concerned with educating students' attention. As Joris Vlieghe (2017) puts it, "the core task of education is precisely to *form* attention, in the sense of disciplining people who are not inclined to sit still for say an hour" (p. 402). While critical educational scholars are likely to balk at the notion of "disciplining" students and "forming" their attention, the purpose of this chapter is to provide a postphenomenological defense of this idea. To do so, I want to discuss *habit-formation*. I first draw on phenomenology and pragmatism to argue that habit-formation constitutes an inescapable aspect of human existence. I then move on to discuss why critical educational scholars are skeptical about educational habit-formation or "discipline." Finally, however, I turn to my own research on digital distraction to argue that, to the extent that contemporary attention is conjoined with tech habits, it makes good sense to talk about disciplining attention.

ON THE IMPORTANCE OF HABIT-FORMATION: PHENOMENOLOGY AND PRAGMATISM

Since its inception, postphenomenology has prided itself in being a theoretical mixture of phenomenology and pragmatism. As Don Ihde (2009) puts it: "Postphenomenology is a hybrid phenomenology. On the one hand, it recognizes the role of pragmatism in the overcoming of early modern epistemology and metaphysics [. . .] On the other hand, it sees in the history of phenomenology a rigorous style of analysis through the use of variational theory, the deeper phenomenological understanding of embodiment and human active bodily perception, and a dynamic understanding of a lifeworld as a fruitful enrichment of pragmatism" (p. 23). Or, as he more succinctly puts it elsewhere, "phenomenology + pragmatism = postphenomenology" (Ihde, 2012: 128). In the rest of this section, I want to explore some points of convergence between these two traditions that I believe are yet to be fully discussed in the context of postphenomenology. More specifically, I want to argue that phenomenology and pragmatism have developed related understandings and vocabularies of "habit" that help us transcend the age-old dichotomy between empiricism and intellectualism.

First, as a sort of historical curiosity, it is interesting to note that John Dewey (1896) and Maurice Merleau-Ponty (2002) used the same candlelight example to emphasize the tight coupling between experience and world. "The light of a candle changes its appearance for a child," Merleau-Ponty (2002) writes, "when, after a burn, it stops attracting the child's hand and becomes literally repulsive" (p. 60). By burning her finger, the child learns that the candle is hot, and this embodied understanding is henceforth present in her perception of the candle as "a-light-that-means-pain-when-contact-occurs" (Dewey, 1896: 360). On the one hand, this deceptively simple example transcends an empiricist account of mechanical reflexes, because, as Nick Crossley (2001) explains, "A child who withdraws their hand from a hot object and thus relieves the pain caused by the heat does not thereby learn to retract their hands from hot objects; they learn not to touch such objects in the first place" (p. 126). On the other hand, the example also challenges an intellectualist account of learning, because what is learned in this situation is not *represented* in the child's mind and added onto experience, it is *presented* in the way that the candle shows up to her (Dreyfus, 2002).

Phenomenology and pragmatism are also united by the importance they accord to habits. "Man is a creature of habit," Dewey (2007) argues, "not of reason nor yet of instinct" (p. 125). Evidently, then, the concept of habit occupies an important middle ground between the Scylla of intellectualism ("reason") and the Charybdis of empiricism ("instinct"). But what does this ground consist of? Merleau-Ponty (2002) provides the answer: "If habit is neither a form of knowledge nor an involuntary action, what then is it? It is knowledge in the hands, which is forthcoming only when bodily effort is made, and cannot be formulated in detachment from that effort" (p. 166). Habits constitute a somatic form of knowledge, an embodied know-how, which is why many phenomenologists use the term synonymously with skill (Dreyfus, 2004). Dewey (2007) concurs with this sentiment and argues that knowledge lives in "muscles, not in consciousness" (p. 177), but Dewey also warns us against discussing habits in terms of skills, because doing so risks portraying habits as mere technical abilities that we call into action at will. Instead, he prefers to discuss bad habits, because in this case it becomes obvious that habits transcend our conscious decision-making.

Phenomenology and pragmatism do not to deny that we sometimes deliberate about what to do or that we occasionally react rather instinctively to things, but they do emphasize that most of our waking moments are spent in the prereflective mode of habit. In fact, our habits are even active when we sleep. "The habit of walking is expressed in what a man sees when he keeps still, even in dreams" (Dewey, 2007: 37). Here, we see the projective quality of habits, the way our habits shape our perception. Furthermore, while habits may be grounded in our bodies, they blur the distinction between body and world: "Habits are ways of using and incorporating the environment in which the latter has its say as surely as the former' (Dewey, 2007: 15). Or, as Merleau-Ponty (2002) puts it: "Habit expresses our power of dilating our being-in-the-world or changing our existence by appropriating fresh instruments" (p. 166). The body is not a fixed entity, but a fleshy network that incorporates and extends over a range of material artifacts. Merleau-Ponty (2002) gives various examples of this phenomenon such as a woman intuitively dodging doorframes when wearing a feathered hat, a blind man skillfully using his stick to navigate, and an experienced driver effortlessly parking his car.

Finally, habits are imbued with normativity. To illustrate this idea, Hubert Dreyfus (2014) gives the example of distance-standing: What constitutes an appropriate distance between two people engaged in conversation cannot be formalized into explicit rules, but because our bodies have such a long history of being socialized into what feels appropriate, some distances simply *feel better* than others, and our bodies are naturally drawn to maintaining such distances. "All habits are demands for certain kinds of activity; and they constitute the self. In any intelligible sense of the word 'will,' they *are* will" (Dewey, 2007: 25). Thus, whenever we meet a person that does not share our sense of personal space (what Seinfeld calls a close talker), we intuitively step back. This happens prereflectively, without us having to think about it. During our upbringing, we are thrown into a web of such normative styles of comportment, which eventually coalesce into a certain understanding of being. "For instance, we train our youth to stand respectfully before their elders, to bow at appropriate moments, not to raise their voices, to use certain forms of address. In all this, they learn to respect, even revere, their parents and elders" (Dreyfus & Taylor, 2016: 116). As a result, habits shape who we are. Or, as Charles Sanders Peirce (1898) puts it: "Man is a bundle of habits" (CP 6.228).

ON THE DANGERS OF HABIT-FORMATION: FOUCAULT AND CRITICAL EDUCATIONAL THEORY

Let us now explore why it is understandable (and indeed laudable) that critical educational scholars view the notion of habit-formation with such pronounced skepticism. One of the main architects of this type of critique is Michel Foucault. As a backdrop to this discussion, recall that Foucault (2003) famously argued that power is not substantial but relational and consists of techniques and strategies used to direct the behavior of others, which led him to define power as *the conduct of conduct*. Related to this idea, Foucault (1980a) also challenged the then-reigning juridical conceptualization of power as a repressive mechanism that prohibits, censors, and refuses, and instead argued that power is "productive" in the sense that it "produces effects at the level of desire" (p. 59). In a way, this analysis makes power much more insidious: Rather than suppressing subjectivity from the "outside," power works by affecting us from the "inside." Finally, Foucault argued that power relations traverse our bodies in ways that bypass our rational capacities. As Foucault (1980b) himself put it, "power relations can materially penetrate the body in depth, without depending even on the mediation of the subject's own representations" (p. 186). Thus, much like phenomenology and pragmatism, Foucault viewed the human body as an important locus of analysis.

In *Discipline and Punishment* (1975/1991), Foucault analyzes a bodily form of power called disciplinary power, which originated in eighteenth century institutions like prisons, barracks, factories, and schools that set out to increase the body's utility while increasing its obedience. A large part of the book consists of detailed instructions for how specific bodies were to be subjected to specific training practices like military drillings, athletic exercises, and pedagogical instructions. In fact, these kinds of institutional training practices are essentially what constitute disciplinary power. "The chief function of the disciplinary power is to 'train,'" as Foucault puts it (p. 170). For our purposes, I therefore propose to interpret disciplinary power in terms of *institutional habit-formation.* One of the institutions in which such habit-formation takes place is school. Here, students are taught to read, write, sit still, and raise their hands before speaking. While such habits may appear benign, Foucault reminds us that there is an unmistakable element of control at stake in their formation. As an example, Foucault provides a lengthy excerpt in which educational reformer Jean-Baptiste de La Salle meticulously describes how students' bodies are to be positioned during the process of writing, right down to the distance to be left between body and table ("two fingers worth"). Evidently, then, writing is not just about expression of thought; it is also about control of the body.

In the first instance, disciplinary power strives for immediate and unthinking obedience. "All the activity of the disciplined individual must be punctuated and sustained by injunctions whose efficacity rests on brevity and clarity; the order does not need to be explained or formulated; it must trigger off the required behavior and that is enough" (p. 166). In the long run, however, prolonged exposure to disciplinary power will result in people eventually coming to "embody" or "incorporate" its norms. "The perfection of power should render its actual exercise unnecessary" (p. 201). When you have acquired the habit of raising your hand before speaking, for example, it *feels wrong* to start talking without raising your hand. You *want to* raise that hand, which renders external control superfluous. This is what Foucault means when he says that power produces effects at the level of desire. In making these arguments, Foucault ultimately inverts Plato's motto and claims that the soul is the prison of the body.[1] As such, Foucault's portrayal gives rise to a rather bleak view of schools as manufacturers of what he calls *docile bodies*. Now, Foucault was by no means alone in this critique of schooling. Around the same time, neo-Marxist scholars began to argue that school is in fact an Ideological State Apparatus (ISA), whose chief function is to reproduce capitalist ideology and distribute the skills needed to sustain this hegemonic world order (Althusser, 1971).

Critical educational theory views schooling as an instrument used by the ruling class to transmit its values, beliefs, and ideology to the working class, and to prepare individuals for their predetermined social roles. "The structure of the educational experience," Bowles and Gintis (1976) wrote, "is admirably suited to nurturing attitudes and behavior consonant with participation in the labor force" (p. 9). Such attitudes include an ability to focus on tedious and repetitive tasks, an indiscriminate acceptance of hierarchy and authority, and a willingness to obey orders. This is what is known as school's *hidden curriculum*. Because critical educational theory perceives educational habit-formation as a means of control, any notion of teachers "helping" students cultivate habits will set its alarm bells ringing. Indeed, critical educational scholars would agree with phenomenologists that habits instill a certain understanding of being, but they would replace a positive (if slightly quaint) term like respect with critical terms like docility and obedience.[2] Colloquially put, we might say that critical educational scholars are worried about students getting "brainwashed" by school, although that term obviously misrepresents the bodily and habitual origins of such indoctrination. In this dialectical model, students either have the choice between subjecting themselves to educational strategies of control or developing tactics that resist them (de Certeau, 1984).

EDUCATIONAL HABIT-FORMATION: FROM DIALECTICS TO HERMENEUTICS

We have now looked at two different accounts of habit, one from a combination of phenomenology and pragmatism (i.e., postphenomenology), the other from critical educational theory. Both traditions agree that habits are embodied and prereflective: We are concerned with a direct bodily responsiveness that does not depend on the mediation of representations, as Foucault puts it. Both traditions would also agree that habits are projective, normative, and constitutive. Note, for example, the similarity between the idea that habits constitute our "will" and the contention that power produces effects at the level of "desire." Having said that, however, the two traditions differ dramatically in their assessment of what this means: While postphenomenology seems quite favorably disposed toward habits, critical educational theory perceives habits as dangerous entities. How are we to make sense of this difference? Well, one option is to argue that the two traditions are simply describing two different things. This understanding has gained a strong foothold in postphenomenology, where Ihde (2002) has distinguished between a Merleau-Pontyian body and a Foucauldian body: "Merleau-Ponty's body is the active body, filled with actional experience, in contrast to the culturally fixed and acted upon body of Foucault" (p. 26). While I have found this heuristic useful, I do worry that pushing its binary logic too far ends up reproducing an unhelpful split between these two thinkers (see also Crossley, 1996).

Foucault has long been critiqued for consigning subjects to pure passivity. Jürgen Habermas (1990), for instance, argues that Foucauldian subjects can only be viewed as "standardized products of some discourse formation" (p. 293), while Charles Taylor (1986) laments how Foucault's notion of power leaves us without "at least the idea of liberation" (p. 92). A popular but for our purposes unhelpful strategy for defending Foucault against these charges is by appealing to the last two volumes of *The History of Sexuality* in which Foucault investigates how people fashion themselves as ethical subjects. As Benda Hofmeyr (2016) puts it: "The self is now no longer considered as the passive product of an external system of constraint and prescriptions, but as the active agent of its own formation" (p. 216). The problem with this strategy is that it cedes the idea that *Discipline and Punish* portrays the self as a "passive product," which does not exactly help our project. For us, it is better to ask why it is so tempting to read the book as structuralist in the first place. Here, I think Foucault's treatment of his source material is at least partly to blame: Disciplinary instruction manuals may have a goal of turning unruly intentionality into mechanical regularity but realizing this empiricist metaphysics is (thankfully) impossible.[3] We should never mistake such ideals for actual reality. While I do not think Foucault himself commits that mistake, his gloomy overall narrative and many verbatim (and comment-free) citations of these manuals certainly do not help readers.

What happens to the concept of discipline when it is divorced from its empiricist ambitions? When its outcome is not mechanical reflexes, but bodily habits? Before we discuss this issue, we should acknowledge two things: First, critical educational theory has aptly demonstrated that there is always power at stake in habit-formation, a fact that is not always evident when postphenomenological scholars like myself draw on the concept. Second, much like STS has done for technology, critical educational theory has shown that school is decidedly "non-neutral": Schools do not just decide what we learn, they also shape who we become. While I wholeheartedly agree with the critical discourse in these respects, I worry that its legitimate *critiques* of educational habit-formation end up becoming an unhelpful *dismissal* of this practice altogether: First, much critical educational theory operates with a reified (and distinctly un-Foucauldian) concept of power in which schools have strategies and power, while students have tactics and resistance (Butin, 2001). This framework correlates with a rigid normativity in which school is portrayed as "the bad guy," which makes it difficult (if not impossible) to see students' and school's interests as in any way aligned. Secondly, critical educational theory often implies that educational habit-formation is dangerous because it originates externally (this is the idea of the self as a "passive product" that we saw earlier).

If there is one thing that Foucault taught us, however, it is that *everything* originates "externally" and that there is no inner sanctum in which to seek refuge. As such, habit-formation is not an illegitimate encroachment on human autonomy, but an ever-present part of our existence. Postphenomenology wholeheartedly agrees with this sentiment. Of course, opposing an externalist view of habits does not give schools a carte blanche to indoctrinate students however they want, it simply challenges the assumption that schools are always on the wrong side of history. Borrowing a distinction from Peter-Paul Verbeek (2013), we might say that this argument moves us from a *dialectical* to a *hermeneutic* understanding in which educational habit-formation goes from being an oppressive force to becoming an inescapable part of schooling. If we grant this argument, it follows that we cannot get rid of educational habit-formation unless we dismantle the institution of school altogether. And even then, the process of habit-formation would just take place elsewhere. So yes, we should be aware that schooling does not "liberate" students from power, but that does not mean we should abolish it altogether. The problem is not *that* schools instill habits in students, but what *kind* of habits we are talking about. Teaching students to write, for example, may involve indelible elements of power and control, but that does not make it pacifying or mean we should stop doing it.

ATTENTION, DISTRACTION, AND TECH HABITS: AN EMPIRICAL CASE STUDY

To get a closer look at educational habit-formation, I now turn to my own research on digital distraction in two Danish high schools or "gymnasiums," as we call them. Over the course of two qualitative research projects conducted in 2013–2014 and 2018–2019, I have ultimately participated in 100 lessons and interviewed 50 students about their use of laptops in school. Of course, one of postphenomenology's key insights is that technologies can be taken up for multiple purposes or stabilities, and this certainly holds true for digital devices like laptops, which have an almost unbounded multistability (Keymolen, 2021). As such, there is no "one way" in which these devices were used and, in addition to becoming distracted, students told me they used laptops to take notes, produce slides, make budgets, perform calculations, share files, search for information, co-author documents, and exchange educational material. Nevertheless, when the topic turned to the off-task use of laptops, students readily and vividly described being "drawn" to distraction (Aagaard, 2015, 2021a).

> *IVAN:* I think it's in my fingers somehow. When I open my computer, I open the internet, then I type "F," then I type "Enter," and then I'm on Facebook. So, you actually have to be a little attentive *not* to do it. I don't know if that sounds a little extreme.

Due to deeply sedimented tech habits that have been built, maintained, and solidified in the course of their everyday lives, the action of logging onto Facebook ("F," "Enter") has become embodied in these students' hands and fingers and now occurs habitually. I call this phenomenon *habitual distraction* (Aagaard, 2015). While nothing in the laptop determines that it be used for distractive purposes, students are habitually inclined to do so because of how they are used to handling these devices. Furthermore, succumbing to this temptation is frustratingly easy since it occurs independently of conscious decision-making ("you actually have to be a little attentive *not* to do it"). As a result, students often become distracted without wanting to or even noticing it. This, of course, does not mean that digital distraction is *always* habitual, but deliberate distraction seems to be limited to off-task activities that involve a specific purpose. Otherwise, distraction mostly occurs on the prereflective level of habits. Students reacted particularly strongly against any kind of perceived boredom or passivity.

> *JACOB:* If the lesson isn't a 100%, like, I guess you can call it "entertaining," but of course it's not always like that, but if it isn't *catchy* or *appealing* to you, then you very quickly log on to Facebook or You . . . No, not YouTube, mostly

Facebook. And then just sit and scroll down. Not really looking at anything, because you've seen it all before, because you did it ten minutes ago, but really just keep scrolling.

AUTHOR: So why do you do it?

JACOB: It's a habit in your hands to push the button and just go in there and look.

I suggest that students' habitual inclination to seek out digital distraction if a topic does not immediately appeal to them can be interpreted as a kind of *mediated impatience* (Aagaard, 2015). An obvious counterargument to this idea is that students being bored by school is an age-old problem that predates the presence of laptops in the classroom. And that is certainly true. To circumvent this objection, I therefore asked students about technology use in settings that they choose more freely: How, for instance, do they use their phones when sitting on the couch with their partners? Around the dinner table with their families? Or in cafés with their friends? While the key devices of their stories shifted from laptops to smartphones, students gave remarkably similar descriptions about how their devices would sometimes distract them and make them act in ways that betray their values like being attentive during conversations. I call this phenomenon *digital akrasia*, which can be defined as the habitual tendency to become swept up by one's digital devices in spite of better intentions (Aagaard, 2020).

These findings show us why we cannot simply collapse habit and skill: While digital distraction doubtlessly requires bodily skills like touch typing (a paradigm case of "knowledge in the hands"), it also seems true that our skillful use of digital devices sometimes makes us do things that *we do not intend to do*. By discussing this phenomenon in terms of (bad) tech habits, we avoid resorting to the pathologizing rhetoric of tech addiction (Aagaard, 2021b). Furthermore, if we accept the idea that digital distraction is grounded in tech habits, it becomes interesting to explore how people cope with this phenomenon. In an educational context, students described how they would set up stumbling blocks like closing their laptops, which transforms distraction from being a "small click" that can occur habitually to becoming an "entire action" that requires conscious decision-making (Aagaard, 2021a). While such technological self-practices are interesting, two things should be noted: Students did not *always* implement such measures and not every measure was implemented by *students*. Sometimes, teachers took charge. And students seemed to appreciate that (Aagaard, 2022).

LAURA: What works best for me is when [teachers] tell us to close the computer when we don't have to use it. Like they do sometimes. Because most of the time, half of us are just sitting there with our computers open on Facebook.

> *AUTHOR:* That seems paradoxical, because you're 19 years old, you can vote in the next election, and you're basically an adult, but you still want someone *else* to control it for you. Can you say something about that?
>
> *LAURA:* Well, for my part, it's about self-discipline: When it comes to distraction, I just don't have it, so I'm really quick to just go on Facebook.

Some students even told me that they would like teachers to enforce the ICT-rules with more consistency. When it came to laptops in the classroom, students actually appreciated adult interventions, rules, and regulations, a presumably uncommon characteristic in the rest of their adolescent lives. In a recent study of device-restricted teaching at the university, we similarly found that university students appreciated how the absence of digital devices in the classroom "forced" them to pay attention during class (Herrman et al., 2021). This result adds to the previous findings in two important ways: First, while the previous findings focus exclusively on high school students, the dynamics of distraction pertain to most (if not all) of us. Thus, if we limit the discussion to teenagers' technology use, we risk trading in an important societal conversation for moral panic about "kids these days." Second, it appears that *having one's technology use restricted can sometimes feel liberating*. For a scientific field like psychology that practically fetishizes the concept of autonomy, this finding is quite remarkable.

FREEDOM THROUGH RESTRICTION: AN ORWELLIAN SENTIMENT?

At this juncture, I hope to have convinced the reader of three things: First, digital distraction is grounded in embodied tech habits that unfold prereflectively (*habitual distraction*). Second, like stepping back when faced with a close talker, we intuitively seek out digital distraction if a topic does not immediately appeal to us (*mediated impatience*). Finally, this dynamic is not confined to the classroom but happens everywhere and sometimes against our will (*digital akrasia*). When we look a bit closer at this research, the normative aspect sticks out: While phenomenology tends to portray the felt normativity of habits as springing from shared social norms (as in the case of distance-standing), digital distraction remains in opposition to the ongoing activities of the classroom. Making sense of this normative discrepancy becomes the crux of our discussion. We have suggested something like the following: Students have developed tech habits that challenge their ability to stay focused when things are not inherently "catchy" or "appealing." Teachers help students resist these tech habits by intervening, taking control, and telling students to close their laptops. By doing so, they are effectively disciplining students' attention. And students seem to appreciate that. Although not talking about freedom in any metaphysical sense, we have argued that this practice constitutes a form of *freedom through restriction*. So, what seems to be the problem?

There is another story to be told about teacher intervention. One that more explicitly involves power. After all, teachers are literally conducting students' conduct. As a result, critical scholars will view the proposed notion of "freedom through restriction" as either hopelessly naïve, or worse, as Orwellian newspeak.[4] In the aptly titled paper "Disciplining the Akratic user," for instance, Chad Valasek (2022) argues that most research on digital distraction is based on an implicit fear of unproductivity and risks becoming a disciplinary device used to normalize technology users. Furthermore, today's hidden curriculum can, indeed, be said to include the goal of turning students into "compliant and 'productive' users of educational technologies" (Selwyn, 2011: 113). Accordingly, in the critical educational framework, students have the choice between subjecting themselves to educational strategies of control (i.e., teacher intervention) or developing tactics that resist them. And the critical framework applauds such resistance. Thus, if schools want to teach you to be attentive, you should resist. Following this line of thought, we might regard digital distraction as a *heroic resistance* against docility and obedience. According to this logic, my endorsement of teacher intervention can be seen as complicit in the capitalist production of docile bodies. I am effectively attacking one of the few sanctuaries left in the ever-tightening grip of disciplinary power.

Before jumping to such conclusions, however, let us engage in a brief thought experiment. In cognitive science, it has become popular advice for educators to schedule "tech breaks" every ten to fifteen minutes for students to look at their phones (see e.g., Wood & Zivcakova, 2015; Guinness et al., 2018). The only way to get rid of temptation, it is said, is to yield to it (Rosen, 2011). To ground this advice, scholars often appeal to the supposed limitations of the adolescent brain: Neurologically speaking, students are incapable of focusing for longer than fifteen minutes at the time (King, 2017). Thus, while critical educational theory and cognitive psychology make for an odd couple, they both believe that leaving students to their own devices is progressive. Taking our starting point in the hermeneutical approach to educational habit-formation, however, let us imagine the long-term implications of adopting this practice: Do we eventually ask cinemas to pause movies every fifteen minutes to allow the audience to check their devices? Should weddings and funerals be over in fifteen minutes, or should such events include tech breaks to keep attendants from getting bored? When taken to its logical conclusion, digital distraction looks a lot less like heroic resistance than capitulation to instant gratification, arguably the pinnacle of capitalist consumerism. In comparison, teaching students to tolerate boredom and resist mediated impatience seems positively liberating.

CONCLUSION

This chapter set itself the thankless task of defending the idea of "disciplining" students and "forming" their attention. To do so, we have tried to rehabilitate the Foucauldian notion of discipline. In doing so, we argued that pointing to the existence (and mechanics) of disciplinary practices should not be considered equivalent to denouncing such practices tout court. And just to preempt one obvious objection to this perhaps controversial argument: By arguing that not all discipline is bad, we are not thereby endorsing the inverse notion that all discipline is good. Paraphrasing Melvin Kranzberg's (1986: 545) famous quip about technology, we might say that discipline is neither good nor bad; nor is it neutral. What we did argue, however, was that, in our specific case, and to the extent that contemporary attention can be seen as conjoined with tech habits, it makes good sense to talk about disciplining attention. As Dewey (1916) puts it: "A person who is trained to consider his actions, to undertake them deliberately, is in so far forth disciplined. Add to this ability a power to endure in an intelligently chosen course in face of distraction, confusion, and difficulty, and you have the essence of discipline" (p. 129). Contrary to what critical educational theory presupposes, it therefore looks like students' and school's interests can, in fact, be aligned.

To summarize our case in postphenomenological terms, we have argued that habit-formation constitutes an inescapable aspect of human existence and that the opposite of educational habit-formation is therefore not autonomy but *different habits*. Some of these habits, like habitual distraction, may be bad. To be clear, such tech habits are not bad because they are "unnatural" or because they alienate us from some authentic state of being. From a postphenomenological perspective, technological mediation *is* the essence of humanity, and we can probably adapt to an existence in which we concentrate in short bursts and turn to technologies if a topic does not immediately appeal to us. No, such tech habits are bad because this is a highly *undesirable* scenario. The critical question is, of course, who gets to decide that. By talking to students, we have seen that, from the students' perspective, 1) being distracted is not always a voluntary choice, 2) this kind of absorption can have negative consequences, and 3) some students actually appreciate having their technology use restricted. This last part is particularly noteworthy: Had it just been teachers complaining about digital distraction, such statements might be dismissed as Luddite or reactionary, but because this critical limit-attitude comes from students themselves, we are thankfully "forced" to have a more nuanced conversation about technologies.

NOTES

1. While references to the "soul" being the prison of the body may sound idealist, Foucault went to great lengths to emphasize the bodily origins of this entity and explicitly warned us that "nothing is more material, physical, corporal than the exercise of power" (Foucault, 1980c: 57f).

2. While such embodied dispositions may resemble Pierre Bourdieu's (1977) notion of habitus, Bourdieu was less concerned with how schools form *new* habits and more interested in how the educational system favors and rewards certain *pre-given* (i.e., middle-class) habits (Bourdieu & Passeron, 1990).

3. Richard Sennett (2006) reminds us that such totalizing discipline is not even found in the military: "In an army, orders modulate as they pass down a chain of command: what the general decrees, the military staff begins to translate into practice, adapting the command to conditions in the field; sergeants, corporals, and rank privates try in their turn to make sense of the command on a particular patch of ground. All obey, but equally, all interpret" (p. 34).

4. While the underlying notion of voluntary passivity may seem ludicrous to critical educational theorists, it is not completely foreign to STS, where Charis Thompson has insightfully described how patients in infertility clinics manifest agency by participating in objectifying procedures (Cussins, 1996).

REFERENCES

Aagaard, Jesper. 2015. "Drawn to Distraction: A Qualitative Study of Off-Task Use of Educational Technology." *Computers & Education* 87: 90–97.

———. 2020. "Digital Akrasia: A Qualitative Study of Phubbing." *AI & Society* 35 (1): 237–244.

———. 2021a. ""From a Small Click to an Entire Action": A Qualitative Study of Students' Anti-Distraction Strategies." *Learning, Media and Technology* 46 (3): 355–365.

———. 2021b. "Beyond the Rhetoric of Tech Addiction: Why We Should Be Discussing Tech Habits Instead (and How)." *Phenomenology and the Cognitive Sciences* 20 (3): 559–572.

———. 2022. "Taming Unruly Beings: Students, Discipline, and Educational Technology." *Technology, Pedagogy and Education* 31 (2): 159–170.

Althusser, Louis. 1971. "Ideology and Ideological State Apparatuses." In *Lenin and Philosophy and Other Essays,* 85–126. New York: Monthly Review Press.

Bourdieu, Pierre. 1977. *Outline of a Theory of Practice*. Cambridge: Cambridge University Press.

Bourdieu, Pierre, and Jean-Claude Passeron. 1990. *Reproduction in Education, Society and Culture.* Second edition. London & Beverly Hills: Sage.

Bowles, Samuel, and Herbert Gintis. 1976. *Schooling in Capitalist America: Educational Reform and the Contradictions of Economic Life.* New York, NY: Basic Books.

Butin, Dan W. 2001. "If This Is Resistance I Would Hate to See Domination: Retrieving Foucault's Notion of Resistance Within Educational Research." *Educational Studies* 32 (2): 157–176.

Crossley, Nick. 1996. "Body-Subject/Body-Power: Agency, Inscription and Control in Foucault and Merleau-Ponty." *Body & Society* 2 (2): 99–116.

———. 2001. *The Social Body: Habit, Identity and Desire.* London: Sage Publications.

Cussins, Charis. 1996. "Ontological Choreography: Agency through Objectification in Infertility Clinics." *Social Studies of Science* 26 (3): 575–610.

De Certeau, Michel. 1984. *The Practice of Everyday Life.* Berkeley: University of California Press.

Dewey, John. 1896. The Reflex Arc Concept in Psychology. *The Psychological Review* 3 (4): 357–370.

———. 1916/1966. *Democracy and Education.* New York, NY: The Free Press.

———. 1922/2007. *Human Nature and Conduct: An Introduction to Social Psychology.* New York: Cosimo Classics.

Dreyfus, Hubert. 2002. "Intelligence Without Representation – Merleau-Ponty's Critique of Mental Representation." *Phenomenology and the Cognitive Sciences* 1 (4): 367–383.

———. 2004. "Merleau-Ponty and Recent Cognitive Science." In *The Cambridge Companion to Merleau-Ponty*, edited by Taylor Carman, and Mark BN Hansen, 26–49. New York: Cambridge University Press.

———. 2014. "The Primacy of Phenomenology over Logical Analysis." In *Skillful Coping: Essays in the Phenomenology of Everyday Perception and Action*, edited by Mark A. Wrathall, 146–167. Oxford: Oxford University Press.

Dreyfus, Hubert, and Charles Taylor. 2016. *Retrieving Realism.* Cambridge, MA: Harvard University Press.

Foucault, Michel. 1975/1991. *Discipline and Punish: The Birth of the Prison.* London: Penguin Books.

———. 1980a. "Body/Power." In *Power/Knowledge: Selected Interviews and Other Writings 1972–1977*, edited by Colin Gordon. New York: Vintage Books.

———. 1980b. "The History of Sexuality." In *Power/Knowledge: Selected Interviews and Other Writings 1972–1977*, edited by Colin Gordon. New York: Vintage Books.

———. 1980c. "Truth and Power." In *Power/Knowledge: Selected Interviews and Other Writings 1972–1977*, edited bei Colin Gordon. New York: Vintage Books.

———. 2003. "The Subject and Power." In *The Essential Foucault*, edited by Paul Rabinow, and Nikolas S. Rose. New York: The New Press.

Guinness, Kendra E., Lauren Beaulieau, and Jacquelyn MacDonald. 2018. „Effect of Technology Breaks on Media Multitasking with College Students." *Behavioral Interventions* 33 (4): 427–439.

Habermas, Jürgen. 1990. "Questions Concerning the Theory of Power: Foucault Again." In *The Philosophical Discourse of Modernity: Twelve Lectures*, by Jürgen Habermas. Cambridge, MA: MIT Press.

Herrmann, Kim Jesper, Katrine Lindvig, and Jesper Aagaard. 2021. "Curating the Use of Digital Media in Higher Education: A Case Study." *Journal of Further and Higher Education* 45 (3): 389–400.

Hofmeyr, Benda. 2016. "The Power Not to Be (What We Are): The Politics and Ethics of Self-Creation in Foucault." *Journal of Moral Philosophy* 3 (2): 215–230.

Ihde, Don. 2002. *Bodies in Technology*. Minneapolis MN: University of Minnesota Press.

———. 2009. *Postphenomenology and Technoscience: The Peking University Lectures*. New York: SUNY Press.

———. 2012. *Experimental Phenomenology: Multistabilities*. Second edition. New York: SUNY Press.

Keymolen, Esther. 2021. "In Search of Friction: A New Postphenomenological Lens to Analyze Human-Smartphone Interactions." *Techné* 25 (3): 354–378.

King, Barbara J. 2017. "Should College Professors Give 'Tech Breaks' in Class?" *NPR*, October 19, 2017. https://www.npr.org/sections/13.7/2017/10/19/558751178/should-college-professors-give-tech-breaks-in-class.

Kranzberg, Melvin. 1986. "Technology and History: 'Kranzberg's Laws'." *Technology and Culture* 27 (3): 544–560.

Merleau-Ponty, Maurice. 1945/2002. *Phenomenology of Perception*. London: Routledge Classics.

Peirce, C.S. 1931–1958. *Collected Papers* (vols. 1–6, C. Hartshorne & P. Weiss [eds.], vols. 7–8, A. Burks [ed.]). Harvard University Press.

Rosen, L. 2011. "The Amazing Power of 'Tech Breaks'." *Psychology Today*, May 16, 2011. https://www.psychologytoday.com/us/blog/rewired-the-psychology-technology/201105/the-amazing-power-tech-breaks.

Selwyn, Neil. 2011. *Schools and Schooling in the Digital Age: A Critical Analysis*. London: Routledge.

Sennett, Richard. 2006. *The Culture of the New Capitalism*. New Haven & London: Yale University Press.

Taylor, Charles. 1986. "Foucault on Freedom and Truth." In *Foucault: A Critical Reader*, edited by David Couzens Hoy. New York, NY: Blackwell.

Valasek, Chad J. 2022. "Disciplining the Akratic User: Constructing Digital (Un) Wellness." *Mobile Media & Communication* 10 (2): 235–250.

Verbeek, Peter-Paul. 2013. "Resistance is Futile: Toward a Non-Modern Democratization of Technology." *Techné* 17 (1): 72–92.

Vlieghe, Joris. 2017. "ICT Literacy: A Technical or Non-technical Issue?" *Foundations of Science* 22 (2): 401–404.

Wood, Eileen, and Lucia Zivcakova. 2015. "Understanding Multimedia Multitasking in Educational Settings." In *The Wiley Handbook of Psychology, Technology and Society*, edited by Larry D. Rosen, Nancy Cheever, and L. Mark Carrier. New York, NY: Wiley Blackwell.

Chapter 12

A Critical Examination of Teaching and Learning in Times of Algorithmic Reasoning

Dan Mamlok

In recent years, there has developed a growing interest in the permeation of digital technology in education and in particular in innovative technological developments, such as artificial intelligence (AI), big data, and machine learning as a means to improve teaching and learning. Governments, Non-Government Organizations (NGOs), and private entrepreneurs push toward harnessing algorithmic rationality as a tool to enhance education (Means 2018). For example, an Organization for Economic Cooperation and Development (OECD) document (Vincent-Lancrin and van der Vlies 2020, 3) opens with the following statement: "In the transition to a digital era, education in G20 countries faces two challenges: reaping the benefits of AI and related technological advances to improve educational processes in the classroom and at the system level." While the development of digital technologies solutions in education holds a great potential for enhancing teaching and learning, it is essential to consider the complex relationship among technology, society, and education (Mamlok 2021). In this respect, philosophy of technology can provide a theoretical conceptualization of the potentials, challenges, and risks as new digital innovations become intrinsic in educational settings.

This chapter will provide insights into various understandings of technology and the ways they can support pedagogical goals in digital times. Particularly, two prominent approaches will be considered: critical theories of technology and postphenomenology. In brief, both approaches have provided a profound contribution to the understanding of human–technology relationships. Andrew Feenberg (2020, 27), whose critical constructivism theory emanates from neo-Marxist critical theory, contends that both approaches share some fundamental tenets, such as "the mutual constitution of technical subjects and objects" and the "rejection of technological determinism." A distinct difference between the approaches pertains to the nexus of technology, society, and the political. From a critical theory perspective, overlooking the political dimensions of how technology is designed, produced, and reproduced merits a caveat. From a postphenomenological perspective, the inclination of critical theories of technology to dystopian predictions distracts one from a central question pertaining to the meaning of being fully human in a mediated technological world. Thus, as Verbeek (2013) argues, the focus should be on ontological, rather than political, understanding. As a point of departure, this chapter will open with a review and complication of postphenomenology. The review will be followed by an examination of a brief analysis of personalized learning as an exemplar of integrating a digital technology solution in education. The second part of the chapter will review Feenberg's critical constructivism. As in the review of postphenomenology, I shall examine the example of personalized learning through a critical constructivist lens. The concluding part will argue that, in addition to the desire to develop an ontological and ethical understanding of technology, a more nuanced epistemological and sociopolitical realization should be considered in a world in which technology has become an intrinsic part of our everyday lives.

POSTPHENOMENOLOGY AND EDUCATIONAL TECHNOLOGY

This chapter deals with the possible meanings of philosophy of technology to education. Indeed, this chapter cannot capture the diverse theoretical perspectives that can help us consider the intricate relations among human–technology and its ramifications on teaching and learning. Thus, this chapter will focus on two prominent approaches: postphenomenology and critical constructivism. While my personal scholarship focuses on critical theories, I find the ontological dimensions of postphenomenology essential for elucidating the influence of technology on educational experience. Given the rapid progress of digital technologies, and the growing interest in utilizing AI in education, it is important to consider both ontological and ethical implications of technology on education. While there is a large body of literature about postphenomenology, there is limited literature that considers the relevancy of postphenomenology to education.

Postphenomenology is a philosophy of technology approach that focuses on human–technology relations. While phenomenology serves as a point of departure to consider human experience, postphenomenologist thinkers tend to "distance themselves from the classical-phenomenological romanticism regarding technology and find a starting point in empirical analyses of actual technologies" (Rosenberger and Verbeek 2015, 9). The analysis of human–technology relations explores the various modalities in which technology develops human relations, experiences, and practices. In addition, postphenomenology starts with looking at certain technologies and the relations that are developed between human and technology as a basis for a philosophical analysis. In this sense, unlike considering philosophy as a starting point to provide meaning to technological artifacts (philosophy and technology), postphenomenology aims to develop a philosophy of technology that combines empirical and conceptual analysis and considers technological artifacts as the point of departure of the analysis (Kaplan 2009; Rosenberger and Verbeek 2015).

Alongside its phenomenological roots, which highlight the mediative aspects of human–technology experiences, postphenomenology is also influenced by American pragmatism. Ihde (2012, 117) points out that, instead of relying merely upon theoretical description, the nonfoundational nature of pragmatism[1] helps us to develop a "vocabulary of practice rather than theory, of action rather than contemplation, in which one can say something about truth." The importance of pragmatism is moving beyond mediation and realizing how technologies develop worldviews, meanings of the world, and human subjectivities (Rosenberger and Verbeek 2015, 9). In other words, examining the practices in which people engage with technology helps us to reveal the various relations and attitudes that we develop toward the world. In this respect, these practices are not merely instrumental or neutral, but rather demarcate intricate lines of relations among humans, technology, and the world.

The human–technology relations, as described by Ihde, can be understood through four main, distinct, but related contours: (1) embodiment relations; (2) hermeneutic relations; (3) alterity relations, and (4) background relations. These four types of relations have been scrupulously discussed in previous works. Yet it seems that what is missed in previous literature is recognizing the meanings of these human-technology relations to education. Thus, following a succinct review of each of the relations, I shall provide concrete examples of technologies used in education.

Embodiment relations refer to incorporated-technology experiences in which the technology becomes an indispensable part of our body. Eyeglasses are a prime example of how our vision is mediated through a technological artifact (Ihde 2009). As one uses eyeglasses, his or her vision is mediated through the lenses, and "our sense of "body" is embodied outward directionally and referentially, and the technology becomes part of our ordinary experience of [seeing]" (Ihde 2009, 42). As pointed out earlier, the discussion about human–technology relations has been greatly developed since Ihde formalized *(human–technology) → world* schema as a basis for describing embodiment relations, which have been explored both by Ihde and by other postphenomenological scholars.[2] The question that assails this chapter is how these relations are, if at all, pertinent to education. Pens, pencils, and microscopes in biology classes are some examples that exhibit *(human–technology) → world* relations (Ihde 2012, 102), where the "machine is "between" me and what is experienced and is in this sense a "means" of experience in the primary focus."

Hermeneutic relations refer to the mediation of information or knowledge through technology. Unlike embodiment relations, where technology has become inseparable from the body and can be described as transparent, the nature of hermeneutic relations involves the interpretation of the technology's auditory or visual signal. For example, safety devices in cars can offer us valuable information through auditory or visual alerts. One needs to be able to differentiate between the different sounds and visual alerts in order to interpret them and respond accordingly. These alerts are particularly useful in situations where we fail to maintain a safe distance or when we attempt to change lanes but are unaware of another vehicle nearby. Ihde (2012) formalizes these relations as *human → (technology–world)*. In education, we can consider tools such as a ruler in geometry class or a thermometer in chemistry class as exhibiting hermeneutic relations.

Alterity relations refer to engaging with "technologies themselves as quasi-objects or even quasi others" (Ihde 2012, 43). Ihde points out that, whereas this relation was initially related to toys or objects, such as ATMs—technological devices that interact with humans—we can think today about interacting with robots. For example, when corresponding with Siri, the robot's ability to reference myriad types of issues, knowledge, and content is mimicked to correspond with humans (Rosenberger and Verbeek 2015; Wellner and Rothman 2020). Educational games, such as Kahoot, require students to utilize their devices to select the correct answer from a set of multiple-choice questions. The answers are represented by shapes and colors and exhibit alterity relations. In a similar manner, personalized learning software provided to students includes dialogue boxes, where students can select the topics they wish to focus on and the exercises they wish to take. The formalization of these relations is *I → technology (– world)*.

Background relations relate to technologies that, despite their presence, are "unattended to [in] the background" (Ihde 2009, 43). Air-conditioning systems, refrigerators, and security cameras are examples of background relations. In education, devices such as classroom projectors and security cameras are often used on a regular basis and exemplify background relations. A timelier example is the prevalence of a teacher dashboard or learning management system that works in the background of the teaching and learning environment. Ihde (2012) schematizes these relations as *I→ technology (world–)*. Although background relations can be understood as an absent entity, these relations constitute a distinct type of experience. As Rosenberger and Verbeek explain:

> Technologies to which we share background relations stand back in our awareness not simply because we have grown accustomed to their usage, but because they quite literally form the backdrop of our experiences. They shape our experiences, protecting us from the elements or keeping our food safely chilled, but do so in ways that do not require direct interaction (Rosenberger and Verbeek 2015, 19)

The premise of each of the four relations is that understanding human–technology–world relations is based on human intentionality. Yet the advents of current technologies, such as smartphones and artificial intelligence, have entailed new modalities of relations that require us to take into account what Verbeek (2008) calls cyborg intentions. He contends that since "Ihde's primary focus is on the *relations* between humans and technologies rather than the intentionalities involved, his analysis tends to blackbox the various forms of *intentionality* involved in these relations" (390). The more technological innovations develop, the more technological intentions should be taken into account as part of human experience. Namely, current technologies (such as AI) challenge the assumption that human intentionally must be the basis for analysis. The immersion of new technologies (such as smartphones and VR) that merge with our bodies creates an environment where "technological background interacts actively with human bodies" (21). "Smart environments "perceive" their users too, and "act" upon them" (Rosenberger and Verbeek 2015, 21). The cyborg relations can be understood through two distinct but related types of relations:[3]

1. Hybrid intentionality: Contemporary developments of technologies have led to a new mode of relations that extends beyond embodiment (such as the case of eyeglasses or earbuds). For example, pacemaker insertions denote a more intimate relation between the individual and the technology. As demonstrated earlier, the first four relations among humans and technology were represented by a dash, since intentionally was primarily assigned to humans. The transition of human–technology relations refers to cases where intentionality is not exclusively human but is partly based also on smart technological decisions. Thus, rather than symbolizing the relations as human–technology, Verbeek (2008) formulates these relations as *(human/technology) → world.*
2. Composite intentionality: Smart technologies, such as smartphones and augmented reality do not merge with the body, but rather with the environment (Wellner and Rothman 2020). As Rosenberger and Verbeek

> (2015, 21) point out, "smart environments "perceive" their users too, and "act" upon them." Smart technology involves a unique type of two-directional relations. This relatively new form of human–technology relations entails an interplay between the subject and the machine. For example, using a sports watch does not simply provide information. Rather, the experience of working out includes the feedback of the watch that dictates, at least to some extent, the nature of experiencing exercise. Thus, in this case, the intentionality of technology takes part in human intentionality. Composite intentionality is represented by the scheme: *human → (technology → world).*

Considering the various relations that have been explored by postphenomenologists, composite relations seem to provide an important insight, considering the push of educational systems to integrate Virtual Reality (VR) and Augmented Reality (AR) tools. These new educational tools can be understood as powerful technologies that immerse the learner in the presented simulations (Maas and Hughes 2020). It is important to note that while VR and AR have not yet been integrated on large scales in educational settings, there are ample pedagogical and administrative solutions that aim to enhance learning processes and support decision making practices in schools and departments of education. A prime example is the hype around personalized learning as an effective means to address the diverse needs of different students. In its essence, personalized learning aims to "allow each learner to gain more power over the learning process and to provide teachers tools to support each individual student based on his or her needs" (Mamlok 2021, 107–8).

In general, personalized learning can be distinguished into two different types: (1) responsive learning and (2) adaptive learning. Responsive learning refers to "an interface to predetermined content, like a hyper-linked menu or a series of digital buttons" (Bulger 2016, 5). This type of personalized learning is designed to support students in attaining standardized goals that are congruent with instrumental approaches to education. Adaptive learning offers the learner a dynamic learning environment, centered on his or her interest, capacities, and goals (Mamlok 2021). The interactive nature of personalized learning tools enables the students to set their own goals, in some cases to choose fields of interest, and to progress at their own pace. Software such as "Teach to One," "Cloud Math," "TenMarks," "McGraw-Hill Thrive," and "Rosetta Stone's Lexia" are some examples of platforms that aim to individualize instruction. Beyond the edifying potentials of these tools, they can be analyzed through a *composite intentionality* perspective. For example, integrative math learning tools involve a degree of communication between the learner and the tool. The learner may be asked various questions by the app, and the decision-making on what question will be shown on the screen, and what lessons (or short explanatory clip) should be rehearsed are made by the software. The intentionality, in this case, is not limited to the learner but entails an interplay (of intentions) between the learner and the technology.

This brief example demonstrates a postphenomenological interpretation of personalized learning. Clearly, this example merely scratches the surface of many other tools that can be used for various practices that take advantage of digital technology tools. But my point in this article is not necessarily to develop an in-depth analysis of each educational tool (and its relations), but rather to consider the sociopolitical ramification of these new technologies on students' experiences and the possible implications on their worldview (*Lebenswelt*) development. While this analysis helps us to better understand the experience of the learner and, more broadly, to consider how this experience can improve learning practices, it glosses over sociopolitical dimensions of educational technology tools and the underlying rationality of their design. Recognizing the ideological and social motivations of technological design is necessary for the development of a more nuanced understanding of the integration of technology in education (Mamlok and Abowitz 2022). As Emejulu and McGregor (2019, 133) claim, "Constructing technology as innocent or neutral misunderstands the social relations of technology and its very real material consequences in our social world." Simply put, I contend that while the contribution of the postphenomenologist perspective to an ontological description of human technology is profound, we ought, in addition, to consider the ethical and sociopolitical meanings of the immersion of new technologies into our lives. This claim is of great importance as we examine the integration of digital technology in education, since those digital technologies shape, at least to some extent, students' worldviews. In this respect, I find Andrew Feenberg's critical constructivism compelling. He rightly notes: "If there is a significant difference [between postphenomenology and critical constructivism] it is in the emphasis critical constructivism places on the role of politics in the process of co-construction" (Feenberg 2020, 27). In the following section, I flesh out the major arguments of critical constructivism and suggest that considering both theories can provide a more nuanced contemplation of educational technologies.

CRITICAL CONSTRUCTIVISM AND EDUCATIONAL TECHNOLOGY

Critical constructivism is based on Feenberg's accounts of critical theory and technology. Around the same period, Don Ihde developed postphenomenology, Feenberg (1991) published his *Critical Theory of Technology*, which has provided a tremendous contribution to philosophy of technology and to those who are interested in critical theories. Considering the dystopian visions of critical theories on technology, such as those of Heidegger and the scholars of the Frankfurt School (i.e., Horkheimer, Adorno, and Marcuse), Feenberg (2010) invites us to carefully consider the interplay among technology, society, and the public. Clarifying this last point requires a brief summary of Feenberg's conceptual framework. In the second part of this section, I shall reconsider the example of personalized learning through a critical constructivist lens.

Feenberg's (1991) critical theory of technology examines the complex relation between technical and social forces in technological design. As pointed out, his work is largely influenced by critical theory (such as Heidegger, Marcuse, and Habermas). Yet, despite the tendency of early critical theory to envision a dystopian perspective of technological development, Feenberg rejects the determinist premise of his predecessors. Technological determinism assumes that the development of technology is independent of social forces and that social transformations emanate from technological progress, regardless of the social, political, and economic conditions. A major problem with determinism is holding the premise that "science and technology are value-neutral products of inquiry, exogenous influences on social life" (Feenberg 2017, 43). Against the deterministic view, social constructivism focuses on the various modalities in which different actors and social groups interpret technical problems and influence technological design. In other words, unlike technological determinism, social constructivism suggests that technological development is intrinsically related to sociopolitical forces.

Examination of technological design, according to Feenberg, cannot be disconnected from the hegemony of technical-managerial rationality that constructs social and political practices. Technological rationality is governed by the neoliberal apparatus, which entails the reproduction of its own rationality. Drawing from Marxist critique, Feenberg (2010, 70) contends that, in modern constellations, the "whole life environment of society comes under the rule of technique." He calls this mechanism "*operational autonomy*," which places decision-makers in a "technical relation to the world, safe from the consequences of their own actions" (Feenberg 2010, 70). This technocratic rationality permeates every aspect of our lives. Alongside the recognition of the hegemony of the neoliberal apparatus on constructing practices, norms, and standards that serve the interests of those in power (such as large corporations), Feenberg (1991) identifies two significant elements in the development of technological design: the technical element and individual technologies. The *technical element* refers to the technical consideration that can be considered relatively independent from social forces. The *individual technologies*, however, demarcate the sociopolitical conventions, the cultural interests, and the historical context in which technology is designed. Thus,

Feenberg (1991, 2010, 2017) contends that the technical code signifies the interplay between the technical and the social considerations.

In the past two decades, Feenberg has responded on several occasions to postphenomenology and articulated his critique, which focused mainly on its sketchy accounts of the sociopolitical dimensions of technological design. Feenberg (2020, 28) points out that the basis of the postphenomenological formulation *human → (technology–world)* rests on "conceiving the subject as embodied and active, and human–world relations as technically mediated." The problem, according to Feenberg, lies in the tendency of postphenomenology to overlook the importance of political, ideological, and social forces in enacting human experiences. Hence, he claims that critical constructivism focuses on the "collective process of transforming technology in which these subjects are engaged" (Feenberg 2020, 29). Unlike the schematic formulation of postphenomenology that is based on the premise that technology constructs the way we interpret the world, Feenberg (2020, 29) reverses this scheme and contends that "it is technology that is interpreted within the world" and that it can be formulated as follows: *human → (world–technology).*

Let's look again at the example of personalized learning. In addition to the analysis of the new relation between the students and the subject matter that I indicated earlier, it is important to consider the ideological and economic powers that propel the integration of these innovative technologies in education. For-profit companies, such as Google, Facebook, and the Chan Zuckerberg Initiative are among the leading entrepreneurs of personalized learning. In addition, the Gates Foundation, a non-government organization, is also a dominant actor in installing digital personalized learning and AI solutions into education. This context is important since it reflects the great impact of transnational corporations on public life, including education. My claim is not against the potential of personalized learning, in and of itself, to enhance learning processes, but I rather argue that we ought to consider the conditions in which those technologies are designed, regulated, and funded and the purposes they serve. Such understanding can be perceived as a *sine qua non* reconsidering the technology design rationality, in which education and teaching and learning tools are not reduced to instrumental goals of education.

Earlier I pointed out the distinction between responsive and adaptive systems. Whereas personalized learning tends to be depicted by technological entrepreneurs in a rhetoric that resonates with the idea of adaptive learning, the common tools and practices used in schools are mainly based on responsive learning systems (Bulger 2016). What is at stake is not whether technology can open new opportunities for teaching and learning, but rather whether current conditions in which educational technologies operate encourage the advancement of pedagogies that challenge teaching practices to move beyond instrumental rationality and standardized goals of education. In other words, rather than reducing the goals of education to training students to reach predetermined standardized knowledge and to gain sufficient competencies for vocation, it is important to nurture a broader concept of education as a common good (Biesta 2010; Ravitch 2011). In this respect, adaptive learning holds the potential to create educational environments that offer students the ability to navigate their learning process based on their interests, capacities, and goals. Yet, under the current neoliberal conditions of standardized-based education, personalized learning and AI education perpetuate the reductive approach to education, rather than carrying its technological potential to transform education.

This brief example is significant since it reflects what Williamson (2013, 221) considers as "algorithmic imaginaries." Akin to Feenberg (2010), who looks at the interplay between the social and the technical, Williamson (2013, 221–222) argues that the instrumental rationality of educational technologies signifies "institutionally stabilized and publicly performed visions of desirable futures that are animated by shared understandings of forms of social life and social order and made attainable through the design of technological projects." In a similar vein, Means (2018, 103) argues that "algorithmic learning systems can serve to naturalize and perpetuate the domination of human beings and the natural world through the conflation of quantification, calculation, measurement, efficiency, and technical mastery with progress."

Feenberg (2020, 29–30) rightly argues that technological design "involves feedback to technologists from users and victims of an initial design released into the world with dire or unexploited consequences." His critical constructivist view suggests that "modernity is characterized by a dominant form of rationality. This rationality is not universal but is context-bound like other aspects of culture" (39). Considering this conviction in the case of personalized learning indicates that the organization of teaching and learning platforms is governed by techno-managerial rationality, which dictates the epistemological assumptions of learning, and the ontological view of education as what Wendy Brown (2005) defines as *homo economicus*.

Earlier in this chapter, I analyzed the example of personalized learning through the concept of composite intentionality, which elucidates how intentionality in personalized learning (and other types of interactive educational technology solutions) is not limited to the learner but also involves the intentionality of technology. Nevertheless, I argue that this analysis misses the sociopolitical contexts that stipulate the technological design of those platforms. In congruence with Feenberg's (2020) reconfiguration of the postphenomenological formula, I suggest that schematizing the relations in personalized learning and in other forms of AI integrative tools can be formulated as *human → (world → technology).*

This schematization of AI in education holds true as we consider the various practices used to evaluate student competencies, such as formative assessments. It should be stressed that my argument is not to devalue the importance of supporting students to achieve high standards and to gain a mastery level of knowledge and skills. Rather, I claim that the current rationality of education overemphasizes the focus on student evaluation and glosses over broader educational goals. As Ravitch (2011, 280) astutely points out, the "biggest risk is in forgetting that test scores are an indicator, not a goal of education. When the indicator becomes the target, we lose sight of other, more important goals." In a similar manner, I argue that the goals of AI tools for education are reductive in nature. As clearly noted in Vincent-Lancrin and van der Vlies:

> AI is for example leading the way to increase completion of quality primary and secondary education, or transform tools such as standardised assessments, allowing to broaden the scope of the skills that can be assessed and increase the relevance of assessments to the skills that will become more important in a world shaped by digital technologies. (Vincent-Lancrin and van der Vlies 2020, 9)[4]

Transforming education can be understood in different ways. Those who advocate for standardized-based education (e.g., high-stake tests, student assessment, etc.), will find the current direction of educational technologies as a promising means to foster such educational goals. However, those who seek to move beyond instrumental goals and perceive education not only as a training facilitator for the labor world but rather as supporting students' growth by allowing them to experience the world in a meaningful way, to explore, and to develop a critical worldview, will find current educational technologies limited. The onus is on us, as a society, to consider how those conflicted interests and views on education are translated toward a transformative perception of education, in its progressive sense.

CONCLUSION

In this chapter, I aimed to examine current trends in AI education by looking at two different philosophical approaches to technology: postphenomenology and critical constructivism. I have argued that postphenomenology provides a profound contribution to articulating how human experience has been shaped and reshaped through technological developments. Exploring the relation between humans, technology, and the world gives insight into how technologies mediate the world and construct human experiences. In this sense, technology is not merely an instrument, but an essential element that constitutes how humans perceive everyday life and how they develop their worldview. Following Feenberg's (2020) critique, I have contended that, beyond realizing the ontological meanings of human–technology relations, it is essential to examine the socio-political dimensions in technological design. This critique is particularly important as one considers the integration of digital technology in education under the neoliberal apparatus. Against the tendency to instrumentalize education and to harness digital technology as an effective means to quantify and standardize education, I have argued for a transformative technological approach that will strive toward a more progressive and humane model of education.

NOTES

1. Foundationalist theories suggest that there are epistemic premises and knowledge structures that are not subject to revision. Peirce, Dewey, and other pragmatic philosophers reject this categorical epistemic conviction (Bernstein 2010).
2. See for example: Ihde (1991, 2012, 2020), Verbeek (2008, 2013), Wellner and Rothman (2019).
3. It is important to note that Verbeek (2008) offers a third type of relations, which he calls mediated intentionality. Building upon Ihde's (1990) schematization of human technology relations, Verbeek (2008) contends that, since those relations are dependent on technological devices (such as glasses, binoculars, microscope), humans' experiences are mediated through technology. This chapter cannot capture in depth Verbeek's analysis, and I have focused on the other two types of cyborg intentionality, since they are more pertinent to the current chapter.
4. While this document is part of working papers that, according to OECD, do not necessarily reflect the official policy of OECD, it is congruent with the general educational goals that appear in other policy papers and is used as an exemplar of the general approach to AI in education.

REFERENCES

Bernstein, Richard. 2010. *The Pragmatic Turn*. Cambridge: Polity.

Biesta, Gert. 2010. *Good Education in an Age of Measurement: Ethics, Politics, Democracy*. Boulder, CO: Paradigm Publishers.

Brown, Wendy. 2005. *Edgework: Critical Essays on Knowledge and Politics*. Princeton, NJ: Princeton University Press.

Bulger, Monica. 2016. "Personalized Learning: The Conversations We're not Having." *Data and Society*. https://www.datasociety.net/pubs/ecl/PersonalizedLearning_primer_2016.pdf.

Emejulu, Akwugo, and Callum McGregor. 2019. "Towards a Radical Digital Citizenship in Digital Education." *Critical Studies in Education* 60 (1): 131–147.

Feenberg, Andrew. 1991. *Critical Theory of Technology*. New York, NY: Oxford University Press.

———. 2010. *Between Reason and Experience*. Cambridge, MA: MIT Press.

———. 2017. *Technosystem: The Social Life of Reason*. Cambridge, MA: Harvard University. Press. https://doi.org/10.4159/9780674982109.

———. 2020. "Critical Constructivism, Postphenomenology, and the Politics of Technology." *Techne: Research in Philosophy and Technology* 24 (1/2): 27–40. https://doi.org/10.5840/techne20202101116.

Ihde, Don. 1990. *Technology and the Lifeworld: From Garden to Earth*. Bloomington, IN: Indiana University Press.

———. 2009. *Postphenomenology and Technoscience. The Peking University Lectures*. Albany, NY: State University of New York Press.

———. 2012. *Experimental Phenomenology Multistabilities*. Albany, NY: State University of New York Press.

———. 2020. "Almost a Critical Theorist . . ." *Techne: Research in Philosophy & Technology* 24 (1/2): 15–26. https://doi.org/10.5840/techne2020319121.

Kaplan, David. 2009. "Introduction." In *Readings in the Philosophy of Technology*, edited by David Kaplan, 7–22. Plymouth: Rowman & Littlefield.

Maas, Melanie, and Jannete M. Hughes. 2020. "Virtual, Augmented and Mixed Reality in K-12 Education: A Review of the Literature." *Technology, Pedagogy and Education* 29 (2): 231–249. https://doi.org/10.1080/1475939X.2020.1737210.

Mamlok, Dan. 2021. *The Great Promise of Educational Technology: Citizenship and Education in a Globalized World*. New York, NY: Palgrave Macmillan. https://doi.org/10.1007/978-3-030-83613-9.

Mamlok, Dan, and Kathleen Knight Abowitz. 2022. "132 Words: A Critical Examination of Digital Technology, Education, and Citizenship." *Technology, Knowledge and Learning* 27, no. 4 (December): 1237–57. https://doi.org/10.1007/s10758-021-09540-3.

Means, Alexander. 2018. *Learning to Save the Future: Rethinking Education and Work in an Era of Digital Capitalism*. New York, NY: Routledge.

Ravitch, D. 2011. *The Death and Life of the Great American School System: How Testing and Choice are Undermining Education*. New York, NY: Basic Books.

Rosenberger, Robert, and Peter-Paul Verbeek. 2015. "A Field Guide to Postphenomenology." In *Postphenomenological Investigations: Essays on Human-Technology Relations*, edited by Robert Rosenberger and Peter-Paul Verbeek, 9–41. Lanham, MD: Lexington Books.

Verbeek, Peter-Paul. 2008. "Cyborg Intentionality: Rethinking the Phenomenology of Human–Technology Relations." *Phenomenology and the Cognitive Sciences* 7, no. 3 (September): 387–395. https://doi.org/10.1007/s11097-008-9099-x.

———. 2013. "Resistance is Futile: Toward a Non-Modern Democratization of Technology." *Techne: Research in Philosophy and Technology* 17, no. 1 (Winter): 72–92. https://doi.org/10.5840/techne20131715.

Vincent-Lancrin, Stéphan, and Reyer van der Vlies. 2020. *Trustworthy Artificial Intelligence (AI) in Education: Promises and Challenges*. Paris: Organization for Economic Cooperation and Development (OECD). https://www.oecd-ilibrary.org

/deliver/a6c90fa9-en.pdf?itemId=%2Fcontent%2Fpaper%2Fa6c90fa9-en&mimeType=pdf.

Wellner, Galit, and Tiran Rothman. 2019. "Feminist AI: Can We Expect Our AI Systems to Become Feminist?" *Philosophy and Technology* 33, no. 2 (June): 191–205. https://doi.org/10.1007/s13347-019-00352-z.

Williamson, Ben. 2013. *The Future of the Curriculum: School Knowledge in the Digital Age.* Cambridge, MA: MIT Press.

Chapter 13

A Postphenomenological-Constructionist Assessment of AI in Education

Toward a Multidimensional Democratic Education

Galit Wellner and Ilya Levin

For a long time, scholars have been trying to understand how to integrate digital technologies into the education system and assess their success. But before these efforts matured, digital technology itself has made a significant step forward, manifested in the emergence of Generative Artificial Intelligence (AI) (e.g., Dall-E, ChatGPT and the like). Obviously, these new technologies challenge the existing educational system, and led educators, scholars, developers—practically everybody—to rethink basic concepts of learning. In this chapter, we wish to lay the theoretical foundations for integrating and assessing AI-based educational technologies (AI-Ed). Our approach is based on the synergies between postphenomenology and constructionism. We believe this combination enables a multifaceted understanding of the impact of AI-Ed on students, teachers, and schools.

Postphenomenology and constructionism can work together thanks to their common logic that rejects the instrumental reference to technology and favors a relational approach according to which humans and technologies co-constitute and co-shape each other (Wellner & Levin, 2023). In other words, both regard technologies as an intrinsic part of what it means to be human. Moreover, they both advocate for a bottom-up approach: for constructionism, it can be regarded as implementing the logic of data-driven machine learning technology, which is close to constructionist historical roots (Doroudi 2022); for postphenomenology, it is framed as a commitment to being "empirical" and contextualize the philosophical understanding in real-world cases. The affinity between constructionism and postphenomenology is not coincidental. Both approaches were developed in the late 1970s, with the advent of personal computers in the homes, and both share the then-new understanding of the interaction between humans and the world around them. They share an awareness of the central role of technology in society in general, and in education in particular. This proximity between constructionism and postphenomenology is the starting point of our chapter.

In his seminal book *Mindstorms: Children, Computers, and Powerful Ideas*, the founder of constructionism, Seymour Papert, draws four elements of an assessment of Educational Technologies (Ed-Tech), based on his experience as a child playing with gears and leanring from them about the world:

> Several qualities contributed to their effectiveness. First, they were part of my natural "landscape," embedded in the culture around me. This made it possible for me to find them myself and relate to them in my own fashion. Second, gears were part of the world of adults around me, and through them I could relate to these people. Third, I could use my body to think about the gears. I could feel how gears turn by imagining my body turning. This made it possible for me to draw on my "body knowledge" to think about gear systems. And finally, because, in a very real sense, the relationship between gears contains a great deal of mathematical information, I could use the gears to think about formal systems. I have described the way in which the gears served as an "object-to-think-with." (1980, viii)

These four qualities match the postphenomenological relations as developed by Don Ihde (Ihde 1979, 1990): The first that marks the gears as an element of "the natural 'landscape'" matches the postphenomenological background relations; the second points to the "world of adults" as a reference point to an "other" thereby leading us to regard the gears in this context as participating in alterity relations through a dialogue between learner and the gears within a "microworld";[1] the third unfolds the role of the learner's body in understanding physical laws and therefore corresponds to embodiment relations; the fourth quality exposes the role of gears as an object-to-think-with thereby participating in hermeneutic relations.

These parallels form the basis of this chapter and will serve as a foundation to model AI-Ed assessment. Our mission is to understand how AI-Ed becomes part of the educational fabric in the twenty-first century and assess its role in learning.

EMBODIMENT RELATIONS: PERSONALIZATION OF THE LEARNING PROCESS

Embodiment relations are those in which technology changes the user's body schema or extends the senses. In an educational context, these relations can occur when using simple technologies such as a pencil that "extends" the hand and enables writing. The learner feels as if her body and the pencil are a combined unit that operates in the world. This applies not only to tools but also to machines and digital technologies. Thus, when the television or the cellphone are integrated into educational processes, they function as an extension of the learner's senses (Wellner 2016), thereby referring to embodiment relations in its wider sense as suggested by Ihde (1990).

In the digital age, embodiment relations do not disappear but rather take on a new form (Wellner 2021b, 2020; Buongiorno 2019). The analysis of embodiment relations in the context of learning in the digital age reveals the importance of the screen when large parts of the learning process take place in front of a screen (Aagaard 2016; Rosenberger 2017b; Vlieghe 2017). The challenge is to "transcend" the glass screen and render the learner's body and senses part of the learning process. It is not sufficient to have a docile body sitting almost motion-less in front of a screen, be it a television, a computer screen, or a cellphone, but to involve the body in such a way that embodiment extends beyond the reference to the body schema and rather refer also to the senses and how they are changed.

The matching concept on the constructionist side is personalization, broadly indicating how EdTech transform the learning process according to the learner's preferences, choices, and body (Papert 1993, 1996). It is the mentioning of the body that constitutes the link to the postphenomenological notion of embodiment. It is central to constructionism as the theory targets not only "thinking with objects" but also "thinking through the fingers" (Papert 1993).

For Papert, personalization begins with simple technologies like the gears he played with. In the opening pages of *Mindstorms*, he emphasizes the ability of pre-digital technologies to adapt themselves to the learner's body (Papert, 1980, 11), making the learner feel like part of that collection of technologies. This feeling is conceptualized as body syntonicity denoting a learning through bodily interaction with the environment as a site of learning technologies.

The entrance of the PC to the classroom marked the beginning of the digital era. It enhanced the personalization of EdTech, especially compared to the mechanical gears, as it can adapt itself to the learner's age, height etc. The PC also transformed embodiment relations as the keyboard and the mouse became part of the learner's body, replacing the pencil and the pen.

Writing was no longer a matter of using the right or the left hand, but of using both hands to type. The next step in assessing embodiment relations would be to examine the effect of the content presented on the screen on the learner's body. The challenge is to identify synchrony between the learner's body movements and the movement presented on the screen (Abelson and DiSessa 1986).

Efforts toward personalization have been further intensified in the age of AI, with the aim of using AI systems to provide a personalized tutoring experience for every learner (Holmes, Bialik, and Fadel 2019). In terms of embodiment, it seems that the challenge remains similar to that of the early digital age—crossing the glass barrier of the screen and revealing how the presented content is linked with bodily schema, senses, and feelings.

In the era of AI, personalization intensifies as AI systems can adapt themselves automatically to the user's traits and preferences. Now embodiment relations can be rethought and reframed to reflect a new balance between humans and AI technologies. In the case of face recognition, for example, we no longer examine changes to the body schema, but rather assess how AI systems extend the digital representation of the human body, what is referred to as algorithmic bodily relations (Wellner 2021a). In this new postphenomenological relation, the process of facial recognition can be described as a relation between a user's face, an AI system that analyzes an image of the face, and a dataset on which the algorithm has been trained. The algorithm extends the face by adding a name to it or—in the more problematic cases—a trait, or ethnicity. It is a digital extension of the user's body, an extension over which the user has no control. In this new relation, the intentionality is therefore reversed so that technological intentionality takes over human intentionality.

In education, such AI systems not only recognize students' faces but also attempt to monitor their facial expressions in order to detect level of engagement and consequently provide more personalized and adaptive learning experiences (Kupriyanov et al. 2020). Whereas the public ethical discussion focuses on the technical limitations of face recognition to correctly identify faces of various genders and ethnicities, or the ability to protect the database from cyber-attacks thereby hampering the students' privacy, there is much less discussion on the more ethically problematic goals of such systems and their propensity to detect feelings.

In summary, the notion of embodiment relations models our connections with a wide range of technologies, ranging from traditional physical technologies to the newer digital technologies, including AI-Ed. Similarly, constructionism's notion of personalization covers gears, software programming languages and AI-Ed. By mixing the two theories, we can better understand the merging of the physical and the virtual and critically analyze AI-Ed. Moreover, this mix can enrich the postphenomenological notion of embodiment relation by pointing to the effect on the world and how the environment changes as a result of the joint activity of the I/learner and the technology/EdTech.

HERMENEUTIC RELATIONS: COMPUTATIONAL THINKING

In hermeneutic relations, the user experiences the world and the technology as a combined entity. This situation is typical of media technologies that promise to show us the world, whether a true representation of it or an imagined scenario. These relations require some forms of reading and consequently of interpretation. In EdTech, hermeneutic relations are probably the easiest to identify. After all, learning assessment focuses on reading, interpretation and meaning generation. For example, learning with a book can be framed as hermeneutic relations in which the learner conceives the world through reading the content of the book. A postphenomenological analysis of reading would highlight the role of the technology that presents the text (which can be alphabetic, numbers or images): is it a printed book? Is it an e-book? (Rosenberger 2017b).

In the age of AI, reading is not sufficient to describe our relations with the world, and we have to take into account the abilities of AI systems to write texts of various kinds, and thus shape our worldview. The "writing relations" (Wellner 2017, 2018) can be regarded as an expansion to the classical hermeneutic relations, and an adaptation to the active role of AI technologies. In these new relations, the focus is on the much-enhanced technological intentionality that is due to the autonomous nature of AI technologies. The new relations also require unpacking the "technology" element into "technology" and "text," in order to represent the now-common situation where we use an AI system to produce text (or draw an image) based on our "prompt," which is a text that interacts with the technology element. Therefore, today we have two types of "texts"—one that is the product of a technological system and another that is our way to instruct the technology how to perform. Both texts have an impact in and on the world. The "text" element also serves as raw material for AI systems in the training phase and affects the systems' results, for example, biases (Wellner and Rothman 2020; Wellner 2021a).

Reading and interpreting are major components of the classical hermeneutic relations and are regarded as active and transformative. Analogously, the constructionist notion of computational thinking (CT) focuses on meaning generation, albeit its goal is to transfer meaning from one context—the computerized environment, to another context—life outside the classroom. CT is conceived as a skill that enables learners to think differently about the world by teaching the computer (Bull, Garofalo, and Hguyen 2020). It was first implemented in the study of programming languages and served to teach mathematical thinking as a way to interpret nature (Lodi and Martini 2021, 887). This interpretation is intended to predict a particular outcome. It is the logic of modern science that uses mathematics to calculate how nature would "work."

Both postphenomenology and constructionism stress the importance of interpretation and both theories examine the role of technology in this process. How does technology participate in the interpretive process? Ihde explains: "hermeneutics should no longer concern itself only with the interpretation of texts but also with the interpretation of technical artifacts, thus expanding the 'interpretive activity' of human beings to a 'thing interpretation'" (Ihde 1998, 8). CT is a prominent manifestation of such "thing interpretation" and the extension of textual interpretation. In constructionism, a thing can be physics and the phenomena it explains. CT led educational scholars to develop a novel teaching method based on spreadsheets, rather than the traditional learning of equations and formulas from textbooks (Gero and Levin 2019). Such a constructionist learning demonstrates how "thing interpretation" operates in a digital world through the mediation of a PC and an appropriate programmable software.

Papert developed the principles of CT as he moved from gears to PCs. These principles were implemented in his 1967 development of LOGO as a programming language intended to teach a new way of thinking. According to constructionism, in coding and debugging programs, students acquire new abstract concepts, just as young Papert had acquired knowledge about the world with his gears. This development fits with the constructionist goal of understanding the laws of nature by designing personal learning environments.

In the world of PC and programming, CT has been a clear manifestation of the hermeneutic relations in constructionist learning. However, ChatGPT's emergence presents a challenge to the basic principles of CT. While CT is focused on programming, ChatGPT allows the learner to create software without using or understanding code but based on text (either written or spoken and then automatically translated into text).

The contemporary discourse on CT as twenty-first-century mode of thinking typically references Papert as its founder (Denning and Tedre 2021; Lockwood and Mooney 2018). However, Papert himself distanced from this idea (Papert 2000), unequivocally emphasizing that although programming played a crucial role in constructionism's initial stages, it was strongly instrumental. He wrote:

> In Mindstorms I made the claim that the ability to program would allow a student to learn and use powerful forms of ideas. It did not occur to me that

> anyone could possibly take my statement to mean that learning to program would in itself have consequences for how children learn and think (Papert 2000, 12).

According to Papert, CTs offered learners a novel, previously unknown method of constructing abstract concepts. However, he did not rule out the possibility that constructionist CT learning would be replaced by another way of learning. Papert's instrumental stance on CT provides a rationale for today's educators and researchers to rethink CT in the age of AI. The challenge is how to integrate and assess in learning processes that use new AI tools like ChatGPT automated programming.

The challenge is even wider and may refer to various usages and contexts in which AI-Ed is integrated into the curricula. It seems that AI-Ed requires students to develop a different set of skills, mostly critical thinking. Instead of regarding the output as an accurate interpretation of the world, they should develop a critical stance toward the output. Reading along is not sufficient. In January 2023, *Wired* Magazine described how ChatGPT is used in this manner in higher education. In the article titled "ChatGPT Is Coming for Classrooms. Don't Panic" (Ceres 2023), one of the examples is brought from a course by Olya Kudina from Delft University where ChatGPT is used as a debate tool to provoke critical thinking. Such thinking was needed because

> the students were dazzled by how quickly the chatbot rendered information into fluid prose—until they read it with a closer eye. The chatbot was fudging facts. When students asked it to back up an argument with citations from scholarly texts, it misattributed work to the wrong authors. And its arguments could be circular and illogical. (Ceres 2023, n.p.)

In Wired Magazine, Kudina does not recommend banning such tools, on the contrary. When used in a critical and appropriate way, they can invoke new ideas or arguments through a collaborative process. Other scholars, like Alexey Semenov from Moscow State University, recommend using AI tools' hallucinations to identify missing pieces in research thereby finding new research directions.[2] Developing the modifications of CT in the age of AI will require both technological and pedagogical efforts. Thus, the future of the constructionist notion of CT remains to be determined.

To sum, the combination of postphenomenological hermeneutic relations and constructionist computational thinking highlights the role of digital technologies in learning processes. For both theories, learning means interpreting the world. And both need to be updated in the face of the AI revolution.

ALTERITY RELATIONS: MICROWORLDS

In alterity relations the technology is experienced as a quasi-other with which the user engages in a kind of dialogue. In these relations, the world withdraws to the background and is left unnoticed. All the attention is paid to the dialogue between the experiencing "I" and the technology. Alterity relations can be identified in simple technologies as well as in AI systems. Whereas a child's doll functions as a quasi-other for the child, although it does not respond, an ATM would be an example of an early digital technology that responds albeit to a well-structured query of the user. It can respond, but in a limited way. As soon as AI systems enter the picture, the responses become richer and more varied. Their autonomy enables a dialogue which is in free form and increasingly resembles a conversation with a human being. It happens with educational robots (Hasse 2020), with Natural Language Processing (NLP) technologies such as voice assistants, and with the more recent conversational interfaces of AI systems like ChatGPT and DALL-E.

Digital technologies, including AI systems, mostly maintain alterity relations through the mediation of a computer screen of various sizes—from television to PC to cellphone, and sometimes even without a screen as in the case of smart speakers like Alexa. In such cases, the screen display functions as a "quasi face," the microphone as a quasi-ear, and the speaker as a quasi-mouth (Wellner 2014; Bergen and Verbeek 2021). The concept of quasi-face enables us to free ourselves from anthropomorphism and identify a basic human capability in technologies and the ways in which they interact with learners. The face metaphor explains how alterity relations are intensified as AI systems become autonomous and may assist in the assessment of learning processes where such technologies are deployed.

A similar form of dialogue with a technology can be found in the famous constructionist turtle (Papert 1980) which moves on the screen according to the learner's instructions. The attention that the learner pays to the turtle led Papert to conceptualize this situation as "microworld." This notion has acquired (at least) three simultaneous meanings (Doroudi 2022): as a model of reality; an environment that is isolated from reality (a "slice"); and as something that enables a reflection on reality. An additional meaning is an "incubator of knowledge" in which learning occurs (Levin and Tsybulsky 2017).

From a postphenomenological perspective, when the learning "I" operates in the microworld, she is in dialogue with the technology, be it the gears, the turtle, or ChatGPT. The "attention" element of the microworld is enriched by the postphenomenological alterity relations that accentuates the human capacity for a dialogue even with "mute" objects like gears and dolls. Combining microworld and alterity relations can thus bring new assessment directions such as studying how students interact within the microworld and examining the dialogues established in these interactions.

In parallel, constructionism can enrich postphenomenology by highlighting how the "conversation" begins already at the setup stage. The rules of the constructionist microworld are determined by the learner, according to her understanding of the world. Being personalized, these rules may differ from those of the real world, thereby allowing the learner to experiment within the microworld. This is the basis for a variety of dialogues with the technology, and through it—with the world. An additional contribution to postphenomenology by the constructionist notion of microworld can be the distinction between the "real world" and the "invented" microworld which functions as a slice of reality, a distinction that is often neglected in the postphenomenological literature.

From a theoretical perspective, a more intuitive positioning for the concept of microworld could have been vis-à-vis the postphenomenological background relations, as they both refer to the world. But a close examination of the concept of microworld reveals an active environment that does not recede to the back. Indeed, recently postphenomenologists pointed to the possibility of an active background that dynamically changes (Aydin et al. 2019). Nonetheless, such a background remains unattended and hence can hardly accommodate the constructionist notion of the microworld. Linking the concept of microworld to alterity relations emphasizes the active role of the technologized background in the digital age and especially after the emergence of everyday AI tools like ChatGPT and DALL-E. Furthermore, the concept of microworld enables us to model learning that is not limited to hermeneutic relations but rather involves a bodily interaction with a quasi-other technological artifact, an interaction that occurs within the microworld.

BACKGROUND RELATIONS: DEMOCRATIZATION OF EDUCATION

In background relations, the technology operates in the back, unnoticed, like air-conditioning and lights in a classroom. The technology becomes "barely detectable" (Ihde 1990, 109) and is positioned as "an absent presence" (111) or "atmospheric" (Ihde 1979, 13). So far, in the three postphenomenological relations described above, the world element attracted little attention, thereby reflecting our tendency to regard the world as mute and even neglect it.

However, in background relations, it is the technology that withdraws to the background and the world becomes "visible." In an educational assessment setting, a successful implementation of background relation means that the technology is unnoticed by the learner. This is the case with AI-based administrative tools that help with the day-to-day management of a school, such as assigning teachers to classes, or scanning students' faces during an exam.

Yet, the technology is not necessarily in the "back stage" of reality. It can be in front of our eyes but still unnoticed. Although unnoticed, the technology's presence can re-color our understanding of a situation and reveal hidden meanings. Staley Kranc (2017) explains that "the background is always technologized, and we are always immersed in it" (27). Moreover, we realize that the world cannot be taken as one unified entity. Rather, "in picturing a world 'out there,' it becomes evident that only part of that world is re-presented, and there is always more" (21). The world as an element in the post-phenomenological relations is fragmented, so that some parts are noticeable, and some are not. Those unnoticed parts share the fate of the technology in background relations, as all these elements populate the unnoticed surroundings. Thus, in the educational setting, technology like (clean) restrooms can impact students' learning abilities, even if it is not *inside* the classroom like the air-conditioner.

This phenomenon of withdrawing to the background is intensified in the digital age in which information and communication technologies "are actually fading away from sight" and "increasingly become an intrinsic part of the material environment in which we live" (Aydin et al. 2019). Being part of the environment, background technologies shape our experience of the world in ways we do not fully recognize or appreciate.

In AI systems, the background is of prime importance as it encompasses the training dataset, which is usually outside the view of the end users, and becomes visible in case of a blunt error. In AI-Ed context, background relations can be relevant for students when their written assignments are automatically assessed by AI-based tools. Likewise, background relations can be spotted in the use of AI to fulfil administrative functions such as class timetabling, staff scheduling, facilities management, finances, etc., or the surveillance tasks such as collecting data on class attendance and assignment submission as well as the effort to automatically identify students at risk.

Background relations represent what Papert terms "the natural landscape" of the gears described at the introduction to *Mindstorm*. The postphenomenological formula models how the technology becomes transparent in the learning process and how it allows the learner to experience the world as if it were not mediated by any technology. The "as if" is an important contribution of constructionism to postphenomenology, because it shifts the focus from the technology to the world/background. Complementarily, postphenomenology can contribute to constructionism by directing researchers to identify the technology which operates somewhere in the background. For both approaches, the background position does not entail passivity.

Furthermore, both theories have recognized that digitalization processes turn the background into a significant player in our everydayness (Aydin et al. 2019; Papert 2020 [1997]). Today the background is more and more active in the presence of autonomous sensors, surveillance cameras etc.

Whereas postphenomenologists are currently developing the ethical argumentation to deal with these technologies (Verbeek 2011; Rosenberger 2017a; Verbeek 2020), constructionists have already taken some steps into the ethico-political domain back in the early 1990s, with the collapse of the iron curtain. Specifically, constructionism offers a normative direction with its concept of democratization, which directs us "to respect children as creators, to enable them to engage in making meaning for themselves through construction, and to do this by democratizing access to the world's most creative and powerful tools" (Holbert et al. 2020, 1; see also McStay 2020). Democratization is manifested primarily in equal access to technology and respect for the learners' autonomy (Holbert et al. 2020, 6–7). These directions can be supported by digital technologies, such as the PC. Whereas learning with textbooks implies a knowledge in the old sense of dictating "objective" ideas, learning with a PC may enable the students to develop subjective meaning, construct their knowledge and eventually freely reveal their individual cognitive abilities (Resnik 2020).

A closer reading of Papert's work reveals a more revolutionary aspect of democratization that is applicable to the entire educational system. It is a shift in perspective from "technology *in* education" to "technology *for* education." For him, the integration of the PC into the classroom should lead to a total reconfiguration at the organizational level and render the system democratic. Papert identifies the hierarchical structure of the educational system not only with that of bureaucratic administrations but also with that of authoritarian political systems (Papert 1991). He contends that this hierarchy is duplicated in the structure of schools and makes the learning experience alienating for students. Moreover, in his view, the existing hierarchical structure of the educational system is at odds with the structure of future organizations into which the PC would be integrated. In the late 1980s, when democratic regimes were emerging in Eastern Europe, Papert drew an analogy between Gorbachev's perestroika as an attempt to correct an archaic hierarchical system of government on the one hand, and the efforts to correct a hierarchical system of education on the other. Papert claimed that both attempts failed because these systems cannot be peacefully transformed, only revolutionized. He linked the political change to the rise of the PC and argued that the then-new technological reality should have led to revolutionary democratic changes in the educational system. Unfortunately, these changes did not happen. Today, this idea of democratization enabled by computerized technologies is perceived as foreshadowing contemporary networked reality (Levin and Mamlok 2021), although Papert explicitly rejected the link between democratization and the Internet.

According to Papert, the integration of the PC into the school system should transform the traditional educational organizations, at the level that he terms as "macro education" (1991), which can be positioned as complementary to the "microworld" which was discussed in the context of alterity relations. Now the focus shifts from the individual learner to the educational system as a "world" that operates and enables the learning processes. As a background, digital technologies such as the PC play a role in creating new forms of non-hierarchical learning, education, and assessment.

To sum, constructionist democratization offers a theoretical basis for the inclusion of politico-ethical considerations in the postphenomenological formula and place them in the "world" component (cf. Verbeek 2020). Consequently, the experiencing "I" and the technology interact not only with the physical environment but also with the social, cultural, and moral aspects of the world around us. Combining democratization into background relations draws our attention to the "structural" elements of the world and its organizational principles. Although originally conceived in the digital age with the advent of the PC, these concepts are not unique to PC or to hyper-connected technologies but rather are relevant to any technology that enables learners to gain the freedom of creativity, equal access to educational resources, and personal self-expression. The challenge today is to ensure these qualities in the age of AI.

FOUR-DIMENSIONAL POSTPHENOMENOLOGICAL ASSESSMENT

We suggest applying postphenomenology as a framework for assessing constructionist learning. The proposed method follows the four basic relations that Ihde framed (1979, 1990), and grade each of them according to the constructionist principles formulated by Papert (1980). Postphenomenology provides the basis for assessment axes as each relation is examined as an axis, all combined into a four-dimensional space. Constructionism establishes the scale, weight, and direction of each axis. Together, they enable a multidimensional analysis of the impact of digital technologies on learners and the educational system, providing a framework for evaluating the prospects of specific technologies in education. Such a unified approach can lead to better integration of technologies into students' learning experiences and

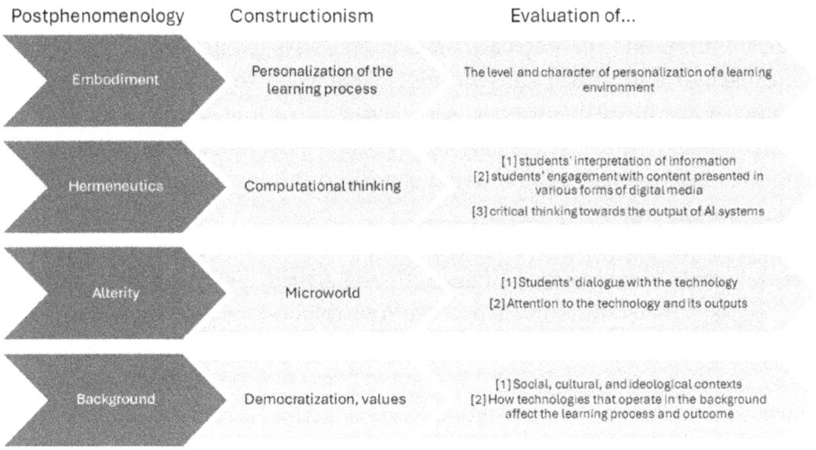

Figure 13.1 Four Axes of Assessment. Image created by author.

benefit the educational system as a whole. Combined, the four axes yield a multidimensional assessment space, as depicted in figure 13.1 and described below.

Embodiment-oriented assessment can identify factors that affect learning by changing the body scheme or the bodily perceptions. Constructionism can be used for measuring the personalization aspects of the combination of "I" and "technology" in embodiment relations in terms of adjustments to the learner's body, senses, and perceptions. Thus, embodiment-oriented assessment can evaluate the level and character of personalization of a learning environment.

Hermeneutics-based assessment can be used to evaluate students' interpretation of information through technology, as well as how they engage with content presented in various forms of digital media. The hermeneutic assessment of constructionist learning focuses on interpretation as a component of computational thinking, but—as mentioned above—this should be updated to the new environment of AI chatbots. A pilot study in this direction examined the differences between data-driven learning and traditional, rule-based programming. The study assessed students' computational perspectives in the realm of ML-based data-driven thinking (Shamir, Wellner, and Levin 2022).

Alterity-focused assessment can evaluate students' interaction with technological artifacts, from mechanical gears all the way to AI chatbots. With constructionism, we can assess how AI-Ed affects students' sense of personal identity and interpersonal interactions. Such an assessment is intended to evaluate students' adaptation to a constructionist microworld as evidenced in the dialogue with the technology.

Background-driven assessment can evaluate learning within specific social, cultural, and ideological contexts and how technologies that operate in the background affect the learning, even when they are unnoticed. For example, it could evaluate how AI-Ed affects various students or how AI systems manage different educational systems worldwide. The constructionist notion of democratization brings values into the analysis and hence a scale for the assessment.

Postphenomenological assessments as a whole can provide a comprehensive view of how technology in general, and AI-Ed specifically, shape learning processes by examining each of the four relations. By understanding how technology is experienced and interpreted by students and how it is situated within broader social and cultural contexts, we can develop more effective educational technologies and practices that better support student learning outcomes.

CONCLUSIONS

This chapter aims to provide a comprehensive analysis of the intersection of postphenomenology, constructionism, assessment, and AI technologies. By examining the correlations between postphenomenology and constructionism, the chapter sheds light on how these two theoretical frameworks can be used together to assess the effectiveness of technology-mediated learning, especially today as AI systems enter the classroom.

The chapter highlights the need for additional studies to fully understand the impact of AI technologies on education, particularly in relation to assessment. The proposed approach demonstrates how the four postphenomenological relations can be used to assess the role of technology in education. Importantly, the chapter suggests that, in the era of AI, the hermeneutic relation in constructionist learning should extend beyond traditional computational thinking. Research in this direction is a particular challenge because of the unpredictability of AI systems.

Overall, the chapter provides a down-to-earth framework for future multidisciplinary theoretical research on technology in and for education. By considering the intersection of postphenomenology, constructionism, and AI technologies, this approach offers a nuanced understanding of the role of technology in education and how it can be effectively assessed. Postphenomenology provides a comprehensive analysis of EdTech in general and AI-ED in particular, as well as a basis for learning assessment based on a multidimensional framework of postphenomenological relations.

NOTES

1. We would like to thank Markus Bohlmann for his constructive suggestion.
2. Based on private conversations.

REFERENCES

Aagaard, Jesper. 2016. "Introducing Postphenomenological Research: A Brief and Selective Sketch of Phenomenological Research Methods." *International Journal of Qualitative Studies in Education* 30 (6): 519–33. https://doi.org/10.1080/09518398.2016.1263884.

Abelson, Harold, and Andrea DiSessa. 1986. *Turtle Geometry: The Computer as a Medium for Exploring Mathematics*. Cambridge, MA: MIT Press.

Aydin, Ciano, Margoth González Woge, and Peter Paul Verbeek. 2019. "Technological Environmentality: Conceptualizing Technology as a Mediating Milieu." *Philosophy and Technology* 32 (2): 321–38. https://doi.org/10.1007/s13347-018-0309-3.

Bergen, Jan Peter, and Peter Paul Verbeek. 2021. "To-Do Is to Be: Foucault, Levinas, and Technologically Mediated Subjectivation." *Philosophy and Technology* 34 (2): 325–48. https://doi.org/10.1007/S13347-019-00390-7/FIGURES/1.

Bull, Glen, Joe Garofalo, and N. Rich Hguyen. 2020. "Thinking about Computational Thinking: Origins of Computational Thinking in Educational Computing." *Journal of Digital Learning in Teacher Education* 36 (1): 6–18. https://doi.org/10.1080/21532974.2019.1694381.

Buongiorno, Federica. 2019. "Embodiment, Disembodiment and Re-Embodiment in the Construction of the Digital Self." *Humana Mente* 12 (36): 310–330.

Ceres, Pia. 2023. "ChatGPT Is Coming for Classrooms. Don't Panic." *Wired*. January 26, 2023. https://www.wired.com/story/chatgpt-is-coming-for-classrooms-dont-panic/ (accessed 11 April 2023).

Denning, Peter J., and Matti Tedre. 2021. "Computational Thinking: A Disciplinary Perspective." *Informatics in Education* 20 (3): 361–90. https://doi.org/10.15388/infedu.2021.21.

Doroudi, Shayan. 2022. "The Intertwined Histories of Artificial Intelligence and Education." *International Journal of Artificial Intelligence in Education* 33: 885–928. https://doi.org/10.1007/s40593-022-00313-2.

Gero, Aharon, and Ilya Levin. 2019. "Computational Thinking and Constructionism: Creating Difference Equations in Spreadsheets." *International Journal of Mathematical Education in Science and Technology* 50 (5): 779–87. https://doi.org/10.1080/0020739X.2018.1501827.

Hasse, Cathrine. 2020. *Posthumanist Learning: What Robots and Cyborgs Teach Us about Being Ultra-Social*. London: Routledge.

Holbert, Nathan, Matthew Berland, and Yasmin B. Kafai. 2020. "Introduction: Fifty Years of Constructionism." In *Designing Constructionist Futures: The Art, Theory, and Practice of Learning Designs*, edited by Nathan Holbert, Matthew Berland, and Yasmin B. Kafai, 1–16. Cambridge, MA: The MIT Press. https://doi.org/10.7551/MITPRESS/12091.003.0002.

Holmes, Wayne, Maya Bialik, and Charles Fadel. 2019. *Artificial Intelligence in Education: Promises and Implications for Teaching and Learning*. Center for Curriculum Redesign.

Ihde, Don. 1979. *Technics and Praxis*. Vol. 24. Boston Studies in the Philosophy of Science ; v. 024. Dordrecht: D. Reidel Pub. Co. https://doi.org/10.1007/978-94-009-9900-8.

———. 1990. *Technology and the Lifeworld: From Garden to Earth*. Bloomington and Indianapolis: Indiana University Press.

———. 1998. *Expanding Hermeneutics: Visualism in Science*. Evanstone: Northwestern University Press.

Kranc, Stanley. 2017. "Picturing the Technologized Background." *Kunstlicht* 38 (4): 20–27.

Kupriyanov, R., A. Uvarov, M. Makarova, P. Unruh, G. Vodopian, and I. Levin. 2020. "Intelligent Tool for Developing Student's Social Skills." In *EDULEARN20 Proceedings: 12th International Conference on Education and New Learning TechnologiesOnline Conference*, 2408–13. IATED Academy. https://dialnet.unirioja.es/servlet/articulo?codigo=8616731&info=resumen&idioma=ENG.

Levin, Ilya, and Dan Mamlok. 2021. "Culture and Society in the Digital Age." *Information 2021*, 12 (2): 68. https://doi.org/10.3390/INFO12020068.

Levin, Ilya, and Dina Tsybulsky. 2017. "The Constructionist Learning Approach in the Digital Age." *Creative Education* 08 (15): 2463–75. https://doi.org/10.4236/ce.2017.815169.

Lockwood, James, and Aidan Mooney. 2018. "Computational Thinking in Secondary Education: Where Does It Fit? A Systematic Literary Review." *International Journal of Computer Science Education in Schools* 2 (1): 41–60. https://doi.org/10.21585/ijcses.v2i1.26.

Lodi, Michael, and Simone Martini. 2021. "Computational Thinking, Between Papert and Wing." *Science and Education* 30 (4): 883–908. https://doi.org/10.1007/s11191-021-00202-5.

McStay, Andrew. 2020. "Emotional AI and EdTech: Serving the Public Good?" *Learning, Media and Technology* 45 (3): 270–83. https://doi.org/10.1080/17439884.2020.1686016.

Papert, Seymour. 1980. *Mindstorms: Children, Computers, and Powerful Ideas*. New York: Harvester Press.

———. 1991. "Perestroika and Epistemological Politics." In *Constructionism: Research Reports and Essays, 1985–1990*, edited by Idit Harel and Seymour Papert. Cambridge, MA: Ablex Publishing.

———. 1993. *The Children's Machine: Rethinking School in the Age of the Computer*. New York: Basic Books.

———. 1996. "A Word for Learning." In *Constructionism in Practice: Designing, Thinking, and Learning in a Digital World*, edited by Yasmin Kafai and Mitchel Resnick, 9–24. London: Routledge.

———. 2000. "What's the Big Idea? Toward a Pedagogy of Idea Power." *IBM Systems Journal* 39 (3–4): 720–29. https://doi.org/10.1147/SJ.393.0720.

———. 2020. "Why School Reform Is Impossible." In *Designing Constructionist Futures: The Art, Theory, and Practice of Learning Designs*, edited by Nathan Holbert, Matthew Berland, and Yasmin B. Kafai, 275–85. Cambridge, MA: The MIT Press. https://doi.org/10.7551/MITPRESS/12091.003.0010.

Resnik, Mitchel. 2020. "Foreword. The Seeds that Seymour Sowed." In *Mindstorms: Children, Computers, and Powerful Ideas*, edited by Seymour Papert, vii–xiv. New York: Basic Books.

Rosenberger, Robert. 2017a. *Callous Objects: Designs against the Homeless*. University of Minnesota Press.

———. 2017b. "On the Immersion of E-Reading (or Lack Thereof)." In *Postphenomenology and Media: Essays on Human–Media–World Relations*, edited by Yoni Van Den Eede, Stacey O. Irwin, and Galit Wellner, 145–63. Lanham: Lexington Books.

Shamir, GIlad, Galit Wellner, and Ilya Levin. 2022. "Postphenomenological Studies of Machine Learning Education in Elementary School." In *EDULEARN22 Proceedings*, 2704–13. IATED.

Verbeek, Peter-Paul. 2011. *Moralizing Technology: Understanding and Designing the Morality of Things*. Chicago: University of Chicago Press.

———. 2020. "Politicizing Postphenomenology." In *Reimagining Philosophy and Technology, Reinventing Ihde*, edited by Glen Miller and Ashley Shew, 141–55. Cham: Springer Nature. https://doi.org/10.1007/978-3-030-35967-6_9/COVER.

Vlieghe, Joris. 2017. "Phenomenology of Digital Media." In *Encyclopedia of Educational Philosophy and Theory*, edited by Michael A. Peters, 1795–1800. Singapore: Springer.

Wellner, Galit. 2014. "The Quasi-Face of the Cell Phone: Rethinking Alterity and Screens." *Human Studies* 37 (3): 299–316. https://doi.org/10.1007/s10746-013-9304-y.

———. 2016. *A Postphenomenological Inquiry of Cell Phones: Genealogies, Meanings, and Becoming*. Lanham: Lexington Books.

———. 2017. "I-Media-World: The Algorithmic Shift from Hermeneutic Relations to Writing Relations." In *Postphenomenology and Media: Essays on Human–Media–World Relations*, edited by Yoni Van Den Eede, Stacey Irwin, and Galit Wellner, 207–28. Lanham: Lexington Books.

———. 2018. "From Cellphones to Machine Learning. a Shift in the Role of the User in Algorithmic Writing." In *Towards a Philosophy of Digital Media*, 205–24. Cham: Springer International Publishing. https://doi.org/10.1007/978-3-319-75759-9_11.

———. 2020. "Material Hermeneutic of Digital Technologies in the Age of AI." *AI and Society* 38: 2159–2166. https://doi.org/10.1007/s00146-020-00952-w.

———. 2021a. "I-Algorithm-Dataset: Mapping the Solutions to Gender Bias in AI." In *Gendered Configurations of Humans and Machines: Interdisciplinary Contributions*, edited by Jan Büssers, Anja Faulhaber, Myriam Raboldt, and Rebecca Wiesner, 79–97. Opladen, Berlin, Toronto: Verlag Barbara Budrih.

———. 2021b. "The Zoom-Bie Student and the Lecturer: Reflections on Teaching and Learning with Zoom." *Techne: Research in Philosophy and Technology* 25 (1): 153–61. https://doi.org/10.5840/techne2021121132.

Wellner, Galit, and Tiran Rothman. 2020. "Feminist AI: Can We Expect Our AI Systems to Become Feminist?" *Philosophy and Technology* 33 (2): 191–205. https://doi.org/10.1007/s13347-019-00352-z.

Chapter 14

Diverging Ethical Concepts

Postphenomenological Analysis through the Lens of Japanese Culture

Tomoki Sakata

This chapter seeks to enrich postphenomenological research by inquiring how technological mediation is perceived in different cultures. The putative human–technology–world relation is not a monolithic phenomenon or theory since each culture, or its ethical principle, determines the relation of technologies to lifeworld (*cultural lensing*). First, two different types of ethics are made manifest through examples: Europe's individual autonomy with an eye to Galileo's telescope and Japan's social conformity apropos of facemasks. In the second section, narratives by Japanese high school students regarding the pandemic-related online learning are empirically scrutinized. The last section philosophically reflects these findings, using two models of the technical mediation. The Japanese version is characterized by the *manipulation* of purpose and the *heteronomy* of technical intervention due to the dominant *we-perspective* of its ethics.

RELATION BETWEEN TECHNOLOGY AND ETHICS

Prior to the main investigation, it may be helpful to illuminate the central issue in more general terms. Postphenomenologists (cf. Ihde 1990, 2011, 2012; Arzroomchilar 2022) often recur to Galileo's telescope to argue for the modern-age technical embodiment. Indeed, this historical event does not only signify a great step forward in science and technology but also in ethics.[1] His efforts symbolize a moral struggle to emancipate reason from the dogmatic teachings of the church (cf. Cassirer 1942–43: 17–19). This ethical principle of autonomy, both in judgment and in performance, is the common foundation of our modern natural science, independent of the fine nuances each philosopher adds, e.g., Cartesian divine substance, Bacon's pragmatism, or Kantian transcendental logic. Science and technology are conceived to liberate our reason from unscientific, mythical thinking and this spirit of the Enlightenment proved still unshakable during the Covid-19 pandemic. In October 2020, the WHO released a video called "Mythbusters" that corrects errors and superstitions about the new virus with the proper, scientific knowledge.[2]

Is this ethics of science or technology universal or culturally imbued? To answer this question, it is germane to distinguish between two aspects of culture, viz., its purely scientific (pragmatic) side and its ethical side (Tylor 1910). In Ihde's pioneering book (1990: 8), we read that anthropologists saw some cultures face "their dying moments before absorption into the high-technology cultures of Western origin." This supposition can be refuted in two ways. The first counterargument is, as Ihde (ibid.: 18) himself argues, that every culture implicates *by definition* technology. However, this only means that mankind as *homo faber* incessantly makes progress. Günther Anders (1980: 433) observes (though critically) that Western science and technology are, divorced from their religious or philosophical origin, totally emulated by Japan. Another way to refute it is to claim that indigenous cultures, judged *ethically*, never became completely extinct in the face of technologization. Let me illustrate this point with the example of a facemask. Boer & Verbeek (2022: 201–203) divide outcomes of this technical device into three categories of "facial expression," "intimacy," and "vulnerability" but fail to see the most important function that it fulfills in Japan, namely, social integration (Nawa et al. 2021).[3] The Japanese wear their masks even where obviously no risk of infecting or being infected is present (e.g., walking alone outside). To stop this nonsense, only the simplest mythbuster suffices, saying that facemasks are only effective in the case of a close contact. While this *scientific* logic is not hard to understand, "social conformity"—the number one motive for the Japanese to wear masks[4]—forces them to this mysterious behavior for fear of social ostracism (cf. Benedict 1947). Evidently, these two aspects are unrelated.

In the postphenomenological discourse, Hasse (2008: 48), in view of Vygotsky's theory, asserts that "artifacts from the very outset are imbued with cultural meanings, which are learned and internalized *not* just in use, but also in *social* use." Japan's example cuts yet deeper into this issue since the very meaning of technological embodiment that is by Western standard defined as an increase in the personal autonomy is *manipulated*. How can a man, as we saw, become independent when the social pressures condition you to wear a mask where he has no scientific reason to do so? The strong force of this *cultural lensing* (parallel to Einstein's gravitational lensing that even refracts the path of light!) must be considered when postphenomenological studies are applied to remote regions.

EMPIRICAL STUDY: JAPANESE STUDENTS' EVALUATION OF ONLINE LEARNING

Purpose and Methodology of the Survey

Postphenomenologists recommend paying close attention to actual practice (Rosenberg & Verbeek 2015: 31). The recent pandemic challenged various actors in an educational context, i.e., learners, teachers, and parents, and there have been numerous empirical studies about these issues (Huber & Halm 2021). But the following empirical study has a very narrow scope (solely the learning experience of pupils) and limited sample (only from two Japanese high schools) because it does not intend to spell out the nation's overall tendency like the PISA study does (to be discussed later). Instead, we use first-hand narratives as truly finger-pointing implications for postphenomenological analysis. If case studies are broadly and philosophically defined, quality can surpass quantity, as Kierkegaard (1882) elucidates the universal religious feeling through just one example of Abraham.

The survey was carried out in close cooperation with two teachers from School A in Tokyo and School B in Toyama in the form of an anonymous online questionnaire (Google Form).[5] It contains four open-ended questions, asking a) the advantages and b) the disadvantages of learning processes, resulting from the shift 1) to online schooling and 2) back to normal schooling (table 14.1). This sort of descriptive or nonparametric question is chosen to let participants express their thoughts freely without any ungrounded presumptions of researchers. The evaluation method resembles "text mining" (Feldman & Sanger 2007) with the only difference that it is undertaken not by

Table 14.1 Questions

1) Concerning the digitalization of learning environment, please describe advantages and disadvantages, comparing between the period before (until March, 2020) and during the Covid-19 pandemic (school years 2020 and 2021).
   - 1.a) Advantages
   - 1.b) Disadvantages
2) Concerning the *normalization* of learning environment, please describe advantages and disadvantages, comparing between the period during the Covid-19 pandemic (school years 2020 and 2021) and the post-Corona (school year 2022).
   - 2.a) Advantages
   - 2.b) Disadvantages

a computer program but by the author himself. Students' narratives are *mined for* relevant keywords and categories to extract collective opinions and tendencies.[6] This manual text mining allows us to penetrate through the language form into the *life form*, to use Wittgenstein's term (1977). Any analyses of natural languages have principally these two dimensions of explicit texts and implicit meanings.[7] Therefore, our first task is a formalization, postulating several rules to evaluate multifaceted comments. Secondly and almost simultaneously, we move from shallow grammar to deep grammar (cf. Wittgenstein 1977: 265, §664) to figure out what is *meant*. The statistics is not the end of the investigation but rather a picture (German *Bild*), in which our life is framed and examined.

Before we move on, there is one additional remark. Our survey took place just once *after* the crisis so that responses about pandemic-related experiences are associated with *older* memories. Cason's (1932) famous study shows that pleasant activities of recent times tend to be more efficiently recalled than unpleasant ones that occurred further back in the past (fading affect bias).[8] More recent research indicates further that such a "positive bias" depends on a person's mental health (Dalgleish et al. 2020). Viewed in this light, personal narratives are entirely vague and unreliable sources. But we should not forget that such discussions on various *biases* are underpinned by a premise that our past experiences are documented like protocols, which a healthy psyche can reiterate accurately but a perverted mind might misinterpret. In contrast to this approach, our methodology supposes that our everyday linguistic communication (documentation) is entirely obscure and must be rediscovered anew at every moment.[9] We are not asking machines to reproduce saved data, which might include unexpected technical glitches. Humans can and must, by narrating their experiences, live again those past moments that can never be reduced to algorithms.

Data and Results: Categorization, Keywords, Frequency

In the evaluation process, pupils' statements were organized into five clusters that were simultaneously used as barometers for the statistics:[10] 1) environment and equipment (both digital and analog), 2) time management, 3) learning motivation and habit, 4) social and emotional experience, and 5) health and lifestyle. In what follows, we expound these aspects, consulting histograms and concrete narratives. To facilitate a comparison, the results from both schools are placed together, though they are separately discussed. Tables 14.2 and 14.3 present the statistic date and figures 14.1 and 14.2 sketch how the advantages and disadvantages are distributed among the categories (i.e., the frequency in percentages) and how these patterns have changed between online and normal schooling.

Table 14.2 Statistical Data School A

		Advantages (%)		*Disadvantages (%)*	
		Online: mid-Covid	*On-site: post-Covid*	*Online: mid-Covid*	*On-site: post-Covid*
#	Category				
1)	Environment and equipment	41.8	8.8	19.8	11.0
2)	Time management	44.0	0.0	1.1	44.0
3)	Learning motivation and habit	17.6	29.7	57.1	8.8
4)	Social and emotional experience	13.2	71.4	52.7	5.5
5)	Health and lifestyle	22.0	22.0	22.0	15.4
#	Tagged keywords				
i)	Commute (通学)	24.2	-	-	22.0
ii)	Lonely, isolation (寂しい、孤独)	-	-	11.0	-
iii)	Fun (楽しい)	-	17.6	-	-
iv)	Friends (友達)	-	39.6	-	-

Table 14.3 Statistical Data School B

		Advantages (%)		*Disadvantages (%)*	
		Online: mid-Covid	*On-site: post-Covid*	*Online: mid-Covid*	*On-site: post-Covid*
#	Category				
1)	Environment and equipment	46.7	2.2	35.6	15.6
2)	Time management	31.1	0.0	0.0	35.6
3)	Learning motivation and habit	22.2	60.0	57.8	22.2
4)	Social and emotional experience	2.2	37.8	13.3	15.6
5)	Health and lifestyle	6.7	2.2	4.4	2.2
#	Tagged keywords				
i)	Commute (通学)	24.4	-	-	22.2
ii)	Concentration (集中) etc.	-	-	35.6	-
iii)	Asking questions (質問)	-	33.3	-	-
iv)	Friends (友達)	-	53.3	-	-

School A (Tokyo)

We begin chronologically with the shift to online schooling (dotted bars in the histograms). The first thing that strikes us are the positive effects in the first and the second category. Many students report that digitalization of learning materials grants them easier access to and better organization of resources. In addition, there is no need to carry books to school and so no need to worry about forgetting them at home. Some say that distance learning is also convenient in the case of sickness or injury. These benefits of technical

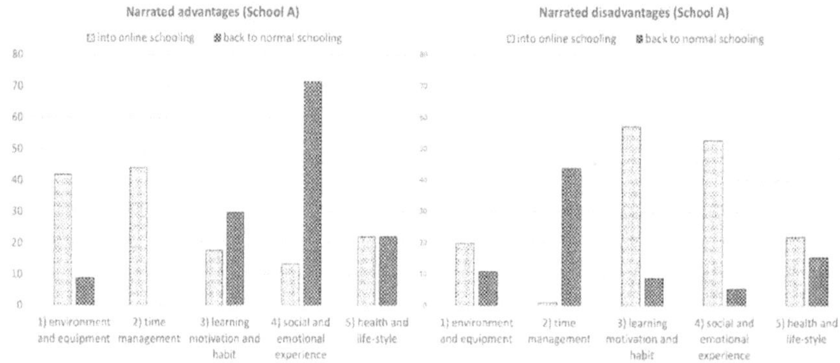

Figure 14.1 Narrated Advantages and Disadvantages School A. Image created by author.

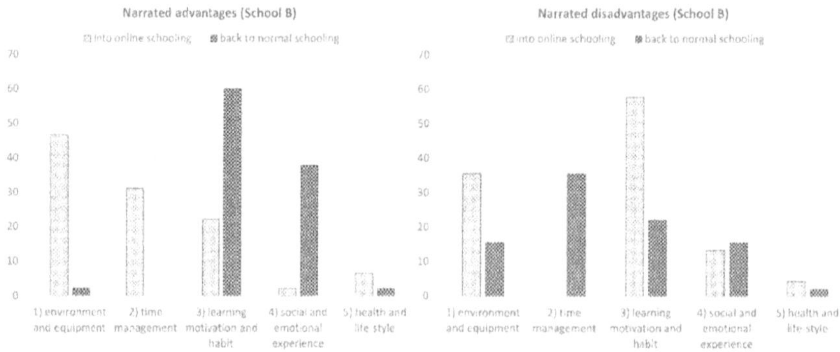

Figure 14.2 Narrated Advantages and Disadvantages School B. Image created by author.

mediation lead directly to the improvement of time management, which seems to be the most attractive feature of remote schooling. Nearly 25% of the students confess that their commute ("通学") through the commotion of Tokyo cosmopolitan areas is extremely stressful. Home schooling is a lucky break to invest more time for other activities such as learning, sleeping, exercise, and hobbies and, to this extent, the category of health and lifestyle, too, is positively affected. But long deskwork also gives rise to physical problems like eye fatigue, backpain, irregular sleeping hours, etc. As regards learning, online classes create more freedom and flexibility because any keywords and topics can be (re-)searched fast and spontaneously. In a similar fashion, communication with schoolmates or distant friends can be done instantly at any given time. However, these welcoming gestures are somehow overshadowed when we turn our attention to the disadvantages of distance school. It is frequently reported that it gives students no strict discipline and structure to keep up with classes. The freedom they get is also a great obstacle they must overcome. Online communication is easy and efficient but prone to be superficial, impersonal, and inauthentic. Another major drawback is that all extracurricular activities (that are placed under the fourth category) are put onto the back burner during the pandemic years. As a result, more than one tenth of the students use explicitly such sentimental terms as "lonely" (寂しい) or "isolated" (孤独).

Next, we assess the resumption of normal schooling (checkered bars in the graphs). Logically enough, all good opinions about optimized time management disappear and metamorphose into harsh critiques about rush hour transportation and so forth. But obviously the most notable change occurs in the feature of socio-emotional life. More than 70% of those questioned describe in a lively manner the pleasure and satisfaction of being in the classroom with their comrades. The keyword *fun (楽しい) is tagged in nearly 18% of the responses and friends* (友) or *classmate* in almost 40%. And even more interestingly, this improvement in sociability positively *impacts* on pupils' learning experience. Although, technically speaking, this *causal* relation cannot be identified in our statistics proper, students themselves confirm that their higher concentration in classes and active participation in discussions are *due to* the support of teachers and other students.[11] Beside this social catalysis, the fixed schedule and environment of the school contribute to re-establishing the habit of constant learning. Concerning the aspect of health and lifestyle, the responses are somewhat ambivalent. It is positive because students can reintroduce in their life regular sleeping hours and exercises through sports classes or afterschool activities. Such routines of usual school life are yet criticized at the same time, for most students get exhausted from fulltime school during the day and private tutoring at night. Nonetheless, emphasis must be placed on some lasting benefits of the technical intervention. Some pupils underline the usefulness of digital learning tools that are provided parallel to the actual class.

School B (Toyama)

Taking a fresh look at the data of School B as a complementary source, there are both similarities and dissimilarities. To begin with, the exact same trend manifests itself in the second category. The burden of commuting to school and related issues are obviously common to both schools situated in a large city. We can also recognize fairly similar patterns in the first, third, and fifth category, while in School B learning motivation occupies the central position rather than social aspects. Our keyword research confirms that here a group of intentional and volitional concepts—*concentration (集中), motivation* (意欲), or *willingness* (やる気)—is employed by about 35 % of the pupils in protest against a lack of them during home schooling. In the usual school environment, deficits of these qualities are quickly compensated mainly because of the increasing opportunities for questions (質問), which are highlighted by more than 30 % of them. And both the advantages and disadvantages of social life are, in comparison to School A, much less mentioned in the answers of School B. However, this difference can be traced back to the difference in *wording*, not in opinion itself since here too we can identify the causal effect of peer support on learning experience. The term *friends* is mentioned by more than half of the respondents as positive factor. Consequently, there seems to be a consensus in both schools that the very definition of school consists in mutual learning by *actually* (not virtually) sharing this process. Nevertheless, our investigation also brings to the fore that several students regard online schooling as preferable, *inter alia*, in order to avoid conflicting situations or peer pressure. In this regard, the technical mediation offers them an escape from social unpleasantries.

To sum up:

1. The greatest advantage of home schooling is to spare the time and stress of the commute, while the efficiency of digital tools is also recognized by many students.
2. Home schooling greatly diminishes both learning motivations and social interactions, which the conventional school setting restores by creating a stable environment of co-learning.

REFLECTION: SITUATIONAL AND HETERONOMOUS PURPOSE OF TECHNICAL EMBODIMENT

To synthesize the last two sections, the following arguments focus first on education technologies and then on the phenomenon of "cultural lensing" that was introduced earlier.

Digitalization of Schools in Germany and Japan

First, let us compare Germany and Japan in terms of digitalization in the field of education. Two OECD reports[12] give us three indications that we analyze more closely here:

1. Neither of these nations was especially prepared for this crisis, although Germany scores higher throughout. According to Huber & Helm (2020: 250–252), German teachers and staff criticize the deficiency of their digital capacity more emphatically than their Swiss or Austrian colleagues. Iwabuchi et al. (2022) show as well that the Japanese school system was neither technically nor pedagogically equipped for the

drastic shift to online teaching. But there were some top-down countermeasures such as Tokyo Smart School Project[13] or GIGA School (Sumimura & Fujimura 2021). In any case, technical conditions were unsatisfactory in both countries.

2. The OECD's survey reveals that only two-thirds of Japanese students have at their disposal a home computer for remote learning (in Germany 92 %). Why this shortage in one of the most technically advanced countries? A plausible explanation can be found in a white paper that deals with the recent development of ICT services.[14] It clearly indicates a decline in the ownership of personal computers, which since 2013 has been roughly *inversely proportional* to that of smartphones. It follows that young Japanese are, on the one hand, highly susceptible to smartphone addiction or overuse (cf. Otsuka et al. 2022) and on the other less prepared for online learning that requires a computer. In Germany, as in any other country, smartphone addiction is often problematized (cf. Montag 2018), but the value of computers for educational purposes seems to be still acknowledged.
3. The last aspect deserves closer attention, for it does not deal with *technical* problems but with the *perception* of young adults. According to the report, Japanese students have lower self-belief in overcoming difficulties in learning (Germany 68 % and Japan 56 %—the second lowest score) and place much less trust in themselves to manage hardships on their own (Germany 85 % and Japan 65 %—the lowest of all).[15] Japanese pupils are hence less independent, i.e., more dependent on others. There is a long history of "shadow education" in Japan, even before this term was coined,[16] in the form of "juku" (塾, private tutoring), as Harnish (1994) points out in the early 1990s. Normally, Japanese children attend, beside their formal education, private lessons in preparation for school exams or entrance exams for higher education. This *external* factor is mirrored in or mirrors the *internal* state of students. Yoshioka et al. (2015) compare adolescents' understanding of science and technology in Japan and Germany and contend, among others, that Japanese young adults are less inclined to recognize them as a source of meaningful knowledge. Some scholars (cf. Kasulis 1995) hold that in Japan technologies are mere practical tools and not rooted in its traditional Weltanschauung. Nakayama (2009), a historian of Japanese and Asian science, dubs this phenomenon "transplantation" of Western technology.

These analyses are of great significance. Postphenomenology claims, just as Einstein did and contrary to Lorentz's theory of material contraction,[17] that the entire condition of our life is affected by the new paradigm of technical mediation, as Ihde (1990: 1) observes that in the Northern Hemisphere "our existence is *technologically textured*." This progress, however, rests on the ideal of autonomy that underlies Western culture including its education. The Japanese do not share this view and instead use technologies randomly and for *ad hoc* purposes (more akin to Lorentz than to Einstein). This inconsistency must be clarified in the framework of postphenomenology.

Philosophical Discussions of Two Types of Ethics

As education represents the moral values of a particular culture, it serves as a pertinent example to assess the seminal role of technology. Even when lifeworld is always or universally mediated by technology,[18] this *de facto* ubiquity does not suffice to justify the *de jure* implementation of it as we have discussed in the introduction. What escapes our notion is our own ethical belief. To get to the bottom of this issue, we pick the brains of a Japanese philosopher, Tetsurō Watsuji (1889–1960), who postulates his ethics against Heidegger, whom Ihde called "the founder of contemporary philosophy of technology" (1990: 19). Watsuji says: "Heidegger characterized human existence by means of the phrase *being in the world* [. . .]. But in his philosophy, the relation between person and person lies hidden behind the relation between person and tools." (Watsuji 1996: 16–17) More accurately, Heidegger's concept of technics presupposes a certain type of intersubjectivity, which is inherited by postphenomenologists to define their human—technologyworld relation.

Watsuji remarks (1996; 13) that in Western philosophy a human being is primarily taken to be an *individual* human. This accusation might be best directed at Heidegger's teacher, Edmund Husserl, who adopts the Cartesian meditative approach and brackets off the ordinary world for the sake of philosophical truth (Husserl 1941). But Husserl's relation to Heidegger is not a simple mentorship but also a sharp rivalry (cf. Harman 2010). Despite Watsuji's protest, Heidegger's being-in-the-world ("In-der-Welt-Sein") requires us both to handle tools masterly and carefully and to treat other persons thoughtfully.[19] Nevertheless, two characteristics of his system stand in conflict with Watsuji's philosophy. First, tools constitute the *first* stage of Heidegger's ontological analysis (cf. §15 et seq.) and concern only the individual existence ("Dasein"), whereas being together with others ("Mitsein") only *follows* it (cf. §25 et seq.). This *methodological* (not ontological) structure betokens an individualistic-technical perspective, which is, for instance, also advocated by Latour's Actor-Network-Theory (ANT).[20] Secondly, both the social and toolmaking aspects are for Heidegger mere checkpoints toward the ultimate source of being—all-embracing *time*—and individuals are historicized within their generation by the force of destiny

("*Geschick*").[21] Although Heidegger stresses the pivotal notion of *betweenness*, as Watsuji does, he refers to the universal *time between* birth and death, which individuals are thrown into.[22] Before this fate, the particularity of tools and other persons fades into the background. As a result, he sacrifices, just like his teacher, our ordinary, *ontic* life for *ontological* manifestation of being, as Yuasa (1987: 77) observes. At variance with this, Watsuji's philosophy is grounded in actual (or act-based) communication in a social and environmental (or climatic) *space between* human and human, with a temporal dimension attached to it.[23] This is evidenced by his demonstration of house-making: according to Watsuji (1961: 8–9), a house must be functional against the climate, viz., it must protect its inhabitants from the heat and cold. Yet this heat or cold is not an objective value that we measure with a thermometer; *we* are the ones that *feel ourselves into* this coldness as a common experience.[24] *Terminus a quo* is always determined by "human relations" or "*act-connections* between person and person like *communication* or *association,*" never by "relations between object and object" (Watsuji 1996: 18).

To summarize, Heidegger's human is, foremost, (ontologically) theorized as *individually* being-in-the-world, while Watsuji observes human beings, (ontically) living in a society. This classification must be considered in the theory of technical embodiment. In the current discourse, it is often taken as given that technologies are immediately related to individuals (cf. Rosenberger & Verbeek 2015). But more sociologically oriented authors (cf. Van Den Eede 2011) raise awareness of various *contexts* of technical mediation. To put it into simple formular, the European model supports:

I – technology → objective world/We → time (as an ontological idea)[25]

In contrast, the Japanese (Watsuji's) version defends:

We – I → technology/climate → space (as ontic reality)

Accidental and Heteronomous Technical Mediation of Lifeworld

It remains to put this theory to the test. Education technologies have revolutionized students' behavior or habits as well as the environment of learning (Aagaard 2017, 2018, 2020). Through digital tools they can act autonomously and interact globally or, as Aagaard (2017) contends, they can integrate ("outside-in") and explore ("inside-out") infinite virtual worlds. This empowerment of individuals through new technologies is not unfamiliar in Japan. Recent research by Kucirkova et al. (2020) shows that Japanese early child education tries out personalized education apps, although there emerges a bundle of problems (paradoxes) such as conflict between personalization and privacy. Even when these authors consider possible cultural differences like individualism and collectivism, the bone of contention is not whether *individualistic use* of technologies can be aligned with social ends because this *I-perspective* is common in the Western schema. Rather, we need to analyze *collective technical intervention*, in which this Western tradition of science and technology might be compromised or jeopardized by the *we-perspective*.

The first implication of our study is that online schooling is very well received as a chance to eliminate laborious commuting hours. Clearly this is just a byproduct of Covid-19 countermeasures for students that are primarily intended to reduce contacts in schools, whereas the use of public transportation is not a necessary condition for going to school. Home offices or teleworks are independent ideas that have been developed since the 1970s as an alternative workstyle to decide freely where to work (Messanger & Gschwind 2016). For our students however, neither such concepts nor the original purpose of technical intervention (i.e., risk management of Covid-19) is decisive (although they are fully aware of them). The purpose of technologies is *accidentally* revised, depending on one's situation. Secondly, individual learning experiences are strongly influenced by social or physical circumstances. Schools generally provide students a place to interact. German researchers (Letzel-Alt et al. 2022) find out that during the pandemic children of various age groups missed social encounters and felt lonely or depressed. But our results show that social factors are not only associated with feelings but also with the perception of technologies such as their purposes. Education is in the West defined as fostering autonomy, as in Kant's "sapere aude!," or, postphenomenologically speaking, healthy "tech habit" (Aagaard 2021). Ihde (2010: 135), for example, emphasizes the "*interrelation*" of humans and technologies: "Even with the pen, I 'use' the pen, but the pen with its material selectivities 'uses' me as well." In Japan, this "I" must be replaced with *we*: the "pen" speaks to *us*, instead of "me" alone. This broadens the concepts of "sedimentation" and "multistability," which are often accentuated in the discourse (cf. Rosenberg & Verbeek 2015: 25–30). The Japanese are accustomed to use only a black pen for official documents. This hermeneutic is segmented and stabilized not in the habit of an individual but in the collective habit as "a black pen for official documents." To reverse the argument, a stable segmentation is not possible if one does not know or cannot see how a tool is employed by other users. Therefore, distant schooling is for this mental frame not only socio-emotionally but also *postphenomenologically unsettling* since students can, being confined in their room, hardly observe how others are operating with machines and programs. This anxiety is caused by deprivation of the "social conformity." Notwithstanding that in our study many students acknowledge the advantages of autonomous learning, these opinions are overwhelmed by critiques of upsetting self-learning and the attractions of fruitful co-learning. Both in theory and in practice, not the *in*dependence but the *inter*dependence of humans constitutes Japanese ethics, which take a grip on the way technologies are arranged as medium to lifeworld.

CONCLUSION

Since the dawn of human history, technologies have helped us to mediate our lifeworld. This postphenomenological principle is not a monolithic concept because each culture perceives technical mediation differently (cultural lensing). The first section elucidated that the notion of science and technology was Janus-faced, being pragmatic and ethical at the same time. The Western tradition is founded in the ethical idea of autonomy, for which technical inventions need to serve. In contrast, Japanese ethics rest on social conformity that maintains their bizarre and contra-intuitive facemask habit. Against this backdrop, we analyzed pupils' narratives about online schooling. On the one hand, home schooling is useful in many respects, especially as an opportunity to do without the unpleasant commute. On the other, it lacks a resilient social and learning environment, which the normal school setting has to offer. These phenomena were then explicated with two models. In contrast to the traditional I–technology–world model, the ego is in the Japanese schema not immediately related to technologies because the we-perspective dominates technical embodiment. The original purpose of technologies can be easily neglected (the manipulation of purpose) and one's tech habit largely conditioned by others' (the heteronomy of technical intervention). This finding can, not only postphenomenologically but also ontologically, shed new light on the technicality of human existence. Ernst Cassirer (1944), whom the young Heidegger challenged in 1929 in a collegial dispute (known as the "Davos Dispute"),[26] maintains that humans are symbolic animals and create various symbolic worlds. In a similar way, Japan's collective approach makes it plausible that technical mediation is of symbolic nature and allows the plurality of meanings (or "multistability"). As a final remark, caution must be exercised not to dogmatize these findings or two models. Cultural lensing effects are only relatively observed and never unique in a literal sense. Japan was just a consummate counterexample to the West to enable the observation we made in this chapter.

ACKNOWLEDGMENT

I owe my thanks to two teachers—Mr. Takahiro Sakata at Senior High School at Komaba, University of Tsukuba, and Mr. Takaaki Taguchi at Toyama Northern High School, Toyama Prefecture—who willingly accepted and carried out my research proposal. My gratitude also goes to the students for their inspiring comments. Although the details of the study cannot be reproduced here, I promise to these collaborators to have treated their contributions with utmost care.

NOTES

1. Compare this to Neil Armstrong's famous dictum: "That's one small step for a man, one giant leap for mankind" (NASA, "Apollo 11"). Technical progress has not only a practical but also a *symbolic* meaning for the mankind.
2. Cf. WHO, "Mythbusters."
3. A cross-cultural study in the United States and Korea also suggests that, in promoting facemasks, the Koreans are more likely to acquiesce in social demands, whereas in America individual freedom must first be guaranteed to introduce the mask requirement on this individual basis. See Chang et al. 2021.
4. Cf. Japan times, "'Everyone else is doing it.'"
5. Since School A disposes of more samples ($N = 90$) than School B ($N = 45$) and also its quality of answers is higher (meaning that students report in a more detailed and careful manner), the latter is regarded as a control sample to detect any particularity of the former. The respondents of School A are roughly 18 years old and those of School B around 17 years old. In both schools, the survey was conducted under supervision of the teacher.
6. In the vocabulary of text mining (Feldman & Sanger 2007), each narrative is seen as a "document," which is "filtered" into one or several "categories" or "labels". For example, a narrative "An easy access to the internet is advantageous and leads to a higher motivation to learn" gets two labels "equipment" and "learning motivation." This method is according to Feldman & Sanger (cf. 67f) "multilabel" (a document can have more than one label/category), "document-pivoted" (each document is assigned to categories, instead of finding all documents that fall under each category), and "soft" (the assignment is assisted by human).
7. Mel'čuk's "Meaning-Text-Model" (2016: 1–4; 34–37) also proposes two aspects of the linguistic operation, namely, having "meaning" (i.e., "something" that must be communicated intersubjectively) and producing "text." The act of speaking is a synthesis of implicit meaning and explicit texts, while "understanding speech" is an analysis where the most appropriate meaning of a text is extracted. In the current chapter, we are not concerned with linguistic problems, but the moot point is that, as Mel'čuk emphasizes (p. 109–110), the operation of language is not "static"

(pure text-analysis) but "dynamic" because behind each text lies its meaning that the speaker wants to express and the listener hopes to capture.

8. Cf. Cason 1932: 84: "The natural preference for the P[leasant] activities, the habit of neglecting and trying to avoid the U[npleasant] activities, the factors of desire, interest, and effort, and the factor of later learning, all favor a relative increase in the efficiency of reproducing activities that are P when they are reproduced and a relative decrease in the efficiency of reproducing activities that are U when they are reproduced."

9. Wittgenstein (1977: 202, §436) quotes (only the latter part of) the following passage from Augustine's Confessions, Book XI (Augustine 2016: 234–235): "dicimus haec et audimus haec et intellegimur et intellegimus. manifestissima et usitatissima sunt, et eadem rursus nimis latent et nova est inventio eorum." ("We say things, and we listen to things: we are understood, and we understand. They are crystal clear and utterly familiar; and then again those same things are all too obscure, and we have to discover them afresh.")

10. Again, my approach was first to obtain (manually) "labels" by "clustering" the documents as they are studied (cf. Feldman & Sanger 2007: 82).

11. This tendency can be identified both in a positive as well as in a negative sense. One positive feedback of the face-to-face class is: "In the environment in which others are present, I can concentrate on learning. Especially, it is fun when I can consult my friends." On the other side, a student claims: "Because the feeling 'I am there (in the classroom)' is lacking, my concentration drops." Although this is not definite proof, it implies a causal linkage between the presence of others and the level of concentration.

12. OECD, "Germany—Country Note"; "Japan—Country Note."

13. Cf. Tokyo Metropolitan Board for Education, "Tokyo Smart School Project."

14. Ministry of Internal Affairs and Communications, „Recent development of ICT service."

15. The chapter "Student self-efficacy and fear of failure" of PISA 2018 (OECD 2019: 189) proves statistically that Japanese students suffer from low self-esteem and the fear of failure.

16. For the origin of this term, see, e.g., Mawer 2015: 135.

17. Epistemologically speaking, Einstein's theory is a more radical shift because the geometry itself is modified whereas Lorentz retains the traditional geometry and lets matter contract *occasionally* (i.e., when objects move almost at the speed of light).

18. Cf. Ihde 1990: 18: "Virtually every aspect of praxis implicates a technology."

19. Cf. Heidegger 2001: 344 (§60): "Resoluteness, as *authentic Being-one's-Self*, does not detach Dasein from its world, nor does it isolate it so that it becomes a free-floating "I". [. . .] Resoluteness brings the Self right into its current concernful Being-alongside what is ready-to-hand, and pushes it into solicitous Being with Others."

20. Cf. Latour 1996: 369: "Whereas social network adds information on the relations of humans in a social and natural world which is left untouched by the analysis, ANT aims at accounting for the very essence of societies and natures. It does not wish to add social networks to social theory, but to rebuild social theory out of networks."

21. Cf. Heidegger 2001: 436 (§74): "But if fateful Dasein, as Being-in-the-world, exists essentially in Being-with Others, its historizing is a co-historizing and is determinative for it as *destiny* [*Geschick*]."

22. Ibid., 426–427 (§72): "Thrownness and that Being towards death in which one either flees it or anticipates it, form a unity; and in this unity birth and death are 'connected' in a manner characteristic of Dasein. As care, Dasein *is* the 'between.'" "

23. This is Watsuji's notion of *yononaka* (世の中) or *seken* (世間), that embraces all three aspects of Welt (world)—the social, climatic, and historical world. Cf. Watsuji 1996: 18.

24. Cf. Watsuji 1961: 8: "That we feel the cold is an intentional experience, in which we discover our selves in the state of "ex-sistere", or our selves already outside in the cold."

25. The human—technologyworld–relation is elaborated both by Ihde as well as by his followers/critics in many directions. But this starting point of the ego seems to be seldom problematized.

26. Regarding this dispute, there are some records and materials available. See Heidegger 1991 and Cassirer 2014.

REFERENCES

Aagaard, Jespers. 2017. "Breaking Down Barriers: The Ambivalent Nature of Technologies in the Classroom." *New Media & Society* 19 (7): 1127–1143. https://doi.org/10.1177/1461444816631505

———. 2018. "Magnetic and Multistable: Reinterpreting the Affordances of Educational Technology." *International Journal of Educational Technology in Higher Education* 15 (4): 1–10. https://doi.org/10.1177/0959354318815766.

———. 2021. "Beyond the Rhetoric of Tech Addiction: Why We Should Be Discussing Tech Habits Instead (and how)." *Phenomenology and the Cognitive Sciences* 20: 559–572. https://doi.org/10.1007/s11097-020-09669-z.

Anders, Günther. 1980. *Die Antiquiertheit des Menschen. Band II. Über die Zerstörung des Lebens im Zeitalter der dritten industriellen Revolution.* München: Verlag C.H. Beck,

Arzroomchilar, Ehsan. 2022. "Some Suggestions to Improve Postphenomenology." *Human Studies* 45: 65–92. https://doi.org/10.1007/s10746-021-09615-1.

Augustine. 2016. *Confessions, Volume II: Books 9–13*. Translated by Carolyn J.-B. Hammond. Loeb Classical Library 27. Cambridge, MA: Harvard University Press.

Benedict, Ruth. 1947. *The Chrysanthemum and the Sward: Patterns of Japanese Culture*. London: Secker & Warburg,

Cason, Hulsey. 1932. *The Learning and Retention of Pleasant and Unpleasant Activities*. New York: Columbia university.

Cassirer, Ernst. 1942–43. "Galileo: A New Science and a New Spirit." *The American Scholar* 12 (1): 5–19.

———. 1944. *An Essay on Man, An Introduction to a Philosophy of Human Culture*. New Haven: Yale University Press.

———. 2014. *Davoser Vorträge, Vorträge über Herman Cohen, Mit einem Anhang: Briefe Hermann und Martha Cohens an Ernst und Toni Cassirer*, edited by Jörn Bohr and Klaus Christian Köhnke. Hamburg: Felix Meiner.

Chang, Hyo Jung (Julie), Seoha Min, Hongjoo Woo, and Jennifer Yurchisin. 2021. "Mask-wearing Behavior During the Covid-19 Pandemic: A Cross-Cultural Comparison between the United States and South Korea." *Family and Consumer Sciences Research Journal* 50 (1): 5–26. https://doi.org/10.1111/fcsr.12416.

Dalgleish, Tim, Caitlin Hitchcock, Jill Newby, Emma Timm, Rachel M. Howard, Ann-Marie Golden, and Willem Kuyken. 2020. "Memory Category Fluency, Memory Specificity, and the Fading Affect Bias for Positive and Negative Autobiographical Events: Performance on a Good Day-Bad Day Task in Healthy and Depressed Individuals." *Journal of Experimental Psychology: General* 149 (1): 198–206. http://dx.doi.org/10.1037/xge0000617.

De Boer, Bas, and Peter-Paul Verbeek.2022. "Living in the Flesh: Technologically Mediated Chiasmic Relationships (in Times of a Pandemic)." *Human Studies* 45: 189–208. https://doi.org/10.1007/s10746-022-09625-7.

Feldman, Ronen, and James Sanger. 2007. *The Text Mining Handbook: Advanced Approaches in Analyzing Unstructured Data*. Cambridge: Cambridge University Press.

Harman, Graham. 2009. "Technology, Objects and Things in Heidegger." *Cambridge Journal of Economics* 34 (1): 17–25. https://doi.org/10.1093/cje/bep021.

Harnish, Delwyn L. 1994. "Supplemental Education in Japan: Juku Schooling and Its Implication." *Journal of Curriculum Studies* 26 (3): 323–334. https://doi.org/10.1080/0022027940260307.

Hasse, Cathrine. 2008. "Postphenomenology: Learning Cultural Perception in Science." *Human Studies* 31 (1): 43–61.

Heidegger, Martin. 1991. *Kant und das Problem der Metaphysik. Gesamtausgabe, I. Abteilung, Band 3*. Frankfurt a.M.: Vittorio Klostermann.

———. 2001. *Being and Time*. Translated by John Macquarrie and Edward Robinson. Oxford: Blackwell.

Huber, Stephan Gerhard, and Christoph Helm. 2020. "COVID-19 and Schooling: Evaluation, Assessment and Accountability in Times of Crises—reacting Quickly to Explore Key Issues for Policy, Practice and Research With the School Barometer." *Educational Assessment, Evaluation and Accountability* 32: 237–270. https://doi.org/10.1007/s11092-020-09322-y.

Husserl, Edmund. 1941. "Phänomenologie und Anthropologie." *Philosophy and Phenomenological Research* II (1) : 1–14.

Ihde, Don. 1990. *Technology and the Lifeworld*. Bloomington: Indiana University Press.

———. 2010. *Heidegger's Technologies: Postphenomenological Perspectives*. New York: Fordham University Press.

———. 2011. "Husserl's Galileo Needed a Telescope!" *Philosophy of Technology* 24: 69–82. https://doi.org/10.1007/s13347-010-0004-5.

———. 2012. *Experimental Phenomenology: Multistabilities*. Second edition. Albany: State University Press of New York.

Iwabuchi Kazuaki, Kouki Hodama, Yutaka Onishi, Shota Miyazaki, Sae Nakae, and Kan Hiroshi Suzuki. 2022. "Covid-19 and Education on the Front Lines in Japan: What Caused Learning Disparities and How did the Government and Schools Take Initiative?" In *Primary and Secondary Education During Covid-19 Disruptions to Educational Opportunity During a Pandemic*, edited by Fernando M. Reimers, 125–151. Cham: Springer Nature. https://doi.org/10.1007/978-3-030-81500-4_5.

Japan Ministry of Internal Affairs and Communications. „Recent Development of ICT Service." Accessed February 12, 2023. https://www.soumu.go.jp/johotsusintokei/whitepaper/ja/r02/html/nd252110.html.

Japan Times. 2023. "'Everyone Else is Doing It': Social Conformity a Top Motive for Japanese to Wear Masks." Accessed February 12, 2023. https://www.japantimes.co.jp/news/2020/08/12/national/social-conformity-masks/.

Kasulis, Thomas P. 1995. "Sushi, Science, and Spirituality: Modern Japanese Philosophy and Its Views of Western Science." *Philosophy East and West* 45 (2): 227–248. https://doi.org/10.2307/1399566.

Kirkegaard, Søren. 1882. *Furcht und Zittern: dialektische. Lyrik von Johannes de silentio.* Translated by H.C. Ketels, Erlangen: Andreas Deichert.

Kucirkova, Natalia, Yuichi Toda, and Rosie Flewitt. 2021. "Young Children's Use of Personalized Technologies: Insights from Teachers and Digital Software Designers in Japan." *Technology, Knowledge and Learning* 26: 535–554. https://doi.org/10.1007/s10758-020-09465-3.

Latour, Bruno. 1996. "On Actor-network Theory: A Few Clarifications." *Soziale Welt* 47 (4): 369–381.

Letzel-Alt, Verena, Marcela Pozas, and Christoph Schneider. 2022. ""I Miss My School!": Examining Primary and Secondary School Students' Social Distancing and Emotional Experiences during the Covid-19 Pandemic." *Prospects* 51: 673–684. https://doi.org/10.1007/s11125-022-09621-w.

Mawer, Kim. 2015. "Casting New Light on Shadow Education: Snapshots of Juku Variety." *Contemporary Japan* 27 (2): 131–148. https://doi.org/10.1515/cj-2015-0008.

Mel'čuk, Igor. 2016. *Language: From Meaning to Text.* Boston: Academic Studies Press. https://doi.org/10.1515/9781618114570.

Messenger, Jon C., and Lutz Gschwind. 2016. "Three Generations of Telework: New ICTs and the (R)evolution from Home Office to Virtual Office." *New Technology, Work, and Employment* 31 (3): 195–208.

Montag, Christian. 2018. *Homo Digitalis: Smartphones, soziale Netzwerke und das Gehirn.* Wiesbaden: Springer.

NASA. "Apollo 11." Accessed February 12, 2023. https://www.nasa.gov/mission_pages/apollo/apollo11.html.

Nakayama, Shigeru. 2009. "Science and Technology in Modern Japanese Development." In *The Orientation of Science and Technology A Japanese View*, edited by Shigeru Nakayama, 114–136. Folkestone: Global Oriental.

Nawa, Nobutoshi, Yui Yamaoka, Yuna Koyama, Hisaaki Nishimura, Shiro Sonoda, Jin Kuramochi, Yasunari Miyazaki, and Takeo Fujiwara. 2021. "Association between Social Integration and Face Mask Use Behavior during the SARS-CoV-2

Pandemic in Japan: Results from U-CORONA Study." *International Journal of Environmental Research and Public Health* 18 (9): 1–12. https://doi.org/10.3390/ijerph18094717.

OECD. PISA 2018 Results (Volume III). 2018. *What School Life Means for Students' Lives.* Paris: OECD Publishing, 2019. https://doi.org/10.1787/acd78851-en.

OECD. 2023. "School Education During Covid-19: Were Teachers and Students Ready? Germany—Country Note." Accessed February 12, 2023. https://www.oecd.org/education/Germany-coronavirus-education-country-note.pdf.

———. 2023. "School Education During Covid-19: Were Teachers and Students Ready? Japan—Country Note." Accessed February 12, 2023. https://www.oecd.org/education/Japan-coronavirus-education-country-note.pdf.

Otsuka Yuichiro, Yoshitaka Kaneita, Osamu Itani, and Yuuki Matsumoto. 2022. "The Japanese Youth Version of the Smartphone Addiction: Scale Among the Youth in Japan: Reliability and Validity Assessment." *International Journal of Mental Health and Addiction* 21: 292–307. https://doi.org/10.1007/s11469-021-00594-z.

Rosenberger, Robert, and Peter-Paul Verbeek. 2015. "A Field Guide to Postphenomenology." In *Postphenomenological Investigations: Essays on Human-Technology Relations*, edited by Robert Rosenberger and Peter-Paul Verbeek, 9–41. Lanham: Lexington Books.

Sumimura, Noriko and Yuichi Fujimura. 2021. "Research on the Support Required by Teachers and School Staffs for Education by Utilizing ICT." *Japanese Society for Educational Technology* 21-2-B4: 112–119.

Tokyo Metropolitan Board for Education. "Tokyo Smart School Project." Accessed February 12. https://infoedu.metro.tokyo.lg.jp/smart_school.html.

Tylor, Edward Burnett. 1920. *Primitive Culture: Research into the Development of Mythology, Philosophy, Religion, Language, Art, and Custom, Vol. I.* Sixth edition. London: John Murray.

Van Den Eede, Yoni. 2011. "In Between Us: On the Transparency and Opacity of Technological Mediation." *Foundations of Science* 16 (2–3): 139–159. https://doi.org/10.1007/s10699-010-9190-y.

Watsuji, Tetsurō. 1961. *A Climate: A Philosophical Study*. Tokyo: Printing Bureau Japanese Government.

———. 1996. *Watsuji Tetsurō's Rinrigaku.* Translated by Seisaku Yamamoto and Robert E. Carter. Albany: State University of New York Press.

WHO. 2020. "WHO's Science in 5 on COVID-19—Mythbusters—16 October 2020." Accessed February 12, 2023. https://www.youtube.com/watch?v=Rr6ABy2Kf_Y.

Wittgenstein, Ludwig. 1977. *Philosophische Untersuchungen.* Frankfurt a.M.: Suhrkamp.

Yoshioka, Ryoei, Takeshi Fujita, Stefan Kaiser, and Yukihiro Komatsu. 2015. "How High School Students and Adults in Japan and Germany Understand Science and Technology." *Journal of Science Education in Japan* 39 (2): 94–103.

Yuasa, Yasuo. 1987. *The Body Toward an Eastern Mind-body Theory.* Albany: State University of New York Press.

Chapter 15

Taking Care for the School in Times of Zoom-Education

A Stieglerian Account

Joris Vlieghe

In this chapter, I develop a postphenomenological account of education in digital times, which is mainly inspired by the work of Bernard Stiegler. The reason for this is Stiegler's original account of the history of technology, and of how evolutions in what count as dominant cultural technologies impact processes of subjectification and hence of education. What it means to educate and to be educated is always technologically mediated and supported. Hence, the shift away from traditional school education, due to the advent and omnipresence of digital technologies, might come with profound changes that demand deeper reflection beyond the platitude that the means we use to educate are but neutral instruments. In this contribution I will particularly focus on the use of conferencing technologies, such as Zoom, that are increasingly replacing traditional classroom teaching, a situation which, in the wake of the COVID pandemic, runs the risk of becoming a 'new normal.' Fundamental pedagogical operations – such as world-disclosure, showing love for one common world old and new generations share, and making the new generation attentive to this world – are at peril. Before developing this argument, I will start with a short methodological clarification for the postphenomenological study at hand. As the notion of attention is the running key idea in this chapter, I will introduce my main methodological tenet with the help of the example of a particular way of becoming attentive, viz. the moment when merely looking at things becomes an act of actually seeing something. It will turn out that seeing is not only deeply embodied, but also deeply shaped by technology.

SEEING VERSUS THINKING: FROM PHENOMENOLOGY TO POSTPHENOMENOLOGY

In this section I want to detail what it means to do research that I regard as 'postphenomenological'. For me the 'post' in the last phraseology refers to the idea that such an approach takes seriously the basic tenets of what is traditionally called phenomenology, but that it also wants to set a step beyond this tradition, especially where phenomenology has not paid enough attention to the way in which technology plays a role in human experience. In that sense there are clear connections between what I propose to do and the school of thought, established by Don Ihde, that goes under the banner of postphenomenology. Even though I share the view that our experience of the world is tightly interwoven with the concrete artifacts we rely on, I will not use Ihde's conceptual schemes and terminology. Instead the main author that will be my guide is the French philosopher of technology Bernard Stiegler, whose main ideas took shape based on a critical elaboration of Husserlian phenomenology. A major reason for turning to Stiegler is that he is also a philosopher of education. More exactly, his reflections on the entanglement of technology and human experience hang closely together with the way he thinks about intergenerational interaction. It is in and through educating youngsters that technologies play a formative (or a destructive) role. Before going deeper into Stiegler's philosophy, I first turn to the issue of why we want to go beyond phenomenology in the first place.

Seeing Further than with Phenomenology Alone

A phenomenological approach could be simply defined as attending to how we experience ourselves, others and the world so as to highlight differences that make a difference and to find a precise language to articulate these differences (cf. Friesen 2011). Consider for instance the difference between thinking and seeing. Throughout Western philosophical history, from Plato to Kant, sight has often been equated with thought or related to it, implying that sight is not fundamentally different from our mental capacity to form representations and ideas about the world. And every often sight was regarded as inferior to thought – as the mere beginning of thinking or a weakened version of true thinking. We have confused 'wiedererkennendes Sehen' (a way of seeing that knows/recognizes) with what seeing is taken as a phenomenon sui generis, a foundational way of relating to reality next to thinking—a 'sehendes Sehen' (a way of seeing that really sees) that has a structure that is quite opposed to thought (Imdahl 1980 as quoted in Kienzler 2020).

Phenomenologically speaking, thinking is experienced as taking distance to the world: ideas come to our mind, which can be checked against a world outside. The outside then proves our thoughts to be true or false. Over and against this, the seeing subject doesn't stand opposed to the world, as seeing is only possible thanks to being fully immersed in the world. Seeing is itself a way of being in the world, of belonging to the world which we see. As Merleau-Ponty (1968, p. 84) famously argues, the one seeing and that what is seen are both part of one and the same 'flesh of the world'. Merleau-Ponty's use of the somewhat enigmatic term 'flesh' does not imply that the fundamental essence of reality is organic. Instead, it signifies that all entities are interconnected through shared qualities resembling flesh, including the ability to touch and be touched. This interconnectedness extends to the capacity to see and be seen. Hence, we can only see the world because we are bodies that are composed out of the same stuff. Moreover, it is vital for the capacity to see that we are able to move about in it. Seeing presupposes the ability to move our eyes (scanning over things and the world saccadicly) and to move our bodies (when approaching the world and things in it from an infinity of viewpoints and angles without ever being able to come to a synthesis of perspectives). This is, again, a completely different logic than the one that structures thinking, which is fleshless and placeless, and which is all about getting at a conclusive synthesis. If we are short of ideas, it makes an enormous difference when we decide to think harder or decide to go for a walk: whilst moving about, ideas will pop up that are not the logically cogent outcome of a combination of existing ideas in our mind. Thinking and seeing are different in essence.

Hence, a camera or a computer cannot possibly 'see' (even though they might be very good at registering and processing visual data), as they are not intertwined with the flesh of the world, nor can they really move about in it. As, again, Merleau-Ponty (1964) says: the human eye is not a camera eye. The eye of the moving body is the place where this body and the world meet and merge. Interesting and convincing as this account may be, it also shows the limits of phenomenology. What remains inconceivable here is the possibility that the very meaning of seeing might change over history, and more exactly due to our dependency on technologies. In the above analysis, seeing appears as a pure, uncontaminated and bucolic capacity (Merleau-Ponty's excellent phenomenological study [Ibid.] is not by accident all based on the work of the painter attempings to portray the world 'en plein air'). Nonetheless, I will defend in the following a divergent position which holds that the increasing use of technologies, such as cameras and smartphones, transmogrifies this very capacity substantially.

Technological Seeing

Ivan Illich offers an interesting take on the changes that our capacity for sight has undergone in human history, and more specifically due to 'the mechanization of the image in photography' (Illich 2005, 164). Before that time, Illich claims, seeing was much closer to the sense of touching, and so it made sense to train the eye according to an ethics of seeing: just the way we do not touch everything (e.g. a priceless painting) or everyone (e.g. a person entirely unknown to us), it made sense to guard one's gaze, to not watch everything and everyone (Op. Cit., 108). This is related to the idea that our gaze actually makes contact with that what we look at, and this can happen in ways that vary from a soft caress to forceful and impudent seizing. With the invention of mechanically reproducible images and the omnipresence of them in our day and age, things have radically altered: 'now the gaze can be conceived as that of the camcorder', Illich says (Op Cit., 164). The eye becomes an arrogant and disrespectful voyeur. Contemporary society seems to have no limits on what should remain unseen or unconsumed. Imaging techniques also afford us to see things with which we could never have been faced in times past, like satellite images of our planet, of which we all earnestly claim that what we do is actually 'seeing' the earth (a perspective sharply contrasting with Merleau-Ponty's account).

The shortcoming of Merleau-Ponty's approach is that it has a particularly one-sided understanding of our bodies as interwoven with the world. Paradoxically enough, one has to assume that the camera and the computer are exceptions in that they seem, according to Merleau-Ponty, excluded from the flesh of the world. This is very implausible: why does what applies to mountains, not apply to my smartphone? Hence, in this contribution I want to stay true to a phenomenological stance but add to it the idea that our reliance on technologies makes a difference that phenomenology also needs to take into account. As mentioned before, I will employ a postphenomenological perspective, I will embrace the idea that all human capacities, including thought, are profoundly influenced by technological mediation, structuring, and conditioning. Since these technological supports evolve over time, it follows that these capacities are contingent and subject to change. More strongly put, the history of humankind is the history of the technologies we have relied on in order to make the world a livable place. What we can (and cannot) do, experience, and think as human beings is fully dependent upon technologies that are the dominant ones at a given time.

THE SCHOOL AS A TECHNOLOGICAL MILIEU FOR TRANSINDIVIDUATION

To the perspective I just described, Stiegler adds three highly relevant points. *First*, Stiegler understands the notion of technology in the largest sense possible. We often take technology to refer to instruments that are driven by fossil or electrical power, and what frequently comes to mind are digital technologies. However, Stiegler's point is that from the very beginning of hominization we have always relied on technologies, as is already demonstrated by the foundational myth of Prometheus and Epimetheus (Stiegler 1998): after his brother has provided all the living creatures with the attributes they need to survive (beaks for the birds, claws for the lions, etc.), there aren't any left for the newly created human beings. To equip these vulnerable and defenseless beings with the means to navigate life, Prometheus defies the Olympic Gods and steals fire, so that metals can be melted and casted into tools. Anthropogenesis is technogenesis. Humanity has depended on the use of tools since its very dawn. A stone axe is thus as much technological as the computer I used to type this text. And more to the point, writing with pen and paper should also be conceived of as a practice that is no less technological. If we want to understand what is at stake when transitioning from predigital school practices to digital ones, it is of the utmost importance to take a technocentric perspective, acknowledging that, for instance, predigital longhand writing is a technology itself.

This is all the more the case, because *secondly*, the protheses (of many different kinds) we rely on define who we, humans, are: for instance, beings that are very much defined by their relationship with their future and past. Telling in this regard is Stiegler's postphenomenological and technocentric rereading of Husserl's theory of retentions (Op. cit.). Retention is not merely memory, i.e. the capacity for recalling the past in the present. Instead, retention is the way in which the past is always co-constituting the present: every state of consciousness, Husserl (1928) claims, is structured and colored by past events we have lived through (e.g. the first time on board of a transatlantic flight will probably feel quite different compared to how an experienced traveler experiences the same span of time). Husserl distinguishes between primary retentions to refer to the effect of past experiences that are still given (e.g. the reader having understood the sentence before this one) and secondary retentions which have to do with bygone experience that is not immediately available (e.g. a prior reading of Stiegler's work, which may make an understanding of this chapter here easier). What Husserl only discusses superficially, but which according to Stiegler (1998) is the most important dimension, is the fact that there always exist external, i.e. material and technological, supports of memory which also and inescapably structure and color our actual experience. This is tertiary retention. It regards past experiences of all our ancestors that are sedimented outside of consciousness in concrete material objects – externalized memory tools. A simple object such as a stone ax contains memory, meaning that we don't have to constantly reinvent our tools from scratch.

It is important to notice here that Stiegler doesn't speak about memory in a figurative sense, as if there is true memory (an internal state of consciousness) supported by an aid that is only called memory metaphorically. On the contrary, it is the other way around: external supports are memory themselves ('hypomnemata') and they are actually a precondition for human consciousness and memory to arise in the first place. Tertiary retention is the most original retention, so to speak. Without monuments that recall past historical events, without calendars which we can access at will, without material bearers such as a piece of paper or a digital notebook to trust our ideas to, etc. we would not have a memory, and thus we would not be who we are. Memory is first of all a feature of the technologies we rely on to become the kind of creature we are.

This is also to say that human beings have no transhistorical essence that expresses itself in each person. Instead, again and again we become who we are through processes of 'individuation', a term Stiegler (2010) borrows from Simondon. Like Prometheus' creatures, we are born without a make-up and always need to become who we are (subjectification) by using the support offered by the 'technological milieu' we happen to live in. This means that individuation which takes place in a milieu shaped by technologies of writing is a different process than individuation in preliterate times. This regards, moreover, also a process of 'transindividuation': we constitute ourselves by taking up the technological legacy of our forebearers to become ourselves, always with slight variations compared to our past (as we have no essence), so as to pass over our own milieu to the generations that come after us. Human history, to Stiegler, is a 'long circuit' during which generations keep going on individuating in relation to artificial memories bequeathed by our ancestors. This history is first and foremost one of – sometimes steadily and at other times rapidly evolving – external supports that shape our humanity in contingent ways.

Expanding on this, Stiegler adds a *third* and most original pedagogical insight. It follows from the last point that every society has a dominant cultural Leitmedium (cf. Böhme 2006), and for a long time this regarded the technology of reading and writing. The inhabitants of the western world are the heirs of a 'culture of the book': as from our earliest education, the world is disclosed to us through (hand)books, and reciprocally one of the first skills we are expected to acquire is how to write (and thus – potentially – to be able to produce books, or book-like texts). In our culture, we don't just consume written or printed texts, but we are also supposed to get the hang of producing 'bookish' texts ourselves.

The purpose of the school is, thus, to introduce newcomers into this Leitmedium. Stiegler's point is that our transindividuation (i.e. that we become a specific sort of human subjects) is due to the school training that is required to read and write accurately (at an early stage of education) and creatively (at later stages). This requires particular routines that we need to come to embody through a long and painstaking process of exercising. We have to learn to write the 26 letters of the alphabet in an acceptable way, which demands much effort. Our bodies need to become 'fully alphabetized bodies', as Kittler (1990) says. Now, in a sense, this elementary training sets the scene for all other learning activities during compulsory education: when we have to learn to get at the meaning of a scholarly text, or to write one ourselves, we need to go through particular exercises which come down to acquiring a certain bodily and mental attention and discipline. It is these exercises that make a school into a school (and which make it differ from other times and places where we can also acquire skills in a spontaneous way, such as learning a new language by living in a family that speaks this language) (Masschelein and Simons 2013). In that sense, the school is not only geared at introducing the new generation in a technological milieu, by making them become fluent and proficient in certain routines. The school is moreover a technology itself. It 'programs' us, Stiegler (2006) says, to become beings who can experience, do, say and think what we can.

Stiegler's account is in line with the so-called literacy hypothesis (Goody and Watt 1968), which comes down to the idea that (alphabetic) writing was not invented by an already brilliant species that, through this invention, made life more comfortable, but that humans are only intelligent thanks to the contingent development of writing technology. To this view, the logical structure of our thinking which is mainly based on lines of argumentation where one proposition follows stringently from the other is rooted in the practice of physically writing one letter behind the other (Cf. Flusser 2011). In view of this we must consider, Stiegler (2010) adds, that with the advent of new and markedly different technologies, and more specifically digital tools that are becoming, or have already become, our new cultural Leitmedium, radically new forms of subjectification might see the light of day. Given that the school, since the invention of writing and reading, has played a key factor in transindividuation (i.e. technologically supported subjectification), the rapid shift from traditional to digital forms of education we are faced with is one of the greatest pedagogical challenges today.

Shifts in Education

Technologies are not just instrumental to education, as if they were just means at the disposal of teachers and students that could easily be swapped for others, without this having an effect on what education is all about. Fundamental changes in our dominant technologies might come with shifts in the very meaning of educating and being educated. One technology, then, comes with different affordances and limitations. It is important to add that all this might sound overly fatalistic and in that sense anti-educational: it could appear that human beings are just the captives or victims of forces that reside outside our own agencies. However, the opposite holds true for Stiegler: it is only thanks to a particular education that we come to possess agency in the first place. Through a long and demanding process of training at school (and, following Kant, for Stiegler this also means to learn how to sit still and concentrate) we constitute ourselves as creatures that are able to be attentive to the world, to think outside of set definitions of the world, to make up our minds and to make reasoned decisions. Aided by a particular school technology, we become beings who desire to lead a life as subjects, not as objects (Cf. Biesta 2022). Stiegler's postphenomenological and technocentric approach is not identical with anti-humanist determinism. Rather, it comes with a much more nuanced and subtle view of what it means to be and to become human.

At the heart of Stiegler's (2010) account is the concern how to ensure processes of transindividuation, and thus of subjectification to go on. Our humanity is at stake. And this means we have to take care, collectively (i.e. as the older generation responsible for the fate of the ones to come), of the technologies we are dependent upon. We should therefore carefully consider the impact of a turn to the unwieldy use of digital technologies, especially in the sphere of education. Stiegler's cautiousness is given in by particular characteristics of the digital which might work against the purposes of education. Surprisingly, Stiegler draws a parallel here between the origins of the school in the Western world, viz. in the Ancient Greek city state, claiming that the school, as a technology of attention, has always been in a fierce competition with inimical societal forces that threaten to capture the youth's attention and to capitalize upon their 'available brain time'. The sophists are a case in point. What is new today, however, is that with the advent and omnipresence of digital tools, attention as such is under threat. Or at least, more and more we loose our capacity for 'deep attention', a term Stiegler (Op. Cit.) takes from Catherine Hayles, to become the slaves of a 'dispersed attention' regime of constant vigilance. The digital tools we so easily embrace are dangerous 'psycho-technologies' that aim at massive control over individuals, and which lead to 'de-individuation'. Instead of supporting processes of (trans) individuation in which people become subjects against the background of an ongoing intergenerational work of initiation, the newest technologies are dangerous devices that are aimed at controlling people and at depriving them to lead meaningful existences as subjects.

A STIEGLERIAN CRITIQUE OF ZOOM-EDUCATION

In the remaining part of this text, I will develop further Stiegler's take on the interconnectedness of technology and education, and also take a critical stance vis-à-vis a too easy acceptance of the digital. However, I will not do this along the lines of a critique of the rise of an attention economy that has installed a complete herd mentality, as Stiegler tends to suggest. Rather, my interest is more pedagogical, and deals with fundamental educational operations that are under jeopardy. More specifically, my focus is on one pedagogical practice, viz. classroom delivery, and on one particular digital technology, viz. 'Zoom-education', which since the COVID pandemic seemed to become a 'new normal'. My focus will thus not be on digitalization as a whole. I will take issue with the ease with which some claim that distance education with the aid of videoconferencing techniques is an acceptable, equally good or even preferable form of teaching and learning compared to traditional, face-to-face school education. However, this is not based on an anti-technological sentiment, but on the contrary on a perspective that accepts our dependency on technology through and through. The future of education depends on the use of adequate technologies, and in view of this, defending the kind of on-screen distance delivery we witnessed over the past years needs to be judged as irresponsible.

I want to stress, once more, that I focus in the following on one specific phenomenon, albeit a most omnipresent and pervasive one, Zoom-education. This by no means implies a generalized anti-digital point of view, let alone technophobia. Taking such a stance would run counter to Stiegler's main idea that we are but who we are thanks to particular technological supports (broadly conceived of). It would also not take seriously the idea that real life education, as it holds for every human activity, is itself supported by technological means, such as blackboards and benches. As indicated, the 'traditional' school itself is a technological milieu. Moreover, given the pharmakon-structure of all technologies, digital ones included, it is more than likely that digital tools can be developed that are educationally speaking defendable and that can offer a good alternative to traditional, predigital forms of teaching. My point is just that what we have introduced as a panacea – and maybe as a development that is here to stay – during the pandemic is not part of digital innovations we should embrace.

The Zoom-bie Seen Differently

From a postphenomenological perspective, Zoom-education has been analyzed recently by Galit Wellner (Wellner 2021) in terms of the figure of the 'Zoom-bie student', a pun on the 'zombies' teachers often find at the other end of the screen, i.e. creatures that are supposed to be (tele)present and to engage in the virtual class, but that appear as 'living dead': for the Zoom-teacher the only sign that there is life at the other side is a black square on a screen. The account I present here differs in three important respects from

Wellner's analysis. First, her concern is mainly with a particular category of students that partake (or better: fail to partake) in online education, viz. those that due to the affordances and limitations of videoconferencing technologies decide to behave in an utterly disengaged manner by not putting on their camera's. My concern, however, is not only with students-turned-zombies, but with students in general and with what it means to attend classes online, i.e. even when they really desire to be taught and don't shy away from putting effort in their own education. Second, Wellner seems to be interested in the effect Zoom-education has on the learning outcomes in young people. I am explicitly not. I will try to show that a confusion between learning and education, especially in the context of digitalization, is something to be carefully avoided. Education comprises of more than learning. And so, in my view, the pervasive rhetoric that we can successfully learn on-line severely threatens the continuation of genuine education – understood intergenerationally and as a matter of transindividuation, as Stiegler wants it. Third, where Wellner sticks to Ihde's postphenomenological framework, looking at the four basic relations Ihde discerned between human, technology and world, I pursue a Stieglerian analysis of Zoom-education.

Attention as an Ecological Phenomenon

If the traditional school, based on face-to-face teaching, is a technologically supported milieu, the question is what it is about this technology that makes education possible in the first place, and why videoconferencing technology gravely impedes this. For Stiegler (2010), the school – as it has materialized in the last centuries – is a particular way of giving shape to space and time, i.e. the bringing together of the new generation with a representative of the old one during a limited time and in a physically shared space, separated from home and yet protected against the public world of adults and their political dreams or commercial interests, so that study for the sake of study becomes possible. The 'old' school operates more exactly as an 'attention machine': it conducts attention towards a thing of study the generations are assembled around. It is important to be precise about the meaning of the notion attention here, because Stiegler points at three different, although interrelated phenomena that lie at the heart of education, understood as a transaction between the generations.

First, attention should be considered from a pedagogical perspective, and hence it should not be taken as an individual and psychological category, i.e. as an inborn and unchangeable capacity (which some possess more and others less), which can be measured under laboratory conditions. On the basis of this kind of research we can conclude that average students can only stay concentrated for about 15 minutes. This way of framing things is often the mindset of those who defend on-line education, because it is supposed to offer more opportunities for immersion and intellectual stimulation. However, when we agree to this view and act accordingly, we give in to a self-fulfilling prophecy and hence abide to a dangerous anti-educational fiction (Rancière 1991). Over and against this, we could also deal with attention not as a fixed given, but as something we have to cultivate, difficult as it is, through exercising (Rytzler 2019). Attention is an ecological category, as Yves Citton defends. With this he makes a double claim. Attention is an attribute of a collective of people: it is not the possession of individual persons, but necessarily always in-between people. Furthermore, the quality and intensity of our common attention also always depends on the material and environmental conditions we are in: the technological context where attention takes place always matters.

In the classroom in which students and teachers are physically present this plays out in the following way. First, it makes an enormous difference whether one is trying to learn in solitude sitting behind a screen, compared to being gathered as bodies in one and the same environment. When a student has lost concentration and is daydreaming, but all of the sudden her peers start jotting down what the teacher is saying, she is immediately brought back to the situation and regains focus. This shows that attention is somehow in-between the people gathered in the classroom (cf. Marin 2021). This is, attention is supported by the physical presence of other students, in so far, the attention others display works as a contagious force that makes one attentive (not unlike the contagious effect of laughter within a group of people that are physically – not digitally – together). Zoom-education, on the contrary, is all about keeping at bay the presence of others during one's learning process. Becoming concentrated and immersed is a purely private responsibility.

Second, attention is also supported by the architectural arrangement one finds oneself in at school. Physical classrooms are conceived of in such a way that it is difficult for students to hide from the gaze of others – peers and teachers – which aids them in staying attentive (even though, obviously, the massive introduction of laptops in schools has critically changed this situation, as teachers no longer have a clear sense whether students are attentive or not for what happens in the classroom or whether they are actually with their minds in the digital realm). School architecture is thus a technological support of attention.

In her analysis, Wellner makes a similar point when she captures the specific feel of being taught online in terms of a felt lack of 'social embodiment', the absence of an ambience which she calls 'buzz' (p. 6). However, she doesn't really go deeper into why it is we need others in order to stay attentive. Moreover, there is another reason why Wellner's argument falls short of getting at the root of the issue. Developing Ihde's conceptual schemes, she defines the Zoom-bie student in terms of a failed merger of an embodied and a hermeneutic relation to the world, i.e. one in which technology has become one with the subject to the extent that it gets to things in the world without being aware of the support (e.g. looking through one's glasses) and one in which we look at the technological tool in order to know something about the world (e.g. looking at a thermometer to find out the temperature in the room). What she implies is that for the engaged on-line student such a successful intertwinement is possible, viz. in this student's experience the world appears on the screen and at the same time the world becomes present, so that she can learn (instead of being condemned to the life of the Zoom-bie).

In my view however, Wellner's analysis misses out on specific technological characteristics of videoconferencing technologies that just make it impossible to really relate to and be exposed to a world. This has been well established by Iain Thompson (2005) in his Heideggerian analysis of the digital university. In a digital learning environment, where the world is disclosed through a screen, the world gets a different ontological aura, Thompson shows. What appears on the screen of our digital apparatuses doesn't have the weight of the things we physically study together in a classroom or via a (hand)book. What we have in front of us is no longer study material, but ephemeral, easily manipulable and qualifiable pieces of information. This is especially the case with recorded classroom delivery, which we can access at the time we find convenient, where we want, at double or triple speed, as many times as we choose, etc. Information can be cut and pasted, as the learner pleases, and – as of recent date – also increasingly automatically generated and endlessly multiplied. With the ever growing use of handheld screens (also in educational contexts) this trend gets reinforced: what appears on the screen as a stream of visual information is simultaneously experienced as being hold in our own hand. Whereas the screen of predigital devices such as the television is an apparatus for looking at something that is so to speak behind it and its user is transported to somewhere else (Carbone 2013), on the screen of a smartphone tactual and visual sense merge (Cooley 2014). Its user literally has the 'world' in the palm of their hand. Reality is just given so as to be privately consumed. Instead of reaching out to a world outside that we can collectively study, students get trapped in a technological milieu that has only resources on offer that reinforce the student and her own perspectives and desires.

This stands radically opposed to the idea that there is a world outside and that education is about introducing the new generation into this world by way of exposure (Arendt 1961). This entails that children leave the sphere of the home and that they come to transcend their immediate lifeworld and the interests they already have, so as to be confronted with a world that is new to them, that might surprise them and sometimes sting them and be at odds with them, that does not yield to their desires and that asks them to start studying it with due care. Only then they become students.

Attention Beyond Learning

At this point in my argument, a second and third meaning of the notion attention comes to the fore (next to attention being an ecological category rather than a psychological one). In line with the Latin etymology of the word, Stiegler (2010) points out that 'at-tendere' means stretching out towards what is not yet given, or stronger: giving heed to something that transcends the sphere of one's own lifeworld. Attention then stands opposed to intention: there is something more at stake that demands that one starts caring for it. If the school is indeed an 'attention machine' as Stiegler wants it, this is bound up with a particular understanding of education that goes beyond mere learning, i.e. the strengthening of one's given position by developing new cognitive powers and skills. According to a logic of learning, the world doesn't really matter. The world appears then as a resource and not as an object of attention and due care. This is the 'world' that Zoom-learners see appear before their eyes: information they can consume and exploitate. The world is no longer something outside that might put demands on them and that might profoundly transform them.

I would therefore like to represent a different perspective on world opening through Zoom than Wellner (2021), for whom education and learning seem to coincide. ZOOM-mediated education implies exactly world-destruction in my view. It locks learners up into the confines of an information solipsism in which a true exposure to the world gets totally excluded. The screen functions not as a window to the world but as a mirror that just increases a sense of self, without really been touched by a world outside and without offering the chance to start attending to it, i.e. to care for something that interrupts our intentional plans and that forces us into a transformation of the self that we already are (Cf. Masschelein and Simons 2013).

Education as Transindividuation

In my view education is an intergenerational interaction in which the continuation and rejuvenation of an already existing world is at stake (Arendt 1961). And here we arrive at the third meaning of attention Stiegler wants to preserve: attending to and caring for the possibility that the older generation can pass on to the new generation what they deem to be of the greatest importance, so that the new generation can actually go on with it and add new beginnings to it. This is, the education of newcomers to our world is taken as being more than learning: they get educated – or in the Stieglerian vocabulary 'individuated' – in relation to a world of meanings that is already there so that they can be transformed by it, start caring for it and renew it. That is, pedagogically speaking, what 'transindividuation' means.

In such a view education is through and through intergenerational: passing on what we find of value and show to the newcomers what is of value, of the things we love, so that they can start to see that there are valuable things to engage with and so that they can rejuvenate a shared world. Whereas videoconferencing technologies may support learning, they cannot support education understood in this sense. For Hannah Arendt (1961), sharing 'love for the world' and showing that and why something is valuable are typically operations that best happen in a direct and physical manner. I think they are seriously endangered when teacher and students appear to each other via a box on a computer screen. For example, when a teacher draws a diagram on the blackboard in a classroom-setting where she physically meets with the newcomers, what she does is not merely presenting information so that students may learn. The care (or carelessness) with which she does this displays (or fails to display) in a most explicit way her love for the subject matter at hand. A teacher who puts away her textbook with diligence, in front of her students' eyes, shows what the value is of the contents in that book, as opposed to the teacher who handles her books roughly. These are small, but not insignificant gestures, which tend to disappear when a talking head appears on the screen of a videoconferencing tool.

These technologies are also highly individualizing (cf. Masschelein and Simons 2013). Of course, seen from a perspective that understands education as mere learning, this doesn't pose any problem, as it doesn't really matter when one is alone or in group when consuming information so as to strengthen one's position in life. But this makes all the difference when it comes to understanding education intergenerationally. When children attend school, they are not primarily addressed as individual learners with their singular needs. They are at school because they are the newcomers to this world. They are there only as the representatives of the new generation – the one with whom we share our love for the world and the one we hope will continue with the world in their unique and new ways. So, it is important for students to leave the sphere of the family (where one is always someone's son or daughter) so as to literally move themselves to a technological milieu that is different and not necessarily comfortable, where they are gathered with people who are sometimes strangers and not the persons one has chosen to be with. To leave the comfort of one's bedroom and the house, and to go to another place to be with peers that are nevertheless strangers is extremely important to facilitate the experience of being the member of a new generation. This experience also highlights the idea that education is not for someone in special, but for everyone (belonging to the new generation). Again, attending a course via Zoom from one's bedroom puts an end to all this. The technological conditions are such that students are forced to behave as atomized individuals who no longer have the feeling to belong to a new generation. This is an education tailored to a personalized customer which excludes the possibility to realize that more is at stake, viz. passing on love for the world and caring for the continuation of this world.

TAKING CARE OF THE SCHOOL

In conclusion, taking up the responsibility for the continuation of a common world and for the possibility of newcomers to go on with the world in unforeseen ways means today, first and foremost, that we take up care for the technologies that inevitably support and give shape to the sphere of education. We must take care after the school in times of Zoom-education. If we take seriously the post-phenomenological and intergenerational account I have worked out throughout this contribution, the most important endeavour for pedagogues today is not to be limited to fostering learning (and thus to stay indifferent vis-à-vis the technologies we rely on and to stay blind to how technology and subjectification are related), but to take up the challenge of thinking about how to deal with educational technologies, and to design these in such a way that fundamentally educational operations are to be supported, even in digital times: disclosing a world that is so valuable that it deserves study and care, teaching the new generation out of love for the world so that they want to add new beginnings to it and supporting attention for the world.

REFERENCES

Arendt, Hannah. 1961. *The Crisis in Education. In Between Past and Future: Eight Exercises in Political Thought*. New York: The Viking Press.

Biesta, Gert. 2001. *World Centered Education. A View for the Present*. London: Routledge

Böhme, Jeanette. 2006. *Schule am Ende der Buchkultur. Medientheoretische Begründungen schulischer Bildungsarchitekturen*. Bad Heilbrunn: Klinkhardt.

Carbone, Mauro. 2013. "La vie mouvementée des écrans." *Ecrans,* 1 (3): 21–33. https://classiques-garnier.com/ecrans-2015-1-n-3-expanded-cinema-la-vie-mouvementee-des-ecrans-en.html.

Citton, Yves. 2017. *The Ecology of Attention*. Cambridge: Polity Press.

Cooley, Heidi Rae. 2004. "It's All about the Fit: The Hand, the Mobile Screenic Device and Tactile Vision." *Journal of Visual Culture* 3 (2): 133–155.

Flusser, Vilém. 2011. *Does Writing Have a Future?* translated by Nany A. Roth. Minneapolis: University of Minnesota Press.

Friesen, Norm. 2011. *The Place of the Classroom and the Space of the Screen. Relational Pedagogy and Internet Technology*. New York: Peter Lang

Goody, Jack, and Ian Watt. 1968. "The Consequences of Literacy." In *Literacy in Traditional Societies,* edited by Jack Goody, 27–68. Cambridge: Cambridge University Press.

Heidegger, Martin. 1978. *Basic Writings*, translated by D. Farrell Krell. New York: Harper Collins.

Husserl, Edmund. 1928. *Vorlesungen zur Phänomenologie des inneren Zeitbewusstseins*, edited by Martin Heidegger. Halle: Niemeyer.

Ihde, Don. 2009. *Postphenomenology and Technoscience. The Peking University Lectures*. New York: SUNY.

Illich, Ivan. 2005. *The Rivers North of the Future. The Testament of Ivan Illich as told to David Cayley*. Toronto: Anansi.

Imdahl, Max. 1980. *Giotto, Arenafresken. Ikonographie, Ikonologie, Ikonik.* München: Fink.

Kienzler, Klaus 2020. *Cézanne, Klee, Kandinsky: Zur Phänomenologie der Kunst des Sehens*. Freiburg: Karl Alber.

Kittler, Friedrich. 1990. *Discourse Networks 1800/1900*, translated by Michael Metteer and Chris Cullens. Stanford, CA: Stanford University Press.

Marin, Lavinia. 2021. *On the Possibility of a Digital University: Thinking and Mediatic Displacement at the University*. Cham: Springer.

Masschelein, Jan and Maarten Simons. 2013. *In Defence of the School. A Public Issue*. Leuven: Education Culture & Society Pub..

Merleau-Ponty, Maurice. 1964. "Eye and Mind." In *The Primacy of Perception*, edited by J. E. Edie, 159–190. Evanston, IL: Northwestern University Press.

———. 1968. *The Visible and the Invisible,* translated by. A. Lingis. Evanston, IL: Northwestern University Press.

Rancière, Jacques. 1991. *The Ignorant Schoolmaster,* translated by Kristin Ross. Standord, CA: Stanford University Press.

Rytzler, Johannes. 2019. "Turning the Gaze to the Self and Away from the Self: Foucault and Weil on the Matter of Education as Attention Formation." *Ethics and Education* 14 (3): 285–297.

Stiegler, Bernard. 1998. *Technics and Time: The Fault of Epimetheus*, translated by R. Beardsworth and G. Collins. Stanford, CA: Stanford University Press.

———. 2006. *La télécratie contre la démocratie. Lettre ouverte aux représentants politiques*. Paris: Flammarion.

———. 2010. *Taking Care of Youth and the Generations*, translated by S. Barker. Stanford, CA: Standorfd University Press.

Thomson, Iain. 2005. *Heidegger on Ontotheology: Technology and the Politics of Rducation.* Cambridge: Cambridge University Press.

Wellner, Galit. 2021. "The Zoom-Bie Student and the Lecturer: Reflections on Teaching and Learning with Zoom." *Techné: Research in Philosophy and Technology* 25 (1): 1–25.

Part III

SUBJECT SUBVERSION

Chapter 16

After Phenomenology?

Digital Subjects and the Relevance of Experience

Anke Redecker

POSTPHENOMENOLOGICAL CHALLENGES: INTRODUCING REFLECTION

Looking at challenges of distance learning during the COVID-19 pandemic that have shown deficits and chances of digitalized schooling, the relevance of postphenomenology for digital learning and teaching can be highlighted. A first step to shape the topic can be the reflection of meaning, extension, and reception of postphenomenology, especially regarding its relation to phenomenology (2.).

After a theoretical discussion of this relation, a practical approach thematizes learning in lockdown scenarios, exemplifying that the experiencing subject—an essential link between phenomenology and postphenomenology—should not be neglected, where the dignity and responsibility of human beings are taken into account (3.). Having a closer look at digital learning strategies the political and ethical relevance of postphenomenology is stressed, especially focusing on power relations of self-guided learning (4.) that can show the affordance of empowering the resistant learner in critical discussions (5.) Finally, a retrospect and outlook round up the reflections (6.).

IS THE PHENOMENOLOGICAL SUBJECT OBSOLETE? A CRITICAL PERSPECTIVE ON POSTPHENOMENOLOGICAL APPROACHES TO PHENOMENOLOGY

We live in times of "postperspectives" in form of attitudes that combine the prefix "post" with problematizing something that was before and is addressed as criticized, transformed, and possibly rejected today. This can be shown with theoretical concepts like postcolonialism (Mecheril 2018), postmodernity (Butler 2012), postdigitality (Jörissen 2017), and finally postphenomenology (Ihde 2009; Verbeek 2011; Rosenberger and Verbeek 2015). Referring to Ihde and Verbeek Rosenberger defines postphenomenology as

> the name of a school of thought in the philosophy of technology that addresses issues of user experience through a practical lens. This perspective brings together ideas from phenomenological and American pragmatic philosophy to develop conceptual tools for the deep description of human–technology relations. (Rosenberger 2020, 1f.)

Overcoming a traditional phenomenology of the Husserlian *ego cogito* as a kind of constructor and dominator of the world (Husserl 1987), postphenomenology rejects a classical phenomenological approach as "a philosophical method that sought to describe 'reality itself'" (Verbeek 2005, 104), while our relation to objects is always a mediated perspective and therefore far beyond knowledge and certainty.

In accordance to postdigitality, postphenomenology stresses that human beings' experiences are not independent from but (pre-)determined by technical, especially digital factors that influence their way of dealing with the world. "Under the postphenomenological perspective, technologies are not conceived as mere instruments in the world that do their jobs in a passive manner. Technologies are understood as active mediators of experience" (Rosenberger 2020, 2). We can see postphenomenology, "one of the central theoretical approaches in the philosophy of technology" (de Boer 2021, 1), in a deep relation to postdigital approaches that problematize "how data is entangled with shifting socio-cultural, political, economic, historic and material orderings and normalisations" (Macgilchrist, 2017, 100; cf. Bettinger and Hugger 2020; Redecker 2020a). Rosenberger points out that "we cannot think of humans as actors, interpreters, and decision makers without considering their relationships to their devices" (Rosenberger 2020, 8). Digitality becomes a special topic in focusing on the technically mediated relation of subject and object. In his e-mail correspondence with Ihde, Kaminski highlights the differentiation of this relation in manifold variations as an essential merit of postphenomenology:

> Classical phenomenology has its merits in the development of a concept of technology that grasps it as an intentional mediator, as the form of the relationship between self and world. This also indicates its weakness. It has little or nothing to say about differences of concrete technologies. This is where

> postphenomenology develops its great potential. Postphenomenology was the first phenomenology to take an interest in the various technologies. And it has developed a rich description of the particular technically mediated experiences. (Ihde and Kaminski 2020, 279)

Concentrating on different practical contexts, postphenomenology rejects fundamental theoretical principles. Due to philosophical developments like postmodernism "fundaments have been challenged as problematic" (Verbeek 2005, 104; cf. Ihde 1993; Rosenberger 2020). The modern relevance of persons is replaced by a postmodern problematized praxeology, no longer highlighting practicing subjects, but their involvement in practices that determine and constitute them. Rejecting fundamental principles, postphenomenology is also related to phenomenology. "Postphenomenology can be viewed as an offshoot of phenomenology that is motivated by the postmodern aversion to context-independent truths and the desire to overcome the radical separation of subject and object, but that does not result in relativism" (Verbeek 2005, 113).

How can we find a meaningful look at postmodernity, postdigitality, and finally postphenomenology beyond relativism? Highlighting postdigitality as a critical reflection of digitality, we can find a focus on something that is still relevant but should be problematized (digitality). Regarding an analogous glance at postmodernity, modernity can be pointed out in a similar position. As shown in Käte Meyer-Drawe's approach to postmodernity in educational theory (Meyer-Drawe 2008), we can come to an interpretation of the prefix "post" that does not address what has been overcome as totally obsolete and irrelevant in today's discussion, but an impulse to discuss its relevance and its chances of transformation. This dialectical approach to the prefix "post"—overcoming and transforming the criticized by highlighting its relevance—can also be found in postcolonial discourses, problematizing that colonial practices, although historically located in the past, are not really over, but still relevant in—even transformed—power relations and the critical glance on them. Thus, postcolonial reflections thematize the ways colonial practices have been constructed. For example, we can find this position in Stuart Hall's *The West and the Rest* (Hall 2019), where he critically highlights a eurocentristic suppression of so-called underdeveloped and subordinated cultures nowadays still struggling for recognition.

In contrast to postdigital approaches, the reflected object (colonization) is strictly damned (Mecheril 2018). We can see this in analogy to radical postmodernism, where the prefix "post" accentuates something that should not be (modern attitudes as a whole). While colonialism is driven by unjustified domination, suppression, and violence, the glance at modernity should be more differentiated to find out what should be rejected and what not. Postcolonialism, postmodernism, postdigitality, and postphenomenology have one essential point in common: rejecting the subject of pure knowledge, self-awareness, and suppressing domination.

Having a differentiated glance on postmodernism, it can be seen as a critical descendant of modernity, still referring to some modern traditions like logical and responsible argumentation, but without claiming fundamental principles. Accordingly, we can shape postphenomenology "as an offshoot of phenomenology" (Verbeek 2005, 113), shedding a new light on phenomenology. It clings to several phenomenological elements like embodiment and human–world–relations while criticizing transcendental argumentation and shaping mediated human–world–relations in reference to new technologies. In postphenomenological approaches, phenomenology is no longer relevant as a philosophical method to find clear truth and certainty, but it can be compared to newer phenomenological concepts highlighting the entanglement of human beings with the world, as Maurice Merleau-Ponty does. Ihde stresses that "it is Merleau-Ponty who inspires Postphenomenology" (Ihde/Kaminski 2020, 274). Merleau-Ponty "no longer addresses the self-aware subject of knowing, grasping the things themselves, but '*relations between human beings and the world*'" (Verbeek 2005, 108). Postphenomenology transforms this concept highlighting the relevance of new technologies. Because of its pragmatic touch (Bohlmann 2022a, 172; Bohlmann 2022b, 25), postphenomenology can benefit from a closer reflection of possible theoretical backgrounds that do not have to be strictly normative. It can consider preconditions we have to take into account if we want to shape postphenomenology reasonably without claiming a fixed final truth in the way these preconditions are described.

If postphenomenology "bases its conclusions on case studies, i.e., on analyzing concrete empirical inter-relations and interactions between humans and technologies" (Ritter 2021, 1502), it has to operate with terms that should be defined and correlated methodically. As "a movement to study technology in a systematic way," that "tries to make sense of technology in its social and cultural context" (Arzroomchilar 2022, 67), postphenomenology should ask how this sense and systematic approach can be constructed and justified. We can refer to the phenomenological subject as an indispensable precondition of experience. "Phenomenological analyses make it clear, that we can only be attentive to something which shows itself, which really concerns and affects us and that we focus on" (Brinkmann 2019, 33). We do not only need that something, but also us, human beings who experience something and deal with it, while being entangled with it. Postphenomenology takes this human-centered but relational idea of experience from classic phenomenology.

> Phenomenology thus overcomes the dichotomy between subject and object, humans and world, by replacing it with a mutual interrelation. Human beings are

> unthinkable apart from a relation to the world, which they continually experience and in which they realize their own existence. [. . .] Human beings are continually engaged with their world, and this engagement precedes any judgment they may have of it. Put another way, it is impossible to speak about the world in the absence of human involvement with it. Reality-in-itself is unknowable, for as soon as we experience or encounter it, it becomes reality-for-us: a world. There exists neither human beings in themselves nor world-in-itself. (Verbeek 2005, 110)

Not only phenomenology but also postphenomenology has various types and aspects. On the one hand, seen as the overcoming of phenomenology—"In light of postmodernism and the linguistic turn, phenomenology seems to be obsolete, a romantic throwback" (Verbeek 2005, 104)—on the other hand, addressed as a "a 'hybrid' phenomenology" (Ritter 2021, 1502), postphenomenology seems to offer a manifold marketplace of meanings and attitudes without a clear and distinct methodology. Referring to Ihde's description, we can shape it as a variable mixture of several theoretical offers.

> Those familiar with postphenomenology know that it is a deliberate modification of versions of classical phenomenology which reject Husserl's egological transcendental (and all quasi-Cartesian leftover aspects), incorporate some aspects of American Pragmatism, and rely upon an interrelational ontology still modeled upon intentionality. (Ihde and Kaminski 2020, 262)

Rejecting the Husserlian approach, postphenomenology denies a world that can be constructed by a pure *ego cogito* beyond the world. The enriched mixture of theoretical elements and directions implies controversies within and in confrontation with further, for example, former orientations, as we can see, for example, in "the 'transcendental/empirical' debate" (Ihde 2022, 852). Ihde criticizes the "conter-movement" of a "transcendental" and "Anthropocene" pushback" (Ihde 2022, 853) with "roots in the older [. . .] largely European philosophies of technology," that he sees "as dystopian, romantic and nostalgic" (Ihde 2022, 853). This is only one perspective on the relevance of phenomenology in postphenomenological concepts; and even phenomenology is not homogeneous, as Ihde points out in his interview with Kaminski:

> Phenomenology has itself had a history of variations—existential, hermeneutic, and now postphenomenology. All these variations reject aspects of Husserl's egological transcendental version of classical phenomenology. (Ihde and Kaminski 2020, 262)

Verbeek accentuates phenomenology as "an important tradition in Continental philosophy in the first half of the twentieth century, but its influence has waned" (Verbeek 2005, 104), while we still find important phenomenological approaches in today's philosophy and pedagogy (Waldenfels 1999; Meyer-Drawe 2008; Brinkmann 2019) that no longer describe themselves on the base of Husserlian egology. Instead matters of perspectivity and positionality, ambiguity, and alienation are common in phenomenological reflections (Waldenfels 2009; Meyer-Drawe 2015). "Human beings never encounter a world in itself, only and always a world for them" (Verbeek 2005, 107)—this seems to be undoubtable for phenomenologists like Waldenfels or Meyer-Drawe in the tradition of Merleau-Ponty (1984, 1986). With Ritter, we can highlight "the evolution of contemporary phenomenology (after Merleau-Ponty)" (Ritter 2021, 1512). Neglecting it, "postphenomenology has gradually diverged from phenomenology. With its inter-relational ontology, postphenomenology now pays attention to relations rather than intentionality" (Ritter 2021, 1512). However, referring to Waldenfels' phenomenology of alienness or Meyer-Drawe's educational theory of bodily experience, we can see postphenomenology in a clear phenomenological tradition, as Kaminski does:

> Postphenomenology is at its core classical phenomenology. Contrary to what the prefix ›post-‹ suggests, it remains within the scope of classical phenomenology and represents an internal differentiation of its ideas. (Ihde and Kaminski 2020, 266)

Kaminski has "the impression that when it comes to the relationship between technology and intentionality, the distance between phenomenology and postphenomenology is minimal" (Ihde/Kaminski 2020, 269) and that "this is an internal differentiation of different technical modes of intentionality within the classical framework" (Ihde and Kaminski 2020, 269). Perception can be seen as a fundamental link between phenomenology and postphenomenology. It accentuates the relevance of the human factor without neglecting the uncertain and (pre-) determined subject.

> The postphenomenological philosophy of technology is phenomenological exactly due to its focusing on experience. Analyzing experience, phenomenology emphasizes the role of intentionality as a "process" through which, to put it as generally as possible, an experiencer gets into, or finds themselves in, a relationship with an experienced world. I deliberately make use of this ambiguous formulation to underline that intentionality does not have to be thought of as based in, or constituted by, the subject. Accordingly, phenomenologists after Husserl have repeatedly criticized his subjectivism and outlined various ways of overcoming it—to do justice to experience itself. (Ritter 2021, 1512)

Here, we can look at the special role of (post-)phenomenological subjects. Thematizing the world we are living in, we are part of what we try to describe, while it is part of ourselves in a kind of entanglement far from clear knowledge. It should be asked, what kind of phenomenology became problematic and which one can be seen as necessary and possibly transformable to fit today's challenges of technical issues like digitality and digital learning. Rejecting traditional phenomenological ideas like Husserl's pure ego cogito beyond the world, does not have to mean rejecting phenomenology and the subject of intentionality as a whole, especially if we remember, that postphenomenology is "an offshoot of phenomenology" (Verbeek 2005, 113). Phenomenology was necessary to find a postphenomenological approach and is still indispensable to explain postphenomenological ideas. In analogy to a differentiated glance on the relation between modernity and postmodernity, we do not have to see postphenomenology in total difference to phenomenology and can look at it as rethinking and possibly transforming phenomenological theses.

That "human access to reality is never direct but always mediated" (Verbeek 2005, 104) rejects the subject of knowledge and certainty, but not the experiencing and reflecting subject finding itself in relation to digitality. If the experience of a mediated world is a typical human one (Rosenberger and Verbeek 2015, 20; Bohlmann 2022, 183), we need an experiencing subject. Human beings can be called subjects who are not obsolete, but ambiguous figures, constitutive and constructed, free and determined, powerful and subordinated at the same time. "The fact that humans are what they are on the basis of their relation to the world does not imply that they are entirely determined by it" (Verbeek 2005, 112). This statement is not only essential for theoretical concepts after the linguistic turn and postmodern interventions, but also for a phenomenology that has reformed itself and to some extent can be labeled as "postphenomenology" (Ihde 1990, 1993).

Not entirely determined, the experiencing and thinking subject can remain a kind of foundation, being necessary, if we try to thematize the relations relevant to the subject. If we do not have a thematizing subject, these relations cannot be thematized. Even if we are entangled with the world, we are only able to problematize this entanglement if we refer to a problematizing mind. The problematizing subject is no longer a knowing but a questioning subject as we can see in Socrates' dialogues (Plato 1988). Socrates knows that he cannot find a final truth, but does not stop to problematize the world he is living in considering contingency and uncertainty. Here we can meet a confused subject finding new orientations in a process of transformative education (Koller 2012), that can also be based on a phenomenological approach (Waldenfels 2011; Meyer-Drawe 2008). As a subject of education (*Bildung*), this subject has to go through and overcome disappointment, astonishment, and aporetic experiences with the aim of learning something qualitatively new. Looking at today's phenomenology we can stress with Bernhard

Waldenfels (2009) that transformative education does not claim to reach true knowledge, but is merged with alienness constructed in the relation between subject and object. Nevertheless, the subject remains essential for this relation without being a subject of independence, certainty, self-awareness, and clear knowledge. Beyond relativism, we need a subject that rejects the domination of an *anything goes* which is in danger of receiving a normative status. In the name of uncertainty, there could be an attempt of justifying everything, even politically and morally unjustified practices. If we neglect the power of logical, senseful, and responsible thinking and acting, of engagement and resistance, we support inhumanity.

PANDEMIC SCENARIOS AND THE POSTPHENOMENOLOGICAL SUBJECT: COMMENTS ON EXPERIENCE

In its empirical direction postphenomenology is "an *applied* phenomenology" (Richardson 2020, 164) offering a basis for a variety of investigations in digital teaching and learning. During COVID-19 pandemic practices of distance learning showed that digital education cannot be understood without experiencing human beings. Qualitative approaches, for example, narrative interviews or group discussions with teachers and students, can deal with the experiencing subject as a co-constructor and not merely an object of empirical studies. In this way, participants do not only show but also reflect their entanglement with digitality. What was fascinating in digital learning at first glance—and finally harmful, or vice versa? Furthermore, not only scientists but also involved teachers and students can reflect limits of problematizing their relation to a digital mediated and transformed world, because even their reflecting can be co-constructed by this mediatization.

The question of how human beings are involved, maybe dominated by digitality, becomes a problem of morality and—as a matter of power relations—a political issue. Here we can highlight the political relevance of postphenomenology (Verbeek 2020, 141; Bohlmann 2022a, 167; Bohlmann 2022b, 29). If schooling is seen "as precisely one salient mode of productive power, which is why there is such an important value in thinking concretely about how best to intervene in how it happens, and how it might be transformed" (Butler 2012, 177), a digital transformation of schooling and the research on it become political. For example, educational data and learning analytics—helpful to shape future education methods and modalities—should lay special importance on sheltering students' privacy and personal dignity. Here postphenomenology meets digital ethics. The relation of privacy and public can also be relevant in learning by video conference, where participants show their personal environment. Private becomes public and vice-versa (Wellner 2021; Bohlmann 2022a, 199). Further power relations can be seen in the way how participants are treated in digital waiting rooms before joining the conference (Bohlmann 2022, 199). They may not get access directly but could be left outside for a long time and feel excluded while others are still communicating inside.

Looking at a postphenomenological approach that addresses technology as a constitutor of the world (Bohlmann 2022b, 28), we can find different ways of constitution. Although there was technology in schools before becoming digital (Bohlmann 2022b, 36), many students and even teachers entered an unknown world, when they faced digital distance learning during the COVID-19 pandemic (Huber and Helm 2020). Several schools were not well-prepared in dealing with technical equipment and digital didactics (Eickelmann et al. 2020), some teachers started to distribute printed worksheets at the beginning of lockdown learning, before this procedure was replaced by digital methods. This transformation was not possible without subjects—involved and influenced, but also engaged. Thus, we should not look at subjects as mere relata, but as experiencing human beings, who reflect relevant relations.

Regarding the heterogeneity of subjects starting distance learning, teachers were not only confronted with learning singularities, but even social aspects, their students' preconditions, their economic, cultural, and social capital (Bourdieu 1983) and further biographic perspectives. Subjects matter—*not although, but because* human beings are interrelated with technology. Looking at aspects of children's social capital we can stress a phenomenological "inter-subjectivity" (Meyer-Drawe 2008) considering that subjects are always socially (pre-)determined by others and co-constructed by them.

> Thus, not only is the other something outside the self, it is also something that can be found within that same self. Only through this pre-given connection does the self emerge, as well as the possibility to interact with one another. While this basic constitution of the body lies the foundation for all communication, it also leads to the fact that the self is never truly and solely with itself. (Breil 2019b, 228)

Digital "inter-subjectivity" is a matter of power and politics—with Ihde we can highlight, that "postphenomenologically we are simultaneously self-and other-aware; hence, already always both individually and socially aware—thus, indirectly political" (Ihde and Kaminski 2020, 283). We live in a social sphere that determines us while—to a certain extend—we have the possibility to shape it, for example, plan regular video conferences to prevent that social distancing becomes social isolation, or reduce technical problems, if interaction is difficult. During the pandemic, students missed their teachers and peers (Joulaei and Zolfaghari 2021) and had to learn without sufficient and adequate support (Helm et al. 2021; Redecker 2022a). Many showed learning deficits (Hurrelmann and Dohmen 2020); and some disadvantaged children did not have the technical devices they needed (Ariyo et al. 2021). Here digitality can improve educational injustice that should be seen as a matter of digital ethics, focusing on the learning subject and how this subject is acknowledged. If "an understanding of moral reasoning [. . .] begins in acknowledging the affective core in the interrelations of things, technologies, and humans, as the basis of a postphenomenological ethics" (Langsdorf 2020, 136), human subjects and their feelings come to the center of interest.

Young learners, not adequately supported by their teachers in the way they deal with their digitally mediated relation to the world, can be left behind. This causes obstacles for themselves, their further biography, and the social world they are entangled with. In this context, we can problematize teaching and learning methods. For example, facing too much unproved research internet content children can be overstressed having to register, choose, and evaluate vast amounts of information (Kirkpatrick 2011, 20). Learning subjects—especially isolated in times of social distancing—can also be restricted by digital communication. Not only "the phenomenon called 'echo chamber,' 'information cocoon,' or 'group polarization'" (Arzroomchilar 2022, 77) shows that "technology is not innocent, it mediates the way we live. The internet, in particular, is no exception of course" (ibid.). With its information overload the internet can lead learners to confusion and disorientation, on the other hand, it seduces people to hold radical positions that avoid sustainable confrontations with the new and strange in transformative education. "Internet in general, and social networks in particular, can function as a barrier in getting exposed to rival views. It deprives people of getting to know other possible worldviews" (Arzroomchilar 2022, 77), which "can ultimately pose a threat to democracy and peace" (ibid.), especially if we take into account, that "an ideological mind could grow more ideological with the help of technology" (Arzroomchilar 2022, 78). With Arzroomchilar, we can illustrate the postphenomenological ambivalence of echo chambers merging autonomy and determination, singularity, and sociality in a postdigital scenario of mediatized communication.

> An "echo chamber" is brought about neither solely by technology nor by humans, rather it is grounded in a heterogeneity of human actors and non-human actors, namely a network. One cannot overlook the effect of featuring the like-minded peers when exploring the effects of the internet. Similarly, one cannot drop the influence of technology either. (Arzroomchilar 2022, 78)

In the internet, learners can find technical constructed and organized, even manipulated information. "Algorithms owned by large corporations often determine the information that people are exposed to" (Ungerer 2021, 560), but even in this case the relation between digitality and human being does not have to mean total domination; "it is evident that outcomes may be unpredictable when vast numbers of users receive this information" (Ungerer 2021, 560). Users and learners matter as subjects of critical experience, discussing and problematizing those scenarios. Taking into account that experience is bodily experience (Breil 2019a, 395), students can become aware of their learning body (Redecker 2023). For example, learning processes can be hardened or hindered because of aching necks and eyes focused on the screen for long times without the possibility to accentuate clearly, in how far this influences the interaction between students and teachers. Postphenomenology meets a phenomenology of bodily experiences that stays sensitive to ambiguity and uncertainty (Breil 2019a, 401). Our body is essential for the way we are in the digitally mediated world and how we respond to this world, while the body is part of it.

DIGITAL (SELF-)SURVEILLANCE: A POSTPHENOMENOLOGICAL GLANCE ON POWER-RELATIONS

If technologically mediated relations between human beings and their world can be seen politically problematic and if postphenomenology highlights this problematic (Bohlmann 2022a, 183), challenges of digital (self-)surveillance can be problematized in postphenomenology—against a typical postphenomenological optimistic outlook on human being's getting around with digitality (Bohlmann 2022a, 172). A first step to enter the world of digital learning can be drill-and-practice-programs (Hoffmann 2020; Jornitz and Leser 2018), where modes of "self-assessment to encourage independence in the learner" (Moisey and Hughes 2011, 421) are refined by algorithms that predetermine and control learning processes. Conquering with their peers, students are entangled with their teachers and other learners in digital inter-subjectivity, always expecting control without knowing who controls whom, when, and how. Using drill-and-practice-programs, visiting learning platforms, reading, and writing posts or checking and providing mails, Foucauldian practices of disciplining, normation, and normalization (Foucault 1995) become digital. Accordingly, Bergen and Verbeek (2021) refer to Foucault, when they highlight practices of self-technology in digital self-surveillance and self-improvement, often driven by gamification, as we can find it in drill-and-practice programs (Jornitz and Leser 2018).

Nikolas Rose, one of the most famous sociologists who deal with the analysis and critique of neoliberalism, rejects "a kind of cybernetics of control" by "mechanisms to fabricate a kind of moral virtuous, self-activating citizen" (Rose 2000, 171). A comparable role of a competitive logic can be problematized in digital learning scenarios (cf. Redecker 2021), where learning outcomes are registered.

> Thought becomes governmental to the extent that it becomes technical. It must connect itself to a technology for its realization: audits, budgets, tests, examinations, assessments, dossiers, types of inscription and calculation, forms of practical know-how and so forth. (Rose 2000, 145f.)

Learners can experience how they deal with digitality as determined and free subjects at the same time. Algorithms give them the opportunity to shape and direct their learning processes, while at the same time, learners are technically ruled and observed. Furthermore, they can realize that they are not able to clearly characterize this relation of surveilling and being surveilled, empowerment, and subordination, because even their self-surveillance can be (pre-) determined by learning programs in the struggle for learning outcomes (Redecker 2020b)—a race that is extremely hard to win for disadvantaged learners.

Addressed as subjects with dignity and not merely seen as postphenomenological relata, learners should have the opportunity to learn by finding and proving reasons in an education (*Bildung*) in contrast to merely internalizing preconstructed content. In creative forms of digital learning like e-portfolios and learning diaries the student at first glance has "responsibility for his or her own content. No longer a passive consumer, the learner is in an ownership role" (Martindale and Dowdy 2016, 129), but even here we have a digitally mediated relation to the world in an entanglement of autonomy and determination, where involved learning subjects suffer from digital observation (Meyer et al. 2011; Allert and Asmussen 2017; Redecker 2022b) in a struggle for profitable learning outcomes.

Where digital practices become more relevant than human beings suffering from impositions of (creative) (self-)surveillance, subjects are not seen in their dignity (Kant 1786). Instead, they have just a prize in the marketplace of multifunctional flexibility, ruled by "an ethics of self-care and self-responsibility, and a battery of market-led rationalities and procedures" (Wilkins 2012, 124). Where digital disregard takes place, human recognition and resistant engagement are indispensable.

CULTIVATING CRITICAL REFLECTION: POSTPHENOMENOLOGICAL RESTRICTIONS AND THE POWER OF RESISTANCE

Postphenomenological approaches that no longer put special emphasis on the autonomous subject (Ihde 2012, 614) should not neglect human dignity, recognition of personality (development), and the responsible and reflecting subject finding and giving reasons. This subject must be seen as—like Kant (1786) points out—and end in itself beyond disregarding instrumentalization. Practices of subjectivation can only be relevant for us when we thematize them, and we have the opportunity to thematize them reflected and critical, recognizing the experiencing subject that should not be seen as "something in the place of which something else, as an equivalent, can also be placed," but an "end in itself" with "inner worth, that is, dignity" (Kant 1786, 77).

Dealing responsibly with postdigitality needs standing its ambivalences in reference to the vulnerable and reasonable, sensitive, and senseful human being in relation to digitality. Not neglecting human determination by technology can be the first step to a resilient resistance taking into account that the criticized cannot be extinguished, but possibly transformed. Thus, the (post-)phenomenological subject is not a knowing, but questioning subject respecting contingency, ambivalences, and ambiguities. Standing ambivalences and cultivating criticism (Mayrberger 2020) in "a space in which risk can be welcomed" (Macgilchrist 2017, 99) is an educational issue. Preferred pedagogical approaches make use of digitality, not neglecting its power, but trying to avoid and reduce instrumentalization by digitality. It is not the issue of pedagogy to copy the procedures of algorithms but to offer chances of learning innovation in dialogic didactics, provoking further questions and proving possible answers.

> Critique moves the conversation forward by raising questions and troubling those previously held assumptions and convictions, including our assumptions about what work the word 'critical' should be doing. It also moves the conversation forward by imagining otherwise. (Macgilchrist 2021, 247)

Interaction between learners and their teachers is essential for pedagogical processes, because we learn from others (Meyer-Drawe 1996, 2013) in confrontation with their ways of dealing with problems and facing their empathy and encouragement focused on supporting learners to find and prove own judgements. Interaction with other learners, for example, in video conferences, is also essential to learn from each other, discussing problems and reasons. If we highlight a world mediated by technology, especially digitality, we should not forget the human factor in learning interaction.

> However, the confrontation with the other in this context does not necessarily have to take place directly, i.e., face to face. Instead, it can also be mediated. Regardless of that, the experience of the other plays a crucial role; one which should not be neglected. (Breil 2019b, 226)

In and beyond digital scenarios critical discussions can be encouraged by teachers realizing their own vulnerability and restrictions, but also their reasonability and responsibility. Facing digital challenges education theory has to focus on the experiencing human being as a critical learner. Learners can show and exercise resistance if they learn to experience and explore digitality (Knaus 2020) and if they experiment and play with it (Allert et al. 2018), cultivating digital creativity that enables them to show, but also hide themselves in different ways and scenarios.

To a certain extent, they can decide and manage where and how they want to take part or not, facing the ambivalence of autonomy and determination. Not using the camera in video conferences can provoke pressure on others by dominating them (Bohlmann 2022, 199). It can also be a way of resistance for those who feel dominated. Similarly ambiguous private communication in video conferences offers another example. While students cannot show resistance usually practiced in scenarios of in-person-learning (whisper to their peers or exchange private news on sheets of paper), they can use digital forms of expression (private chat), undetected by most of the conference participants, which can be seen as an act of digital resistance (Bohlmann 2022, 199).

REFLECTING AND RECOGNIZING A POSTPHENOMENOLOGICAL APPROACH TO PEDAGOGY: RETROSPECT AND OUTLOOK

The discussion started with an attempt to shape the relation between postphenomenology and phenomenology, highlighting the experiencing subject as a link between both. This was exemplified with digital learning during the COVID-19 pandemic, where especially disadvantaged children suffered from the entanglement of digital challenges, social preconditions, and lacking pedagogical support. Here we should not only focus on postphenomenological relata in technically mediatized constellations, but on human beings with dignity and responsibility, being harmed by impositions such as excessive (self-)surveillance. Facing these challenges in the context of the political and ethical relevance of postphenomenology, pedagogical responsibility can concentrate on supporting and empowering the critical learner, finding and discussing reasons without neglecting digital determination, contingency, and human vulnerability.

Considering recent phenomenological concepts in the tradition of Merleau-Ponty, postphenomenology can discuss them to find an ethical and pedagogical approach to digital education. This would be a kind of transformative education thematizing phenomenology. Elaborating on this approach is a future issue of postphenomenological reflections in a critical and constructive glance at phenomenology.

REFERENCES

Allert, Heidrun, Michael Asmussen, and Christoph Richter. 2018. "Formen von Subjektivierung und Unbestimmtheit im Umgang mit datengetriebenen Lerntechnologien—eine praxistheoretische Position." *Zeitschrift für Erziehungswissenschaft* 21: 142–158.

Ariyo, Esther, Michal Amurtiya, M., Lydia O. Yemisi, Ariyo Oludare, Orgunsanmi Ololade, Ajike Patience Taiwo, Lasode A., Abolanle Olukemi, and Daniel Oguniyi. 2022. "Socio-demographic Determinants of Children Home Learning Experiences During COVID 19 School Closure." *International Journal of Educational Research Open* 3: 100111. https://doi.org/10.1016/j.ijedro.2021.100111.

Arzroomchilar, Ehsan. 2022. "Some Suggestions to Improve Postphenomenology." *Human Studies* 45: 65–92. https://doi.org/10.1007/s10746-021-09615-1.

Bergen, Peter and Peter Paul Verbeek. 2021. "To-Do Is to Be: Foucault, Levinas, and Technologically Mediated Subjectivation." *Philosophy & Technology* 34: 325–348.

Bettinger, Patrick and Kai Uwe Hugger. 2020. *Praxistheoretische Perspektiven in der Medienpädagogik*. Wiesbaden: Springer.

Bohlmann, Markus. 2022a. *Bildung—Philosophie—Digitalisierung. Eine Curriculumtheorie*. Berlin: Springer.

———. 2022b. "Distale und proximale Technologien der Krise im Distanzunterricht. Critical Theory of Technology und Postphänomenologie." In *Technologien der Krise. Die Covid-19-Pandemie als Katalysator neuer Formen der Vernetzung,* edited by Dennis Krämer, Joschka Haltaufderheide, and Jochen Vollmann, 21–44. Bielefeld: transcript.

Bourdieu, Pierre. 1983. "Ökonomisches Kapital, kulturelles Kapital, soziales Kapital." In *Soziale Ungleichheiten*, edited by Reinhard Kreckel, 183–198. Göttingen: Schwartz.

Breil, Patrizia. 2019a. "Haltung. Ein Entwurf mit Merleau-Ponty." In *Leib—Leiblichkeit—Embodiment. Pädagogische Perspektiven auf eine Phänomenologie des Leibes*, edited by Malte Brinkmann, Johannes Türstig, and Martin Weber-Spanknebel, 389–403. Wiesbaden: Springer.

———. 2019b. "Altering Identities. Possibilities of Understanding Identity in Phenomenological Pedagogy." *Argument* 9: 225–235.

Brinkmann, Malte. 2019. "Embodied Understanding in Pedagogical Contexts." In *Leib—Leiblichkeit—Embodiment. Pädagogische Perspektiven auf eine*

Phänomenologie des Leibes, edited by Malte Brinkmann, Johannes Türstig, and Martin Weber-Spanknebel, 21–36. Wiesbaden: Springer.

Butler, Judith. 2012. “Gender and Education.” In *Judith Butler: Pädagogische Lektüren*, edited by Norbert Ricken, and Nicole Balzer, 15–28. Wiesbaden: Springer.

de Boer, Bas. 2021. “Explaining Multistability: Postphenomenology and Affordances of Technologies.” *AI & Society* 38: 2267–2277. https://doi.org/10.1007/s00146-021-01272-3.

Eickelmann, Birgit, Kerstin Drossel, and Melanie Heldt. 2020. “Vorteile digital fortgeschrittener Schulen in der Pandemie-Zeit.” *Schulmanagement* 51 (3): 28–31.

Foucault, Michel. 1995. *Discipline and Punish: The Birth of the Prison.* New York: Vintrage.

Hall, Stuart. 2019. “The West and the Rest: Discourse and Power.” *Essential Essays.* Volume 2., edited by David Morley, 141–184. Morley: Duke University Press.

Helm, Christoph, Stephan G. Huber, and Tina Loisinger. 2021. “Was wissen wir über schulische Lehr-Lern-Prozesse im Distanzunterricht während der Corona-Pandemie?—Evidenz aus Deutschland, Österreich und der Schweiz.” *Zeitschrift für Erziehungswissenschaft* 24: 237–311.

Hoffmann, Ilka. 2020. “Die Corona-Pandemie als Katalysator für Schulreformen?” In *„Langsam vermisse ich die Schule ...“. Schule während und nach der Corona-Pandemie, Die Deutsche Schule, Beiheft 16*, edited by Detlef Fickermann and Benjamin Edelstein, 95–101. Münster: Waxmann.

Huber, Stephan G., and Christoph Helm. 2020. “COVID-19 and Schooling: Evaluation, Assessment and Accountability in Times of Crise—Reacting Quickly to Explore Key Issues for Policy, Practice and Research With the School Barometer.” *Evaluation and Accountability* 32: 237–270.

Hurrelmann, Klaus, and Dieter Dohmen 2020. *Das Deutsche Schulbarometer: Corona-Krise verstärkt Bildungsungleichheit.* https://deutsches-schulportal.de/stimmen/das-deutsche-schulbarometer-hurrelmann-dohmen-corona-krise-verstaerkt-bildungsungleichheit/.

Husserl, Edmund. 1987. *Cartesianische Meditationen. Eine Einleitung in die Phänomenologie,* edited by Elisabeth Ströker. Second edition. Hamburg: Meiner.

Ihde, Don. 1990. *Technology and the Lifeworld. From Garden to Earth.* Bloomington: Indiana University Press.

———. 1993. *Postphenomenology.* Evanston: Northwestern University Press.

———. 2009. *Postphenomenology and Technoscience: The Peking University Lectures.* Albany: Suny Press.

———. 2012. “Peter-Paul Verbeek and a Postphenomenological Mirror. Book Symposium on Peter-Paul Verbeek’s *Moralizing of Things*.” *Philosophy of Technology* 25: 610–615.

———. 2022. “Postphenomenology, the Empirical Turn and ‘Transcendentality’.” *Foundations of Science* 27: 851–854. https://doi.org/10.1007/s10699-020-09741-6.

Ihde, Don, and Andreas Kaminski. 2020. “What is Postphenomenological Philosophy of Technology?” In *Autonomie und Unheimlichkeit. Jahrbuch Technikphilosophie*

6, edited by Alexander Friedrich, Petra Gehring, Christoph Hubig, Andreas Kaminski, and Alfred Nordmann, 261–287. Baden-Baden: Nomos.

Jörissen, Benjamin. 2017. *Subjektivation und ‚ästhetische Freiheit' in der post-digitalen Kultur. kubi-online*. https://doi.org/10.25529/92552.421.

Jornitz, Sieglinde, and Christoph Leser. 2018. "Mit Antolin punkten oder: Wie sich mit dem Leseförderprogramm der Bock zum Gärtner macht." *Pädagogische Korrespondenz* 57: 55–73.

Joulaei, Hassan, and Zahra Zolfaghari. 2021. "Covid 19, School Closures, and Its Complications." *International Journal of School Health* 8 (1): 1–2.

Kant, Immanuel. 1786. *Groundlaying toward the Metaphysics of Morals*. Second edition. Riga: Hartknoch.

Kirkpatrick, Denise. 2011. "Flexibility in the Twenty-First Century. The Challenge of Web 2.0." In *Flexible Pedagogy, Flexible Practice. Notes from the Trenches of Distance Education*, edited by Elizabeth Burge, Chère Campbel Gibson, and Terry Gibson, 19–28. Edmonton: AU Press, Athabasca University.

Knaus, Thomas. 2020. "Technology Criticism and Data Literacy: The Case for an Augmented Understanding of Media Literacy." *Journal of Media Literacy Education (JMLE)* 12 (3): 6–16.

Koller, Hans-Christoph. 2012. *Bildung anders denken. Einführung in die Theorie transformatorischer Bildungsprozesse*. Stuttgart: Kohlhammer.

Langsdorf, Lenore. 2020. "Relational Ethics: The Primacy of Experience." In *Reimagining Philosophy and Technology, Reinventing Ihde*, edited by Glen Miller and Ashley Shew, 123–140. Cham: Springer.

Macgilchrist Felicitas. 2017. "Backstaging the Teacher: On Learner-Driven, School-Driven and Data-Driven Change in Educational Technology Discourse." *Kultura—Społeczeństwo—Edukacja* 2 (12): 83–103.

———. 2021. "What is 'Critical' in Critical Studies of Edtech? Three Responses." *Learning, Media and Technology* 46 (3): 243–249.

Martindale, Trey, and Michael Dowdy. 2016. "Issues in Research, Design, and Development of Personal Learning Environments." In *Emergence and Innovation in Digital Learning*, edited by George Veletsianos, 119–141. Edmonton: AU Press.

Mayrberger, Kerstin. 2020. "Praxistheoretisch informierte partizipative Mediendidaktik." In *Praxistheoretische Perspektiven in der Medienpädagogik*, edited by Patrick Bettinger, and Kai-Uwe Hugger, 61–85. Wiesbaden: Springer.

Mecheril, Paul. 2018. "Orders of Belonging and Education: Migration Pedagogy as Criticism." In *Migration. Changing Concepts, Critical Approaches*, edited by Doris Bachmann, and Jens Kugele, 121–138. Berlin: de Gruyter.

Merleau-Ponty, Maurice. 1984. *Das Auge und der Geist. Philosophische Essays,* edited and translated by Hans Werner Arndt. Hamburg: Meiner.

———. 1986. *Das Sichtbare und das Unsichtbare—gefolgt von Arbeitsnotizen*, edited by Claude Lefort, translated by Regula Giuliani and Bernhard Waldenfels. München: Fink.

Meyer, Torsten, Kerstin Mayrberger, Stefan Münte-Goussar, and Christina Schwalbe (Eds.). 2011. *Kontrolle und Selbstkontrolle. Zur Ambivalenz von E-Portfolios in Bildungsprozessen.* Wiesbaden: Springer.

Meyer-Drawe, Käte. 1996. "Vom anderen lernen. Phänomenologische Betrachtungen in der Pädagogik." In *Deutsche Gegenwartspädagogik*, Bd. II, edited by Michele Borrelli and Jörg Ruhloff, 85–98. Baltmannsweiler: Schneider.

———. 2008. *Diskurse des Lernens.* München: Fink.

———. 2013. "Lernen braucht Lehren." In *Pädagogische Reform. Anspruch—Geschichte—Aktualität*, edited by Peter Feuser, Wolfgang Beutel and Jürgen John, 89–97. Seelze: Friedrich.

Meyer-Drawe, Käte, and Johanna Schwarz. 2015. " (Uber-)Sehen, (Uber-)Horen, Zuschreiben. Lernende im Licht und im Schatten der Aufmerksamkeit." In *Pädagogische Erfahrung. Theoretische und empirische Perspektiven*, edited by Malte Brinkmann, Richard Kubac, and Severin Sales Rödel, 125–145. Wiesbaden: Springer.

Moisey, Susan D., and Judith A. Hughes. 2011. "Supporting the Online Learner." In *The Theory and Practice of Online Learning*. Second edition, edited by Terry Anderson, 419–439. Edmonton: AU Press.

Platon. 1988. "Menon." In *Platons Sämtliche Werke.* Band 2, 8–42. Hamburg: Rowohlt.

Redecker, Anke. 2020a. "Vom quantified zum qualified Self. Machbarkeitsmythen und Bildungschancen des Digitalen." In *Vom E-Learning zur Digitalisierung—Mythen, Realitäten, Perspektiven*, edited by Reinhard Bauer, Jörg Hafer, Sandra Hofhues, Mandy Schiefner-Rohs, Anne Thillosen, Benno Volk, and Klaus Wannemacher, 55–68. Münster: Waxmann.

———. 2020b. "Communication and Control. Scenarios of Digital Learning." In *Communication and Learning in an Age of Digital Transformation*, edited by David Kergel, Birte Heidkamp-Kergel, Ronald C. Arnett, and Susan Mancino, 141–151. London: Routledge.

———. 2021. "Von faulen Eiern und fleißigen Kreativtitäsprediger*innen—oder: Wie das Wiesel weiterlernt. Impulse für die digitale Bildung." *merz* 5: 114–126.

———. 2022a. "Dialogische Didaktik gegen den digitalen Drop-out. Pandemische Interaktionsszenarien und die Zukunft des schulischen Online-Lehrens." *Ludwigsburger Beiträge zur Medienpädagogik* 22: 1–13.

———. 2022b. "Kreativität unter Kontrolle—das E-Portfolio zwischen Prozess- und Produktorientierung." *Medienimpulse* 60 (2): 1–35.

———. 2023. "Die Abgründe der Anwesenheit und die Anmaßungen der Abwesenden. Zur Phänomenologie leiblichen Lernens in digitalen Räumen." In *Digitalisierte Lebenswelten—Bildungstheoretische Reflexionen*, edited by Marc Fabian Buck and Miguel Zulaica y Mugica, 27–46. Heidelberg: Metzler.

Richardson, Ingrid. 2020. "Postphenomenology, Ethnography, and the Sensory Intimacy of Mobile Media." In *Reimagining Philosophy and Technology, Reinventing Ihde*, edited by Glen Miller and Ashley Shew, 159–174. Cham: Springer.

Ritter, Martin. 2021. "Philosophical Potencies of Postphenomenology." *Philosophy & Technology* 34: 1501–1516.

Rose, Nikolas. 2000. "Governing Liberty." In *Governing Modern Societies*, edited by Richard V. Ericson and Nico Stehr, 141–176. Toronto; Buffalo, NY: University of Toronto Press.

Rosenberger, Robert. 2020. "On Variational Cross-Examination: A Method for Postphenomenological Multistability." *AI & Society* 38: 2229–2242. https://doi.org/10.1007/s00146-020-01050-7.

Rosenberger Robert, and Peter Paul Verbeek, eds. 2015. *Postphenomenological Investigations. Essays on Human-Technology Relations.* Lanham: Lexington Books.

Verbeek, Peter Paul. 2011. *Moralizing Technology: Understanding and Designing the Morality of Things.* Chicago: The University of Chicago Press.

———. 2015. "Postphenomenology." In *What Things Do. Philosophical Reflections on Technology, Agency and Design* by Peter-Paul Verbeek, translated by Robert Crease, 99–119. Pennsylvania: The Pennsylvania State University Press.

———. 2020. "Politicizing Postphenomenology." In *Reimagining Philosophy and Technology, Reinventing Ihde*, edited by Glen Miller und Ashley Shew, 141–155. Cham: Springer International Publishing.

Waldenfels, Bernhard. 1999. *Vielstimmigkeit der Rede. Studien zur Phänomenologie des Fremden 4.* Frankfurt/M.: Suhrkamp.

———. 2009. "Lehren und Lernen im Wirkungsfeld der Aufmerksamkeit." In *Umlernen. Festschrift für Käte Meyer-Drawe*, edited by Norbert Ricken, Henning Röhr, Jörg Ruhloff, and Klaus Schaller, 23–33. München: Fink.

———. 2011. "In Place of the Other." *Continental Philosophy Review* 44: 151–164.

Wellner, Galit. 2021. "The Zoom-bie Student and the Lecturer: Reflections on Teaching and Learning with Zoom." *Techné: Research in Philosophy and Technology* 25 (1): 1–25.

Wilkins, Andrew. 2012. "Commodifying Diversity: Education and Gorvernance in the Era of Neoliberalism." *Human Affairs* 22: 122–130.

Chapter 17

Asking Educational Questions about Technology

Håkon Jakobsen Aaltvedt

Digital technologies are becoming increasingly present in education across all levels, often driven by educational policies. Themes of technology and digitalization are increasingly permeating educational discourse within both practical, public, and political spheres. As such, it seems appropriate that a growing portion of educational research is focusing on digital technologies within education. However, several scholars (Oliver 2011; Hamilton and Friesen 2013; Aagaard 2017; Pischetola 2021) are critical to the tendency in both educational research and political discourse to approach educational technologies in terms of instrumentalism and/or technological determinism[1]; arguing that both approaches represent simplistic, and mutually exclusive understandings of human–technology relations—but which nonetheless often co-exist in what Aagaard (2017) calls "an uneasy armistice." Also, Selwyn (2016), Biesta (2016), and Adams and Turville (2018) are critical of a tendency in both the political discourse and educational research of viewing education more or less only in terms of learning, and learning efficiency, approaching educational technology solely as *learning instruments,* whose purpose is to make learning more effective.

In this chapter, the tendency to approach educational technologies with an unnuanced understanding of technology on the one hand, and education on the other, are addressed and framed as "two blind spots" in research on educational technologies. It is argued that this "double blindness" leads to a rather limited understanding and approach to educational technologies which is characterized as *asking technological questions about learning.* To remedy this limited approach, a combination of postphenomenology and Gert Biesta's educational philosophy is suggested. A brief exposition of selected concepts in both frameworks is given—pointing out some conceptual similarities, which make them compatible, and some differences that make them complementary—and it is argued that the combination of the two, puts us in a better position to (1) illuminate the two "blind spots" of educational technologies, and (2) instead of asking *technological questions about learning,* puts us in position to ask *educational questions about technology.*

INSTRUMENTALISM, TECHNOLOGICAL DETERMINISM, AND TECHNO-OPTIMISM

Technological determinism is the view that technology determines societal development and human affairs, and comes in both optimistic varieties (technologies make human lives better), and pessimistic varieties (technologies make human lives worse).[2] The optimistic variety seems to be most prominent in the discourse on educational technologies. Although hard technological determinism of the optimistic kind seems rare, we can often come across softer variants; either in the form of techno-solutionism (Pischetola 2021) or what we can more generally call "techno-optimism."[3] *Instrumentalism* is, in a sense, the opposite of technological determinism; it is the view that technology has no determining force on human affairs; technologies are but neutral tools, serving human ends. In the combination of instrumentalism with the soft technological determinism of techno-optimism, technology is often conceptualized as an important motor and agent of educational change, transformation, and innovation (techno-optimism, soft determinism), and the users—though understood as ultimately in control, and responsible for how the technologies are affecting educational practice (instrumentalism)—are at the same time conceptualized as either "barriers" (Pischetola 2021, 55) or promoters of this wider technological transformation of education. This can lead to what Aagaard calls the "paradox of educational technology," in which the technology is praised when "something good happens," and the students and/or teachers are blamed when "something bad happens" (Jesper Aagaard 2017, 1129).

To illustrate this combination of instrumentalism and techno-optimism, I will give some examples from the OECD report "Digital Education Outlook 2021: Pushing the Frontiers with AI, Blockchain, and Robots" (OECD 2021). The report represents a blend of both research and policy agendas, coming from an intergovernmental organization with considerable influence on educational policies across and beyond the 83 countries participating in PISA.[4] In the editorial, the OECD's Director for Education and Skills, Andreas Schleicher, writes that "[t]echnology and AI are not magic powers, they are just extraordinary amplifiers and accelerators that add speed and accuracy. AI will amplify good educational ideas and good practice in the same way it amplifies bad ideas and bad practice" (OECD 2021, 4). Technologies are "extraordinary amplifiers and accelerators," but in terms of educational ends, they are neutral; they will amplify an accelerate whatever task you put them to; good or bad. In discussing possible ethical challenges with educational technology, technology is also rendered as ethically neutral, with Schleicher postulating that "[w]hile technology is ethically neutral, it will always be in the hands of educators who are not neutral" (OECD 2021, 4), placing the ethical responsibility squarely in the hands of the (human) educators. This instrumental understanding of technology as neutral instruments is coupled with an understanding of technology as an agent of change and educational transformation: "Artificial intelligence, robots and blockchain are poised to transform how we teach, learn, and run schools. The technology is ready; are we?" (OECD 2021, 4). The technology is ready, it has the potential to transform education, but for it to happen, educators have to adapt and reinvent themselves to accommodate the technology: "What is clear is that for robots, classware, predictive analytics and the like to work effectively will require reinventing the role of teachers" (OECD 2021, 4). So, although technology is conceptualized as neutral vis-à-vis educational practice and ethics, it is certainly not conceptualized as neutral in terms of educational transformation; rather, it is seen as a prime mover; leading the way of educational innovation and transformation, whereas educators must reinvent themselves to adapt to the technological transformation. Commenting on the educational experiences with digital technologies during the COVID-19 pandemic, this becomes even more clear, as the digitalization of education due to the constraints of the pandemic is understood as catapulting educational innovation *forward,* in a linear progressive motion:

> [T]he experience has catapulted education systems, traditionally laggards when it comes to innovation, years ahead in what would have been a slow slouch towards smart schooling (. . .) The pandemic has picked up our education systems and hurtled them at light-speed from the 19th to the 21st century. (OECD 2021, 3–4)

Later in the report, the question of whether schools and universities should resist or embrace digitalization is (in my interpretation; rhetorically) posed—and answered as such: "While formal education systems should empower everyone to enjoy, access and learn from all the knowledge and experiences that have been developed by humanity, education should be more than a museum" (OECD 2021, 38).

The discussions about *how* technology is "poised to transform" educational practice, revolve almost exclusively around *learning*,[5] emphasizing how technologies promises to enhance the *efficiency* of learning. Technologies can enhance learning efficiency indirectly, by enhancing *teaching efficiency,* as smart technologies promise to "enhance the effectiveness of teaching and learning for better student learning," and classroom analytics promises to improve learning effectiveness by "providing more effective teaching" (OECD 2021, 23). Also, technologies can amplify student learning efficiency more directly, by way of personalised learning; either by providing ". . . all students with the appropriate curriculum or task, and scaffold them within a task, based on a diagnosis of their knowledge and knowledge gaps" (OECD 2021, 23), and even more directly, by keeping students engaged by way of technological monitoring solutions that ". . . identify their affective states during learning and nudge them towards re-engagement when they seem to disengage" (OECD 2021, 23).

TWO BLIND SPOTS IN APPROACHING EDUCATIONAL TECHNOLOGIES

The examples from the OECD outlook illustrate two blind spots that are common in both policy and research on educational technologies; one blind spot toward *technology*, the other toward *education*. The technological blind spot is the result of assuming and approaching technology as *neutral*, blinding oneself to the intricacies of how technologies co-shape human–world relations in non-neutral ways. This is of course the exact area of interest of postphenomenology, illustrated by the concept of technological mediation.[6] Postphenomenology assumes that technologies are *not neutral*, and the concept of technological mediation provides a framework for studying *how* technologies co-shape human perception, interpretive frameworks (Ihde 1990; Verbeek 2005), action, practices (Verbeek 2005, 2016), and normative frameworks (Verbeek 2011) in non-neutral ways. The other—*educational*—blind spot is connected to the tendency to reduce education to matters of *learning*; measuring and evaluating education in terms of the efficiency and effectiveness in producing (pre)defined learning outcomes (Biesta 2016a, 2010, 2022). In being oriented toward educational efficiency and effectiveness (of learning), this kind of approach assumes a kind of normative neutrality toward educational purpose and toward *what* is learned, blinding it to (1) how education functions and has impacts on *other* domains than those explicitly formulated in terms of learning outcomes, and (2) how both learning and education in general always have a normative dimension; how they are *never neutral*. Both the technological and the educational blind spot are conceptually related; they both assume some kind of neutrality, making them blind to how both education and technology shape human–world relations in non-neutral ways.

ASKING TECHNOLOGICAL QUESTIONS ABOUT LEARNING

In combination, this "double blindness" leads to an understanding of educational technologies as *neutral learning instruments*, which then are approached and evaluated in terms of their capacity to amplify the effectiveness and efficiency of producing (pre)defined learning outcomes.[7] Conceptualizing educational technologies as neutral learning instruments conceals how educational technologies impact other educational domains than those of learning, and how they actively co-shape our educational practices, our understanding of education, and even our educational intentions, purposes, and ends. This "double blindness" and alleged neutrality makes it hard to see and question the normative aspects of both education and technology, and risks implicitly promoting certain normative views without knowing or acknowledging it.

Though technology is treated as neutral in some aspects, the OECD report also illustrates a form of techno-solutionism or techno-optimism; treating educational technologies as potential solutions to educational problems—which in OECD's framing most often are problems related to *learning efficiency*—and as having the potential to *transform* education for the better. Combined with the societal push for digitalization, education can either try to resist technological transformation, risking becoming a "museum," or it can embrace technological transformation by adapting to it, or even reinvent itself to accommodate it, and thereby keep up with the technological progress and transformation of the rest of society. In this kind of techno-optimist framing, technology itself becomes the main reason and motor for educational change and transformation, whereas the task of educational practice is to adapt and transform itself to cater to society and technology. In other words, technology is driving the educational agenda; *technological* questions are framing *educational* questions, which, as shown, are more or less reduced to questions of *learning*. This approach, then, amounts to *asking technological questions about learning.*

In the following, I will put postphenomenology—emphasizing Verbeek's framework of technological mediation—in dialogue with Biesta's educational philosophy, and explore how these two approaches can be combined to provide a conceptual framework for (1) illuminating the blind spots of technology and education, and study how educational technologies co-shape education in non-neutral ways, and (2) as a framework for *asking educational questions about technology,* (rather than asking technological questions about learning).

POSTPHENOMENOLOGY: TECHNOLOGICAL MEDIATION AND MULTISTABILITY

Postphenomenology draws on the phenomenological concept of intentionality; the understanding that humans are always directed toward, and in relation to the world, with the most basic unit of analysis not being humans in isolation, but rather, *human–world relations*.[8]

In developing a postphenomenological philosophy of technology, Ihde (1979, 1990)[9] combines this phenomenological insight with the acknowledgment that humans to a large extent perceive and relate to the world with and through technologies, conceptualizing technologies as *mediators* of human–world relations; expanding the unit of analysis to *human–technology–world relations*. Ihde initially focused on how technology mediates human–world relations in terms of perception; how technologies shape the way in which humans perceive the world, and he differentiates between the mediation of microperception (bodily, sensory perception) and macroperception (cultural, interpreted perception) (Ihde 1990, 29; Verbeek 2005, 122–123). Drawing on Latour and actor-network theory, Verbeek (2005) develops Ihde's postphenomenological concept of technological mediation further, adding to Ihde's hermeneutical concept of mediation of *perception*, the existential concept of mediation of *action,* and fleshes out a "postphenomenological vocabulary" centered around technological mediation: The hermeneutical perspective of mediation of perception (developed by Ihde) is oriented toward how technologies shape how the world appears to humans, by *transforming* perception (microperception) and interpretive frameworks (macroperceptions), structured in the way of *amplification/reduction* where "some interpretive possibilities are strengthened while others are weakened" (Verbeek 2005, 195). The existential perspective of mediation of action (developed by Verbeek) is oriented toward how "humans appear in the world"; and how technologies mediate human existence "by giving concrete shape to their behavior and the social context of their existence" (Verbeek 2005, 195). Existential mediation of action *translates* human action and behavior structured in the way of *inviting/inhibiting*; "some forms of involvement are fostered while others are discouraged" (Verbeek 2005, 195).

With the concept of technological mediation, postphenomenology avoids instrumentalism by pointing out how technologies are not neutral means to human ends, but instead mediators that shape how we perceive and act in the world; and thus co-shape human ends. Postphenomenology also avoids technological determinism, as the concept of mediation is coupled with the concept of *multistability* (Ihde 1990); pointing out that the function, effect, or *stability* of technologies are not determined by the technology alone, rather, it is a product of both the particular technology, the user and its use-context

(Rosenberger 2014).[10] The possible stabilities are not infinite though, and the technological stability can't be totally determined by the human user either. The range of possible stabilities thus constitutes a *degree of freedom* for the human user; an existential space of possibility in human–technology–world relations.

With the concepts of technological mediation and multistability, we see that neither are humans determining (the stability of) technologies, nor are technologies determining humans; instead, both are co-shaping each other. In addition to absolving the dichotomy between instrumentalism and determinism, postphenomenology is also pointing to something more fundamental in human-technology relations. Technological mediation is not just affecting the human user and the technology itself, more fundamentally; technological mediation is shaping *human–world relations.* Through technological mediation, the human subject and the world object (not just the technology) "are brought into existence in a specific way, and both subjectivity and objectivity acquire specific shape" (Verbeek 2005, 130). Put another way, both subject and object are *co-constituted* through technological mediation; mediation is not an intermediary between a fixed subject and object, but "an origin" of both (Verbeek 2012). As such, postphenomenology represents a kind of process philosophy and a relational ontology, oriented toward *becoming* or *emergence* through *relations,* rather than the being or essence of entities (postphenomenology explicitly rejects cartesian subject–object dualism, and the essentialism and transcendentalism associated with Husserl's phenomenology).

MEDIATION OF MORALITY AND EXISTENTIAL NORMATIVITY

Whereas the postphenomenological concept of technological mediation can be approached as a descriptive framework, Verbeek (2011) also pushes the concept of mediation into the normative domain ("mediation of morality") making the case that technologies are not neutral in a normative or moral sense. In shaping human perception, understanding, action, and practice, technologies are morally "charged" and morally significant, both in the way of shaping concrete human action in ways that alter their moral significance, but also in the way of shaping the very normative frameworks we use to interpret the moral significance of those actions. As human lives are deeply infused with technologies, to the point where it is hard, if not impossible, to imagine human life without technologies; technologies have always already shaped our world and ourselves to the point where it is not possible to evaluate technology from *outside;* from a technology-free, or "purified" human standpoint. And so "ethics is a thoroughly hybrid affair" (Verbeek 2011, 88), and we must evaluate technology from *within* (the hybrid reality of human–technology relations). Verbeek proposes the concept of technological mediation as a framework for describing how technologies play a part in shaping human lives, so as to get a better view "from inside"—and be better positioned to evaluate the moral significance of technologies. As such, it is still a descriptive framework (describing how technologies mediate human lives in ways that *can* be interpreted as morally significant), and postphenomenology does not provide any explicit ethical standards against which technologies can be evaluated. But, arguably, Verbeek's project nonetheless has a certain normative direction, what de Boer, Hoek, and Kudina (2018) refer to as *existential normativity*. This existential normativity has to do with human freedom and responsibility in relation to technology. In discussing human freedom vis-à-vis technologies, Verbeek draws an analogy to Foucault's conception of freedom vis-à-vis power:

> The freedom of the subject does not consist in being liberated from power but in interacting with it. One becomes a subject not by securing a place outside the reach of power but by shaping one's subjectivity in a critical relation to it. (Verbeek 2011, 73)

If we substitute "power" with "technology," we see that the freedom comes from "within" our relation with technology; in interacting with, and critically relating to it, rather than being liberated *from* technology, securing freedom from "outside" of technology. This (positive) freedom from within is what Verbeek's ethical framework is *directed* towards; it is a conceptual framework for better understanding how technologies mediate our existence, and thereby getting into a position to have a *critical relation* to technologies. In terms of freedom, this is crucial, because it is "[o]nly when making the mediations explicit, we can develop a free relation to it" (Verbeek 2014, 84). With freedom comes (moral) responsibility, and Verbeek suggests that

> . . . [w]e can get actively involved in how these technologies have an impact on us. By critically examining how technologies help to shape situations of choice and frameworks of interpretation, it becomes possible to take responsibility for one's technologically mediated agency. (Verbeek 2014, 84)

Verbeek's framework of technological mediation can be interpreted as directed toward illuminating—and thereby coming in better position to also *realize*—human freedom and responsibility in human–technology relations. As such, Verbeek's postphenomenological ethics has a normative direction; an existential normativity that implicitly puts value on human freedom and responsibility. This kind of existential normativity complements Gert

Biesta's educational philosophy, which also has a notable existential normative directionality in which the realization of human freedom, by way of subjectification, is understood as an essential educational purpose.

THE EDUCATIONAL PHILOSOPHY OF GERT BIESTA

Biesta has over a large body of work covered many different educational and philosophical topics, drawing on a wide variety of sources. Though a bit eclectic in his sources and general approach, there is arguably a consistency in the overall direction of his work, maybe best captured in his latest book; *World-Centered Education: A View for the Present* (Biesta 2022). Here, Biesta positions his educational philosophy and agenda within the paradigm of "existential education," in contrast to what he calls the paradigm of "education as cultivation," which he argues, represents the dominant understanding of education today. In framing these two paradigms, both his critical and constructive agendas are neatly captured, as much of his work oscillates between critiquing the dominant paradigm of "education as cultivation" and constructing an alternative framework for understanding education as "existential education."

Throughout his work, Biesta associates with, builds on, and develops his educational philosophy in dialogue with the continental tradition of educational studies (associated with the traditions of Pädagogik, Didaktik, Bildung, and Erziehung), and with continental philosophy in general; with reference to both philosophers associated with the phenomenological tradition (e.g., Heidegger, Levinas, Arendt, Marion) and poststructuralist traditions (e.g., Foucault, Derrida, Rancière).[11] Also worth mentioning is his engagement with philosophical pragmatism, both in terms of epistemology and methodology (Biesta 2020a) and his recurring engagement with Dewey's educational philosophy (Biesta 2016a, 2010, 2013, 2017, 2022). This amalgamation of the continental traditions of phenomenology, post-structuralism, and American pragmatism bears some general resemblances with the amalgamation of postphenomenology—which also draws on phenomenology, post-structuralism, and pragmatism—and can explain some of the conceptual similarities between the two frameworks. I will expand on these similarities in the following section.

Although drawing on the continental tradition of educational studies, Biesta rejects the "philosophical humanism" associated with (parts of) this tradition, and formulates the agenda of his educational philosophical project as exploring ". . . what will follow if we overcome the humanist foundations of modern education" (Biesta 2016a, 4).[12] Biesta frames "philosophical humanism" as

> . . . the idea that it is possible to know and articulate the nature or essence of the human being, often accompanied by the assumption that such knowledge can provide the basis for subsequent action in such domains as politics and education. (Biesta 2014, 16)

Biesta rejects humanistic essentialism (as does postphenomenology), arguing that the problem with essentialist, humanistic education is that it (pre) defines a human essence, nature, ideal, or a certain *subjectivity,* in which education becomes a project of *cultivating* this (pre)defined human subjectivity. In rejecting the possibility of knowing and articulating the essence of the human being, trying to define it, then, becomes an arbirtary and possibly a dangerous endevor, illustrated by questions like: who should define it? how to define it? who/what counts as properly human and who/what does not? Biesta argues that for education to "overcome" philosophical humanism, it needs to

> . . . treat the question of what it means to be human as a radically open question, a question that can only be answered by engaging in education rather than as a question that needs to be answered before we can engage in education. (Biesta 2016a, 5)

But, essentialistic or not, education will always be a deeply normative project; always defined in relation to values, questions, and ideas about what is good for the human individual, society, and the world. As such—in studying and discussing education—one needs to acknowledge and have a vocabulary for addressing the normative dimension of education; to be able to critically assess, address, and discuss educational purpose. This brings us to one of Biesta's starting points and major criticisms of mainstream educational discourse, which is directed toward what he calls the "learnification" of education, in which the "language of education" is substituted by a "language of learning" (Biesta 2005, 2010). He argues that "learning" is a "rather empty process-term" (Biesta 2022, 43)—empty in terms of educational content and purpose—and that it is "basically an individualistic concept" (Biesta 2010, 18), substituting a relational understanding of education as built upon educational relationships, to a project of facilitated (mass) individual learning. As such, the language of learning tends to invite questions about learning effectiveness and efficiency at the expense of questions ". . . about educational content, purpose, and relationships (. . .) or assumes that the answer to these questions is already clear and decided upon" (Biesta 2022, 43).

BIESTA'S THREE DOMAINS OF EDUCATIONAL PURPOSE

To counter the tendency within educational research and discourse to reduce education to matters of learning—and to provide a richer language for addressing the complexity and normativity of education—Biesta suggest a tripart conceptualization of educational function and purpose; that of qualification, socialization, and subjectification (Biesta 2010). The domain of qualification encompasses the educational function and purposes of qualifying students for life and participation in work-life and society in general. In this domain, *learning* of course plays an important role, as qualification most often entails learning particular knowledge, skills, and competencies. But importantly, the teaching and learning of particular knowledge and skills are not value-neutral enterprises—they are always permeated with a certain view of what kind of knowledge and skills are *valuable*, both for the individual and for society. Qualification, therefore, always implies some kind of *socialisation*, the educational domain which ". . . has to do with the many ways in which, through education, we become part of particular social, cultural and political 'orders.'" (Biesta 2010, 20). While qualification and socialization can't be separated in practice, it is important to distinguish the two analytically, so as to illuminate how education (and learning) is always promoting certain norms, values, and worldviews, either as explicitly formulated goals, or implicitly as side effects of learning and educational practice, and to be able to engage critically with these normative aspects of education.

But if we conceptualize education *only* in terms of qualification and socialization, we run into a fundamental problem, because education then becomes hard—or even impossible—to differentiate from what Biesta interchangeably refers to as indoctrination, reproduction, management of objects, or *cultivation*. The point being that what is truly *educational* about education, disappears from view. Thus, Biesta introduces a third educational function and domain of purpose, so as to be able to distinguish education from cultivation, that of *subjectification*.

SUBJECTIFICATION IN BIESTA'S EDUCATIONAL PHILOSOPHY

For Biesta subjectification has to do with ". . . the interest of education in the subjectivity or 'subject-ness' of those we educate" (Biesta 2013, 4), as opposed to being "object[s] of educational intervention" (Biesta 2020b, 89). Subjectification is about letting the students react and respond to qualification and socialisation in unique ways, establishing a dialogue between the generations instead of a one-way conversation, giving room for unique subjects to emerge; and opening up a space for critique, change, and renewal. As such, subjectification can be understood as the opposite of socialization, reproduction, or "education as cultivation," in which education *cultivates* in its students certain human qualities or subjectivities, so as to adapt or fit in with existing "social, cultural, and political orders." Whereas cultivation is about getting from A to B, so to speak, subjectification must in principle be open-ended, allowing for, or even *summoning* or *encouraging* something unique, new or *other* (than "B") to emerge; unique subjects that are not reducible to, or not (only) determined by, "social, cultural and political orders." Subjectification, then, is at odds with cultivation, prediction, predefinition, and control, it is instead conditioned on (at least some degree of) unpredictability, indefinability, and therefore also entails *risk,* what Biesta refers to as "the beautiful risk of education" (Biesta 2013). But, importantly, although subjectification is conditioned on some degree of freedom vis-à-vis that which has power, influence, or control over us, this is not to say that subjectification therefore should happen in isolation, *outside* the social (and material) world, so to speak. It is also not to say that humans can become *anything*, because the social and material world always exerts some force and puts limits on what humans can become (Biesta 2022). Instead, drawing on Arendt and Levinas, Biesta argues that subjectification is a thoroughly *relational* matter, it is only in relation to the world (with its social and material forces and limits), and in relation to, and in responding to, who or what is *other*, that human subjectivity comes into play. On the one hand, the relationality of subjectification speaks to its limits, as subjectification is limited by that which one responds to, and on the other hand, the relationality of subjectification points to the conditions of possibility of subjectification, in that *otherness,* or plurality, are the relational conditions which make subjectification possible. As such, part of the educational task in terms of subjectification, is to provide a "worldly space (. . .) of plurality and difference." (Biesta 2016a, 100)

BIESTA'S TWO PARADIGMS OF EDUCATION

Returning to the two paradigms of education which Biesta (2022) frames as "education as cultivation" and "existential education," we are now in a better position to understand the difference between the two. Biesta critiques both the "new language of learning" and "philosophical humanism" for rendering "education as cultivation," and thus as ill-equipped to see, evaluate, and discuss that which makes education *educational.* What makes education *educational,* that is, what makes education to something more, or other, than socialization and reproduction, is its interest in subjectification; the interest in providing an educational and *existential* space of unpredictability and possibility in which unique subjects can emerge. It is important to note here that Biesta is not arguing that education should avoid all forms of "cultivation"—qualification and socialization are important and necessary functions and domains of the purpose of education—but for education to be "worthy of its name" (Biesta 2022, 76), it needs to *also* have an existential orientation and an interest in subjectification, which Biesta (2022) argues, "the prevailing description of educational reality that can be found in contemporary research, policy, and practice"—the paradigm of education as cultivation—is not equipped to articulate or appreciate.

Connecting these arguments to the title of this chapter, *asking educational questions*, then, means to ask questions, not only about qualification and/or socialization, but also about subjectification. Following this line of thought, asking *educational questions* (about technology) is distinctly different from asking *technological questions* (about learning). Whereas educational questions (about technology) need to be open—in the sense that the outcome of education should not be predefined (protecting the space for subjectification)—*technological questions about learning* are typically closed questions (of learning or cultivation). They are closed in the sense of being premised on a predefined educational output, typically in terms of measurable learning outputs; asking questions about how technologies most effectively can help to achieve said output. This approach is suffering from the aforementioned "double blindness," ill-equipped to see or envision substantial educational possibilities and pitfalls of technologies within education.

CONCEPTUAL SIMILARITIES BETWEEN POSTPHENOMENOLOGY AND BIESTA'S EDUCATIONAL PHILOSOPHY

I will very briefly summarize some conceptual similarities between Biesta's framework and postphenomenology:

- Both approaches are phenomenologically and relationally oriented, with the most basic meaningful unit of analysis not being human subjects in isolation, but rather, human–world relations.
- Both approaches reject essentialism and are process-oriented, focusing on *emergence* and *becoming* (through relations), rather than essences and fixed entities.

- In much the same way as the postphenomenological concept of multistability provides a conceptual antidote to technological determinism, so does Biesta's concept of subjectification provide an educational antidote to social reproduction/determinism in education.
- In much the same way as Verbeek's ethical framework is directed toward illuminating the existential space of possibility needed to realize human freedom and responsibility in and through human–technology relations (existential normativity), so is Biesta's framework directed towards illuminating, and protecting the existential space needed for human freedom and subjectification in and through education.

BRINGING POSTPHENOMENOLOGY AND BIESTA'S EDUCATIONAL PHILOSOPHY TOGETHER

In combining postphenomenology with Biesta's educational philosophy, we are better positioned to illuminate the two mentioned blind spots in educational policy and research on educational technologies: First, postphenomenology can shed light on how educational technologies are not neutral in education by providing a framework for investigating how technologies mediate and shape educational perceptions, interpretations, intentions, and actions. Through the concept of multistability, postphenomenology also helps us to see how technologies might have multiple stabilities, including other stabilities than those intended. Second, Biesta's framework can illuminate how educational practices—with or without technologies—are not neutral and not reducible to matters of learning, by showing how educational function and purpose are always multidimensional, operating in (at least) the three domains of qualification, socialization, and subjectification. Also, Biesta's framework provides a distinctly *educational* approach; counterbalancing the paradigmatic understanding of "education as cultivation," with the perspective of "existential education," and counterbalancing the tendency within policy and research to conceptualize *technology* as a prime mover and agent of change and transformation, emphasizing instead *subjectification* as a source of transformation and renewal. As such, Biesta's framework can provide a distinctly *educational* and *critical* directionality to postphenomenological studies of educational technologies, which arguably is complemented by the existential normativity and directionality in Verbeek's concept of technological mediation (of morality). In combining postphenomenology and Biesta's educational philosophy, then, we are well-positioned for *asking educational questions about technology.*

ASKING EDUCATIONAL QUESTIONS ABOUT TECHNOLOGY

In line with Biesta's critical and constructive agendas, we can ask educational questions about technology that are both critical and constructive. Combining Biesta's critique of "learnification" and "education as cultivation" with the postphenomenological concepts of technological mediation and multistability, we can ask whether educational technologies amplify the "learnification" of perceptions in and of education, and whether they invite actions and practices that are oriented toward learning and learning efficiency. And if so, what other perceptual and interpretive possibilities are reduced, and what other actions and practices are inhibited? Biesta's framework suggests that a strong focus on learning efficiency and qualification in education also entails a relatively strong socialization; both in the way of the prioritization of learning outcomes—in which some type of knowledges and skills are valued, and others devalued—and in the way of encouraging educational practices conducive to higher learning efficiency—which often entails a strong focus on outcome, predictability, standardization, testing, control, and competition. Here, it is interesting to ask what particular (and dominant) stabilities educational technologies take on in the domain of socialization. Referring to what OECD (2021) highlights as promises of educational technologies, how will technologies of *personalized learning*, *adaptive learning*, and *behavioral nudging* that "keep the students engaged," transform perceptions and translate practices in education relevant to the domain of socialization? Will it amplify the ability and urge for predictability and control, and invite practices conducive to "nudging" and manipulation? Will it amplify perceptions and invite practices that individualizes students, rather than provide a "worldly space" for subjectification?

If technological amplification of "learnification" and learning efficiency is coupled with amplifications in the domain of socialization, will this threaten to tilt the overall educational balance more toward socialization and *cultivation,* at the expense of subjectification, which is conditioned on openness, unpredictability, risk, and trust on the one hand, and on plurality and difference on the other hand?

These critical questions are of course framed at a conceptual, philosophical level, and at the surface, they may look a bit techno-pessimistic and "Heideggerian"—with technology *framing* education toward *managing* its students as *objects* or *resources* ("standing reserve"), rather than subjects.[13] But the postphenomenological perspective can help us from falling into this kind of techno-pessimism. First, postphenomenology's empirical orientation suggests that these critical questions should not only be investigated philosophically, but also empirically; what is the "state of the actual" of particular educational technologies in particular educational settings? In other words, these conceptual questions represent starting points for investigation, not end points. Second, postphenomenology rejects both technological transcendentalism, essentialism, and determinism, and instead emphasizes technological *multistability*—which in this case means that educational technologies do not with necessity determine education in one way or another; rather, each particular technology represents a range of potential stabilities and existential possibilities. In light of Verbeek's ethical framework and *existential normativity;* educators, students, and researchers should not try to escape or liberate themselves from technology; rather, they should actively engage and experiment with technologies in education, so as to be able to critically evaluate different stabilities and mediations, and illuminate the existential space of possibility and freedom in those human–technology–world relations. This complements Biesta's notion of subjectification, and invites questions about how educational technologies can invite actions and practices that encourage and provide the existential space needed for subjectification. How can educational technologies amplify perceptions and invite actions and practices that foster plurality and difference, and invite educational encounters with what and who is *other*?

These tentative questions—both critical and constructive—serve to illustrate, at a conceptual level, the kinds of *educational questions* that can be pursued, both in educational practice and research, and provide a much-needed counterbalance and supplement to the plethora of *technological questions* being asked *about learning.* I also hope to have shown how the combination of postphenomenology with Gert Biesta's educational philosophy can provide a framework for asking *critical* educational questions about technology without falling into techno-pessimistic determinism, and *constructive* educational questions about technology without falling into instrumentalism and/or techno-optimism.

NOTES

1. These terms are explained in the next paragraph.
2. See Feenberg (2009) or Kaplan (2009, xvi–xvii) for more detailed definitions of technological determinism and instrumentalism.
3. Peter Königs (2022) defines a techno-optimist as someone who "…firmly believes that technology (or a technology) will make the world a much better place" (Königs 2022, 2), and Danaher (2022) points out that although "techno-optimism is compatible with both determinism and non-determinism (…) there may be a tendency for techno-optimists to lean into determinism" (p. 10).
4. The number of countries that participated in OECD's Programme for International Student Assessment (PISA) in 2022, according to the OECD homepage, as of

the 28th of Febuary, 2023. https://www.oecd.org/pisa/aboutpisa/pisa-2022-participants.htm.

5. In addition to making the administration and management of data more efficient.

6. I will elaborate on technological mediation in more detail later.

7. Either directly, or indirectly; focusing on how technologies impact learning efficiency through mediating variables such as teaching, engagement, motivation, learning environment, etc.

8. "Human–world relations" is a postphenomenological formulation (Ihde 1990; Verbeek 2005). It points to how intentionality is not just approached in terms of consciousness (Husserl), but in terms of humans as embodied (Merleau-Ponty) and embedded in the world (Heidegger), also reflecting postphenomenology's rejection of Cartesian subject–object dualism in favor of a relational ontology.

9. Only later coined as "postphenomenology."

10. Although a hammer is designed for (the stability of) hammering nails, it can also take up other stabilities, for example, paperweight, prying tool, or murder weapon, depending on the user and use-context.

11. It is a bit crude to lump all those philosophers in the two categories of phenomenology and post-structuralism; that's why I chose to phrase it as "associated with."

12. Biesta draws on Heidegger, Levinas, and Foucault's criticism of humanism.

13. Following this line of thought to the extreme, one could frame educational technologies as "instruments of cultivation," *only* capable of amplifying/inviting "education as cultivation," while reducing/inhibiting educational practices that encourage subjectification. This kind of technological determinism is of course not in line with postphenomenology, and I would argue that this is neither line with Biesta's project either, as he nowhere, to my knowledge, frames educational technologies in this kind of deterministic techno-pessimistic way.

REFERENCES

Aagaard, Jesper. 2017. "Breaking Down Barriers: The Ambivalent Nature of Technologies in the Classroom." *New Media & Society* 19 (7): 1127–43. https://doi.org/10.1177/1461444816631505.

Biesta, Gert. 2005. "Against Learning. Reclaiming a Language for Education in an Age of Learning." *Nordic Studies in Education* 25 (1): 54–66. https://doi.org/10.18261/ISSN1891-5949-2005-01-06.

———. 2009. "Good Education in an Age of Measurement: On the Need to Reconnect with the Question of Purpose in Education." *Educational Assessment, Evaluation and Accountability(Formerly: Journal of Personnel Evaluation in Education)* 21 (1): 33–46. https://doi.org/10.1007/s11092-008-9064-9.

———. 2010. *Good Education in an Age of Measurement: Ethics, Politics, Democracy*. Vol. 5. Interventions : Education, Philosophy & Culture. London: Paradigm Publishers.

———. 2013. *The Beautiful Risk of Education*. Interventions: Education, Philosophy, and Culture. Boulder: Paradigm Publ.

———. 2014. "Cultivating Humanity or Educating the Human? Two Options for Education in the Knowledge Age." *Asia Pacific Education Review* 15 (1): 13–19. https://doi.org/10.1007/s12564-013-9292-7.

———. 2016a. *Beyond Learning: Democratic Education for a Human Future*. 1st edition. New York: Routledge. Kindle.

———. 2016b. "ICT and Education Beyond Learning." In *Digital Expectations and Experiences in Education*, edited by Eyvind Elstad, 29–43. Rotterdam: SensePublishers. https://doi.org/10.1007/978-94-6300-648-4_2.

———. 2017. *The Rediscovery of Teaching*. New York: Routledge. https://doi.org/10.4324/9781315617497.

———. 2020a. *Educational Research: An Unorthodox Introduction*. London: Bloomsbury Academic.

———. 2020b. "Risking Ourselves in Education: Qualification, Socialization, and Subjectification Revisited." *Educational Theory* 70 (1): 89–104. https://doi.org/10.1111/edth.12411.

———. 2022. *World-Centred Education: A View for the Present*. London: Routledge. Kindle.

Boer, Bas de, Jonne Hoek, and Olga Kudina. 2018. "Can the Technological Mediation Approach Improve Technology Assessment? A Critical View from 'Within.'" *Journal of Responsible Innovation* 5 (3): 299–315. https://doi.org/10.1080/23299460.2018.1495029.

Danaher, John. 2022. "Techno-Optimism: An Analysis, an Evaluation and a Modest Defence." *Philosophy & Technology* 35 (2): 54. https://doi.org/10.1007/s13347-022-00550-2.

Feenberg, Andrew. 2009. "What Is Philosophy of Technology?" In *International Handbook of Research and Development in Technology Education*, edited by Alister Jones and Marc J. de Vries, 159–66. Brill: Leiden, Netherlands. https://doi.org/10.1163/9789087908799_016.

Hamilton, Edward, and Norm Friesen. 2013. "Online Education: A Science and Technology Studies Perspective / Éducation En Ligne: Perspective Des Études En Science et Technologie." *Canadian Journal of Learning and Technology / La Revue Canadienne de l'apprentissage et de La Technologie* 39 (February). https://doi.org/10.21432/T2001C.

Ihde, Don. 1979. *Technics and Praxis: A Philosophy of Technology*. First edition. Vol. 24. Boston Studies in the Philosophy and History of Science. Dordrecht: Springer Netherlands : Imprint: Springer.

———. 1990. *Technology and the Lifeworld: From Garden to Earth*. Bloomington: Indiana University Press.

Kaplan, David M. 2009. *Readings in the Philosophy of Technology*. Lanham, MD: Rowman & Littlefield Publishers, Incorporated.

Königs, Peter. 2022. "What Is Techno-Optimism?" *Philosophy & Technology* 35 (3): 63. https://doi.org/10.1007/s13347-022-00555-x.

OECD. 2021. *OECD Digital Education Outlook 2021: Pushing the Frontiers with Artificial Intelligence, Blockchain and Robots*. Paris: Organisation for Economic Co-operation and Development. https://www.oecd-ilibrary.org/education/oecd-digital-education-outlook-2021_589b283f-en.

Oliver, Martin. 2011. "Technological Determinism in Educational Technology Research: Some Alternative Ways of Thinking about the Relationship between Learning and Technology." *Journal of Computer Assisted Learning* 27 (5): 373.

Pischetola, Magda. 2021. "Re-Imagining Digital Technology in Education through Critical and Neo-Materialist Insights." *Digital Education Review* 40 (December): 154–71. https://doi.org/10.1344/der.2021.40.154-171.

Rosenberger, Robert. 2014. "Multistability and the Agency of Mundane Artifacts: From Speed Bumps to Subway Benches." *Human Studies* 37 (3): 369–92. https://doi.org/10.1007/s10746-014-9317-1.

Selwyn, Neil. 2016. *Is Technology Good for Education?* Cambridge: Polity Press. Kindle.

Turville, Joni, and Catherine Adams. 2018. "Doing Postphenomenology in Education." In *Postphenomenological Methodologies: New Ways in Mediating Techno-Human Relationships*, edited by Jepser Aagaard, Jan Kyrre Berg Friis, Jessica Sorenson, Oliver Alexander Tafdrup, and Cathrine Hasse, 3–25. Lanham, Maryland: Lexington Books.

Verbeek, Peter-Paul. 2005. *What Things Do: Philosophical Reflections on Technology, Agency, and Design*. University Park: Pennsylvania State University Press.

———. 2011. *Moralizing Technology: Understanding and Designing the Morality of Things*. Chicago: University of Chicago Press. Kindle.

———. 2012. "Expanding Mediation Theory." *Foundations of Science* 17 (4): 391–95. https://doi.org/10.1007/s10699-011-9253-8.

———. 2014. "Some Misunderstandings About the Moral Significance of Technology." In *The Moral Status of Technical Artefacts*, edited by Peter Kroes and Peter-Paul Verbeek, 75–88. Philosophy of Engineering and Technology. Dordrecht: Springer Netherlands. https://doi.org/10.1007/978-94-007-7914-3_5.

———. 2016. "Toward a Theory of Technological Mediation: A Program for Postphenomenological Research." In *Technoscience and Postphenomenology: The Manhattan Papers*, edited by J.K. Berg, O. Friis, and Robert C. Crease, 189–204. London: Lexington Books.

Chapter 18

Speaking of Education's Technologically Mediated Character(s)

Jan Peter Bergen

> [. . .] it is from the same causes and by the same means that every virtue is both produced and destroyed, and similarly every art; for it is from playing the lyre that both good and bad lyre-players are produced [. . .] For if this were not so, there would have been no need of a teacher, but all men would have been born good or bad at their craft [. . .]. (Aristotle, NE 1103b7-14)

Education is, by its very nature, a mediated practice. In between the student and that about which they are to be educated stands a teacher, ideally not as a hindrance but as a guide, a facilitator, a challenger, an instructor, and a selector of topics, materials, discussions, and exercises; as a mediator. In so doing, teachers focus students' attention on certain aspects of the world in specific ways, and invite and disinvite, reward, and even penalize, some ways of acting. However, such a mediating role is not restricted to teachers. The material environments and artifacts with(in) which education unfolds likewise mediate educational practices.

One particularly striking account of the role of technological artifacts in education comes to us from Michel Foucault. As part of his analysis of modern techniques for the exercise of authority, Foucault (1977) describes in appreciable detail how in the modern school,[1] the design of classrooms, timetables, logbooks, benches, slates, pens, bells, and other material aspects of educational practice were integral to techniques of discipline. From the design of classrooms focusing attention on (and establishing the authority of) the teacher to the use of writing implements in imbuing in students the supposed virtues of obedience and efficiency through the "instrumental coding of the body," material artifacts co-shaped students into "docile bodies." Of course, our goals for student development have shifted significantly since (Walker, Roberts, and Kristjánsson 2015), but Foucault's analysis nevertheless prompts us to be mindful of the role that technologies play in contemporary education, especially as more and new technologies get introduced.

Given this, one central question to be answered is *what things do*, and what their mediating role is in educational practice. Indeed, if technological artifacts play a formative role, understanding it becomes paramount. In turn, this necessitates a conception of technological artifacts that can do that role justice. One such philosophical outlook can be found in postphenomenology (Ihde 1990; Verbeek 2005), which aims to take technological artifacts seriously by focusing on the active role they play in our lives, mediating our experiences of the world and our actions in it. Over the years, a considerable "postphenomenological vocabulary" (Verbeek 2005) has been developed with which to analyze the mediating role of technology in human–technology–world relations, a vocabulary that has seen some, albeit limited, application to educational contexts (e.g., Voordijk and Vahdatikhaki 2020; Bildik and Altun 2022; Kennedy 2022; van Kraalingen 2022; Lepe-salazar and Cortes-alvarez 2023). This indicates that postphenomenology has considerable potential here, even if it remains to be fully explored.

At least equally important is the question of *what we want things to do* in education. That is, the fact of technological mediation in education raises the normative question of what those mediations *should* help to achieve. An answer focusing on the (effectiveness of) the acquisition of knowledge and skills would at least *prima facie* be in line with the technological mediation of perception and action. However, this would amount to ignoring an important area of existing overlap between education and postphenomenology; between educational goals in terms of character development (e.g., Walker, Roberts, and Kristjánsson 2015) and existing proposals for ethics of technological mediation (e.g., Dorrestijn 2012; Verbeek 2011), that is, *a focus on the shaping of moral subjects*.

This chapter explores this overlap, taking contemporary "character education" and the ethics of technological mediation as its starting points. In so doing, it identifies a conceptual gap in the postphenomenological vocabulary on the technologically mediated shaping of subjects that would need to be filled if postphenomenological analysis is to do justice to the role of technology in the processes of character education. Lastly, it proposes to bridge that identified gap with concepts from virtue theory.

LEARNING TO BE GOOD: CHARACTER EDUCATION

Education is meant to teach concrete knowledge and practical and intellectual skills but, as a primary source of socialization, it is also much more than that

(Savater 2001). Students also learn to become members of their respective societies; they are shaped as people, as (moral) subjects. Some educational philosophies make this part of the stated goals of the educational process, for example, so-called "character education." This type of education—what's in a name—focuses on the cultivation of character, or "a set of personal traits or dispositions that produce specific moral emotions, inform motivation and guide conduct" (The Jubilee Centre for Character & Virtues 2022, p. 7). This explicit focus on character in thinking about education, while dating back at least to antiquity, had waned during modernity. And yet, despite that, there has been a resurgence of interest in it recently (Curren 2010), due in no small part to a rejuvenation of ancient philosophical traditions focusing on (the ethics of) character in the second half of the twentieth century (e.g., Foot 1978; MacIntyre 1981) and a reorientation of the legitimating principles of education in the twenty-first century toward pupils' *flourishing* (Walker, Roberts, and Kristjánsson 2015). How to achieve such flourishing, however, is up for debate.

In current scholarship on flourishing-oriented character education, there are two main approaches: one based on positive psychology, and the other based on (neo-)Aristotelian theory. The former usually conceives of flourishing as perceived or subjective happiness and aims to cultivate meaningful and positive emotions (Han 2014), whereas the latter has a more ethically substantive notion of flourishing, that is, as the highest human good that is constituted by good character traits called *virtues* (with bad traits called vices) that should be cultivated through practice and education. Examples of such virtues include courage, temperance, generosity, justice, truthfulness, and curiosity. These are excellent traits of character, habits, or dispositions, either moral or intellectual, that lead the possessor to perform good actions and lead a good life. Being so deeply embedded, they involve "emotions and emotional reactions, choices, values, desires, perceptions, attitudes, interests, expectations and sensibilities. To possess a virtue is to be a certain sort of person with a certain complex mindset" (Hursthouse 2006). Importantly, such virtues are not innate, they must be learned from a young age through practice and education, with the formation of good habits being a prerequisite for developing the right reflective capabilities later on in life (Curren 2010; Peters 1981). Unsurprisingly then, character education of the neo-Aristotelean sort "includes all explicit and implicit educational activities that help young people to develop positive personal strengths called virtues" (The Jubilee Centre for Character & Virtues 2022, p. 7).

In what follows, we will assume an Aristotelean outlook on character education because it has appreciable benefits over its positive psychology counterpart in the context of this chapter. First, Aristotelean virtues, being stable and engrained traits of character, are broader and more constitutive of one as a subject than positive emotions are and thus closer to discussions of how education shapes us *as* subjects. Second, virtues are constitutive of a good life rather than instrumental to it and, thus, also more central to what kind of subject we should be. Third, Aristotelean virtue ethics recognizes the intimate link between individual character and living well together with others, that is, good societies need virtuous citizens[2] and this requires general, accessible character education (Curren 2010). Lastly, character traits, when possessed separately and unreflectively, are unlikely to lead to good actions in a variety of situations. Virtuous persons need some way to "regulate" how the virtues come into play in specific instances. This is why Aristotelean accounts of character include a virtue called *phronesis* or practical wisdom, which is "nothing less than the glue that holds the Aristotelean character system together and makes it unique" (Kristjánsson 2016, p. 530). So, if the goal is to think through the role of technology in character education and shaping students *as moral subjects*, the Aristotelean account of character education seems to hold the most promise.

All of this tells us nothing, however, about the role of technology in such virtue-oriented character education, and even less about the overlap with postphenomenology. Before engaging with the latter in the fourth section, however, the next section explores the different links that have been identified between virtue and technology.

TECHNOLOGY AND VIRTUE

The idea of a connection between virtue and technology is, of course, not new. Different conceptions of that relation exist in the literature, which can be distilled into three "strands": virtues-for-technology, technology-contra-virtues, and technology-for-the-virtues.

The first and most well-developed, *virtues-for-technology*, acknowledges that the increasingly technological nature of our lives requires us to develop the virtues necessary for such technological lives. On the one hand, this means that engineers and designers require the necessary virtues to fulfill their role responsibly (e.g., Harris 2008; Bowen 2009; Schmidt 2014; Bergen and Robaey 2022). But this need for virtue does not stop with engineers. As MacIntyre (1979) notes, "there are necessary limits to our predictions about the future of technology [. . .] we will have to design people with all those traits [. . .] necessary for living in an unpredictable environment" (p. 5), meaning that members of society more broadly also need to cultivate the appropriate "technomoral virtues" (Vallor 2016) and/or "cyber-virtues" (Dennis and Harrison 2020) to prepare for ever-changing technological lives. While this strand of literature strengthens the call for contemporary character education, it adds little to thinking through the role of technology in shaping the character(s) it advocates for.

The second strand, *technology-contra-virtues*, does explicitly recognize the role of technology on virtues, albeit negatively. That is, technology could be detrimental to the cultivation of virtue, risking "moral deskilling" (Vallor 2013) as a result of (interacting with) technologies. For example, the virtues of empathy, self-control, humility, honesty, and perspective are said to be threatened by new social media technologies (Vallor 2016). So, if technology poses such risks of moral deskilling, shouldn't we be wary of its inclusion in (character) education? Not necessarily.

At least some of the authors that highlight the risks of moral deskilling also indicate the possibility of moral *re*skilling through technology (Vallor 2015). Interacting with and through artifacts could also contribute to the cultivation of virtue, hence the third strand: *technology-for-the-virtues*. Ironically, while this would be the most fruitful strand for understanding and planning the role of technology in character education, it is arguably the least philosophically well-developed. Sure, some philosophers acknowledge and encourage the idea of technology contributing to the cultivation of virtue (Vallor 2015, 2016; Dennis and Harrison 2020), yet how it is to do so remains philosophically underexplored despite some design methods having explicitly taken the cultivation of virtue as a goal for the designs produced (e.g., Pohlmeyer and Desmet 2017; Reijers and Gordijn 2019).

Both philosophical and design approaches to technology-for-the-virtues require a thorough understanding of *what things do* in such processes to be successful, a question that is the central focus of postphenomenology. As such, one might expect postphenomenology to already be equipped to answer it. However, the extent to which postphenomenology can be useful here depends in part on the extent to which it can speak specifically to what technologies do *in cultivating virtue or shaping moral character* since that is what they are supposed to do. This in turn depends on how well the postphenomenological vocabulary can capture the idea of moral subjectivity or character and how it develops in interaction with technology. As the following section shows, however, postphenomenology is still lacking in this regard.

WHAT THINGS DO TO US: POSTPHENOMENOLOGY AND SHAPING SUBJECTS

While postphenomenology focusses on the question of what technological artifacts do, it does not conceive of them in isolation, as "entities in themselves." Instead, "[w]hen they are used, they help to establish relations between users and their environment [. . .] In so doing, technologies also help to shape the 'subjectivity' of their users and the 'objectivity' of their world" (Rosenberger and Verbeek 2015, p. 19). Postphenomenology thus has a "relational ontology," with Verbeek even claiming that these relations are not between pre-existing subjects and objects (2005), but rather, "[w]hat the world 'is' and what subjects 'are' arise from the interplay between humans and reality" (2011, p. 15). As such, the technologically mediated constitution and shaping of subjects is baked into the theoretical foundations of postphenomenology.

One example of this can be found in Verbeek's analysis of obstetric ultrasound (Verbeek 2008, 2011). Verbeek describes the active role that ultrasound plays in shaping experiences and practices surrounding pregnancy, as well as what the subjects involved are and are supposed to do. Not only does the way ultrasound visualizes the fetus on the screen invite the perception of the fetus as an individual person and patient, but it also mediates the way prospective parents are to experience and understand themselves and relate to one another. This includes the mother being seen less in unity with the fetus, but rather as its environment, and it sets up parents as active decision-makers because the ultrasound allows for the detection of abnormalities during pregnancy that would have gone unnoticed before the use of obstetric ultrasound.

What obstetric ultrasound does in this analysis, that is, shaping perceptions, practices, and subjects, raises ethical questions. However, according to Verbeek, answering them by applying "standard" ethical theory may be impossible because the "actively mediating role of ultrasound in moral interpretations, decisions, and practices is at odds with the humanist orientation of mainstream ethics" (2011, p. 27). That is, Verbeek contends that the ethical questions raised by obstetric ultrasound cannot be dealt with if we do not take postphenomenology's relational ontology and the hybrid character of subjects and objects seriously. What postphenomenology needs if it is to have an ethics consistent with its ontology, then, is a hybrid notion of subjectivity that is constituted and shaped by its mediated interactions with the world but also has the agency necessary to shape itself in responsible, technologically mediated ways. Dorrestijn (2012) and Verbeek (2011) have sought inspiration for such a hybrid notion of subjectivity in the work of Michel Foucault.

Subjectivation as an Ethics for Technological Mediation

In his early work, Foucault reconceptualized subjectivity as formed through processes of *subjection*, that is, of structures of power molding our ways of life, the ways we think, speak, and act, and what we are and are supposed to be (Foucault 1973, 1977). On such a view, subjectivity is not the self-evident given that it was for modern humanism, but rather something that is imposed on us through so-called "technologies of power," which—as the example of the school in the introduction already indicated—include the use of technological artifacts. Such subjectivity is hybrid, impossible to separate from the very (technologically mediated) processes that constitute and shape it through the exercise of power. This formative influence of technology on our perception, our actions, and even our subjectivity makes subjection at least *prima facie* compatible with postphenomenology's idea of subject–object co-constitution.

However, does this hybrid notion of subjectivity through subjection allow for an ethics that could do justice to the phenomenon of technological mediation? After all, does the idea that we cannot escape the influence of "technologies of power" not deprive us of the freedom necessary for ethical thought and action? Possibly, if subjection was all there was to subject-formation. However, according to Foucault, that is not the case. In his later work (e.g., 1986, 1990), he turned to the subject of freedom and ethics, and their relation to subjectivity. In "The History of Sexuality," Foucault finds inspiration in antiquity (especially the Stoics), proposing to revive the ancient practice of "care of the self" through the employment of "technologies of the self," the latter being "intentional and voluntary actions by which men not only set themselves rules of conduct but also seek to transform themselves" (Foucault 1990, p. 10). That is, he salvages a relational notion of freedom that does not lie outside of technologies of power, but one that can be established in practice only within and in relation to them, a process called *subjectivation* or subjectification. This freedom does allow for an ethics, albeit an unfamiliar one; one of freedom as a situated way of life, an *ēthos*: "a way of being and of behavior" which revolves around "extensive work by the self on the self" (Foucault 1997, p. 286). That work on the self takes the form of *askēsis*, a training of the self by oneself through ascetic practices or spiritual self-discipline.

Dorrestijn (2012, 2017) and Verbeek (2011) have incorporated these notions into an ethics for technological mediation: rather than resisting the formative influence of technology, living well with technology means to shape one's life and *oneself* in relation to technological mediations, that is,, by "finding a relation to these mediations, incorporating them in our existence, human beings can further shape and stylize their moral subjectivity" (Verbeek 2011, p. 82), where this shaping is an important responsibility of the subject[3] (p. 87). Moreover, both Foucault and postphenomenologists have recognized that technological artifacts can play an active role in these "technologies of the self." For example, Foucault (1997) describes the role of letters, notebooks, and diaries in practices of self-care, where they act as "functional devices that would enable individuals to question their own conduct, to watch over and give shape to it, and to shape themselves as ethical subjects" (Foucault 1990, p. 13). Similarly, Bergen and Verbeek have explored (post)phenomenologically the way To-Do list apps can support processes of self-shaping (Bergen and Verbeek 2021). This approach to technologically mediated subjectivity already contains relevant connections to this chapter's central questions concerning the role of technology in character education: both agree that shaping ourselves is an important part of living well, that technology plays an important role in the process of that shaping, and that such a process needs a foundation in reflective and subjectivity-oriented practice(s); all of these aspects seem compatible with the technologically mediated cultivation of virtue.

And yet, the question of *what aspects/characteristics/qualities about the subject are to be shaped* through "technologies of the self," be it virtues or otherwise, is not prescribed by postphenomenological theorizing. Instead, this question is left tantalizingly open by Dorrestijn and Verbeek, falling back on Foucault's general identification of the four central aspects of practices of the self (Foucault 1990): the *ethical substance* (the part of the subject taken as the "prime material" of their moral practices), the *mode of subjection* (one's relation to the rule/moral system they wish to live by and how they recognize themselves as obliged to live by it), the *ethical work* or self-practices through which the subject is shaped, and the *telos* (the mode of being to which the subject aspires). While this "Foucaultian fourfold" presents a general model of self-formation as adopted by some postphenomenologists, it is purposely light on specifics, including on the ethical substance. And herein lies an important difference between postphenomenology and character education: whereas character education explicitly focuses on virtues as the ethical substance to be cultivated through practice, the theory of technological mediation generally lacks such explicit normative commitment.[4] Still, the generality of the fourfold means that it *is* in principle compatible with thinking specifically about the technological mediation of virtue cultivation, even if that involves more normative commitment than postphenomenologists have generally allowed themselves. However, that possibility *in principle* does not mean that we have the conceptual resources for doing so *in practice*. That is, what concepts does the postphenomenological vocabulary offer that could help us here? Beyond the reflections on subjectivation referenced above, the set of useful concepts is limited, but not empty: the closest theoretical fit for further talk of virtue can be found in existing postphenomenological reflections on *habit and sedimentation.*

The Force of Habit: Habituation and Sedimentation in Postphenomenology

As already introduced in the first section, virtues are dispositions or *habits* of character that will elicit (morally) good (re)actions from those that exhibit them. And while postphenomenology is generally light on normative commitment (and thus not generally in the business of judging certain habits as good or not), it does have a theoretically developed notion of habit that may be helpful in connecting postphenomenology to character education.

This postphenomenological understanding of habit builds on the notion of *sedimentation,* referring to how cumulative past experiences inform and shape current ones. Although sedimentation has a long history in phenomenology, the specific version that has found its way into postphenomenology stems from Maurice Merleau-Ponty, for whom sedimentation has an explicitly embodied character:

> [m]y act of perception [. . .] takes advantage of work already done, of a general synthesis constituted once and for all, and this is what I mean when I say that I perceive with my body or my senses, since my body and my senses are precisely that familiarity with the world born of habit, that implicit or sedimentary body of knowledge. (Merleau-Ponty 2002, p. 277)

While already somewhat recognized in Don Ihde's work (e.g., 1998, p. 121), sedimentation as a specifically *post*phenomenological notion was developed later by Robert Rosenberger. Explicitly retaining Merleau-Ponty's focus on sedimentation as effecting experience and the central role of the body therein (Rosenberger 2012), Rosenberger proposed that we understand sedimentation in the context of human–technology relations as referring "to the strength of the habituation that characterizes a particular human–technology relation. That is, any human–technology relation whose features are accompanied by a strong force of habit is one that is deeply sedimented" (Rosenberger 2014, p. 376). Through repeated embodied interactions with a technological artifact, one becomes deeply accustomed to certain ways of interacting with it (and thus, to the kinds of embodied perceptions and actions that that way of interacting invites). That is, those interactions sediment into "bodily perceptual" habits.

Given the embodied nature of sedimentation, it is not surprising that many examples concern so-called "embodiment relations" with technological artifacts (cf. Ihde 1990), such as the use of hammers, eyeglasses, cell phones, binoculars, car driving, and facial prosthetics (Rosenberger 2012, 2014, 2021; Yaron, Widdershoven, and Slatman 2017). This also invited reflection on the connection between sedimentation and "transparency," referring to how an artifact in embodied use tends to disappear into the background of a user's awareness depending on how habituated that way of using the artifact is for the user (Ihde 1990). However, the phenomenon of sedimentation is not limited to embodiment relations. As Rosenberger and Verbeek (2015) indicate, other human–technology relations are also dependent on sedimentation.

They give the example of an expert reading an fMRI result where years of training and experience in reading these kinds of displays have sedimented into bodily perceptual habits that make "the image's meaning [appear] all together in a perceptual gestalt" (p. 25). As such, sedimentation does not just lead to transparency. Its effects are more general: "[s]edimentation provides the pre-perceptive context that enables our current perception to occur with immediate meaningfulness" (idem). As a consequence, the successful use of artifacts generally depends on some level of habituation and automaticity that is the outcome of "bodily and perceptual training" (Rosenberger 2021, p. 30).

Sedimentation, while it makes the use of technologies possible, thus blinds us to the specific ways in which those technologies mediate our lives. But that is not all. Despite its constant influence on our perceptions and actions, sedimentation is itself somehow transparent, not usually present to conscious thought. It operates under the surface, pre-structuring our engagements with the world around us. This led Jesper Aagaard (2021b) to propose the notion of the "*passive subject*." He argues that while approaches from STS, such as Actor-Network Theory, have successfully argued that technological artifacts are much more active than often given credit for, it has failed to give due deference to the extent to which "most human behavior is unmediated by thought and occurs on a prereflective bodily level," that much of human behavior—including technology use—is best described in terms of habit, or an "immediate and prereflective inclination to act in certain ways due to familiarity with that type of situation" (p. 15). This "passive subject" is not to be read mechanistically or in terms of helplessness, Aagaard warns, but rather as a middle ground between human voluntarism and technological determinism[5]. To him, this is the value of Rosenberger's account of habit: that it understands habits as "*neither* choices *nor* reflexes, they are *both* self-initiated *and* involuntary" (p. 17). As Rosenberger further interprets,

> [t]his kind of habituated action is thus "passive" in the sense that [it] is happening with a level of automaticity coming from within the user, as the result of the user's own past experiences, and yet without an explicit act of will on the user's part. (Rosenberger 2021, p. 31)

In sum, postphenomenologists have posited that sedimentation and habits pre-structure our technologically mediated perceptions and actions in ways that we are generally unconscious of. This account conceives of habits as automated ways of responding to situations. *Prima facie*, this seems at odds with the more conscious, reflective, even critical account of subjectivation presented above. And while both could be valuable in their own right for discussing the role of technology in character education, there is an apparent gap here that might hinder such analyses; a gap further specified below.

The Gap between Habituation and Subjectivation

The previous section introduced a general postphenomenological account of habits that are historical (i.e., acquired through sedimentation/repetition), bodily perceptual, non-neutral, passive, unreflective, tied to specific human–technology relations, and not characterized by moral valence. Being such, can they function as the "ethical substance" for a process of technologically mediated subjectivation?

On the one hand, there is some promising overlap. For one, habits are developed by processes of sedimentation, and subjectivation likewise requires (ascetic) practice. Moreover, habits can be worked on as part of subjectivation. On the other hand, there seem to be significant incompatibilities between the postphenomenological accounts of subjectivation and of habit. First, subjectivation is presented as an engaged, conscious, and critical process, whereas habits are seen as passive and unreflective. Second, subjectivation is supposed to result in changes that shape one's life and oneself generally, not just vis-à-vis a specific technology. Last, Subjectivation calls specifically for shaping our *moral* subjectivity, but there is little in the account of habits as presented that indicates that they are moral in nature.

Some postphenomenological reflections have already mellowed the latter two incompatibilities somewhat. First, Lewis (2020) combines "the temporal concepts of sedimentation [. . .] as well as the future potentiality of technologies in order to better understand the intrasubjective transformation caused by technologies" (p. 128), proposing that sedimented habits "continue to influence us even when we are not materially using those technologies" (p. 135). Focusing mainly on how technologies can frame experiences (the "technological gaze"), framings we internalize through habituation, Lewis recognizes that bodily perceptual habits developed in a specific human–technology relation can transcend that relation and influence future situations. However, this is still far from the encompassing character habits that character education, and arguably a habit-oriented process of subjectivation, would aim to cultivate. Secondly, Aagaard (2021a) and Rosenberger (2021) have at least recognized the moral relevance of habits. Aagaard argues for speaking of "tech habits" when discussing the phenomenon of technology addiction while arguing that the latter represents "bad habits" and good ones might be trained. Rosenberger in turn identifies how certain perceptual habits can blind us to real injustices. They thus show that habits are not only non-neutral but are morally relevant. What they do not do, however, is grant habits themselves (moral) valence. That is, such habits might be instrumental to achieving good outcomes, but give us little reason to assume they could be constitutive of the good—something that, again, character education and a habit-oriented process of subjectivation would endorse.

As such, despite these proposals, a gap remains between postphenomenological accounts of reflective and general subjectivation and unreflective, largely technology-specific habit. How to bridge it? At least for the context of technology's role in character education, I propose that this postphenomenological notion of habit only accounts for a limited subset of possible habits—unsurprisingly, those that are usually linked to specific human-technology relations. However, we could bridge the gap with "habits" of a different sort; the sort of "habits" that includes virtues.

BRIDGING THE GAP: THE SORT OF HABIT THAT IS VIRTUE

> "[. . .] by doing the acts that we do in our transactions with other men we become just or unjust, and by doing the acts that we do in the presence of danger, and being habituated to feel fear or confidence, we become brave or cowardly. [. . .] It makes no small difference, then, whether we form habits of one kind or of another from our very youth; it makes a very great difference, or rather all the difference." (NE II.1)

The gap identified above is not unique to postphenomenology: it corresponds quite well with differing emphases of moral education in the past: "[o]n the one hand there is an emphasis on habit, tradition and being properly brought up; on the other hand there is an emphasis on intellectual training, and on the development of critical thought and choice": a clash between habit and reason that, at first glance, constitutes a "paradox for moral education" (Peters 1981, p. 45). However, virtue-oriented character education need not accept this framing of character education as paradoxical, as Aristotelian virtue theory connects habit, upbringing, critical thought, and choice through *hexeis* (singular: *hexis*).

Aristotle explains that the virtues are examples of *hexeis,* and *hexis* has often been translated as "habit," through the Latin term *habitus*. Despite this, it would be a mistake to equate a modern conception of habit with Aristotelean *hexis* (Lu 2015). That narrow, modern conception of habit is succinctly described by Austin as follows, which , in turn, implies the absence of an action's purpose or reference to consciousness (Malikail 2003).

> Once a habit is established, if given a particular trigger the behavior is *automatic* [. . .] Yet the more something is done automatically, the less it is done consciously and voluntarily. The modern concept of habit therefore picks out a mechanistic pattern of response, generated by multiple repetitions of identical actions. Habits, therefore, lead to *unthinking* and *nonvoluntary* action. (2017, p. 24 my emphases)

This notion should sound somewhat familiar: not only is it close to its colloquial use, but it is also rather similar to the postphenomenological notion of habit at the basis of the "passive subject": passive, automatic, and leading to actions without a corresponding act of will. So, if *Hexis* is *not* like the modern or postphenomenological notion of habit, what is it like instead? As the fact that it has been translated by quite different notions including habit, disposition, and state of character indicates, there is no modern English word that neatly captures its meaning. Instead, we need to explore what Aristotle tells us about *hexis* and its characteristics.

Hexis is a type of quality of the soul, a characteristic of a person of a specific kind. So what characterizes this "kind"? For one, a *hexis* is a characteristic that is stable, long-lasting and difficult to change, a kind of "second nature." This makes *hexeis* like "branches of knowledge and the virtues" different from mere "conditions" like being healthy or sick,[6] cold or hot (Cat. 8b28-9a4). Another characteristic of *hexeis* is the fact that they need to be developed, through habituation (*ethos*) and/or education (with the development of moral virtues more reliant on the former, and intellectual virtues more on the latter): "[. . .] some men become temperate and good-tempered, others self-indulgent and irascible, by behaving in one way or the other in the appropriate circumstances. Thus, in one word, states of character arise out of like activities" (NE 1103b19-22). This makes *hexis* different from some other qualities of the soul, for example, "capacities" that might occur naturally like the ability to hear sound or feel cold, or certain natural talents and intuitions (NE II.5). A third characteristic of a *hexis* is that it structures how we experience and engage with situations. Possessing a *hexis* leads those possessing them to be sensitive to the relevant aspects of a situation to which the *hexis* corresponds, feel specific sensations like emotions and relate to those in specific ways (Hursthouse 2006). So far, so similar to postphenomenological habits being sedimented into the subject through habituation and providing the pre-preceptive context for experience and action. However, for all these similarities, there are also vital differences.

First, *hexis* is not as "static" as habit, in the sense that it is (a) not restricted to the situation(s) of habituation and (b) does not result in the same behavior in different circumstances. Sure, we cultivate *hexeis* by engaging in like acts, but the resulting states of character apply beyond it, are "paradoxically dynamic rather than static" (Lockwood 2013, p. 22). That is, they are enduring, but manifest themselves very differently and more or less strongly in different circumstances. Generosity may be a stable *hexis* (of the kind that is moral virtue) for someone, but how and to what extent it manifests depends heavily on the circumstances. The same holds for other *hexeis* like truthfulness, curiosity, or creativity. Second, just like *hexis* is not static, it is also not mainly passive. Instead, it is characterized by an intricate "intertwinement of activity and passivity" (Oele 2012), of acting and being affected. While *hexeis* do pre-structure how we are affected by or experience a situation, they also shape how we engage in it in a way that "involve[s] a kind of active holding of oneself in some respect" (Lu 2015, p. 200). This makes *hexeis* different from, say, emotions, which Aristotle characterizes as "passive experiences and their occurrence does not imply any trait of character. What is responsible for whether these emotions occur in accord with reason, or in opposition to it, is states of character" (EE 1220b11). This connection to reason leads to a third difference: *hexeis* are not merely automatic or unreflective. Rather, they are engaged with reason, either because (a) they are intellectual in nature or (b) because they are receptive to and harmonious with reason, like moral virtues are to practical wisdom (NE 1102b20-1103a10). This also explains how *hexis* can be dynamic rather than static: it is guided by wisdom (if said *hexis* is virtuous). This intertwining of moral and intellectual virtue adds to them being engaged and active, since moral virtues are *hexeis* concerned with choice (NE 1106b36) and practical wisdom one concerned with action (1140b4-5). Lastly, *hexeis* have valence. In the *Metaphysics*, Aristotle states that *hexis* is "a disposition according to which that which is disposed is either well or ill disposed, and either in itself or with reference to something else" (Met. 1022b5-15). In other words, *hexeis* are good or bad with reference to something. In the case of artistic or technical production (*techne*) the corresponding *hexis* would be good or bad in reference to that which is produced. Moral and intellectual virtues on the other hand are good with reference to the person that possesses them, how they act and "hold" themselves and whether they experience the right emotions, pleasures, and pains. And given that the virtues are constitutive of the highest human good that is flourishing, Aristotle's ethics provides a basis for judging certain *hexeis* as virtues and others as vices.

To cultivate good *hexeis* is demanding, which is why Aristotle calls for general character education (starting with developing good habits) from a young age (Curren 2010). But how does this solve the aforementioned "paradox" between habit and reason in moral education? It is because good habits and moral virtues need to be cultivated before one can properly engage in the further cultivation of reason/practical wisdom (NE 1095b2-5). Small children may not be capable of practical wisdom, but they can already be educated in good habit. Thus, "they can and must enter the palace of Reason through the courtyard of Habit and Tradition" (Peters 1981, p. 52). Once *hexis* develops based on such habits, however, habit and reason are not to be conceived of as separate. Instead, we should strive to have them harmoniously intertwine in virtuous *hexis*. Likewise, if we are to take character as the "ethical substance" for a postphenomenological ethics of subjectivations, the gap between subjectivation and the narrow notion of habit can be bridged. At least in the context of character education, it could be bridged by *hexis* by focusing on those habituated, morally valent, reflective, and engaged qualities of our subjectivity. Just like for Aristotle, developing good "technological habits" (narrow as they may be) might be part of shaping our moral subjectivity, but would not amount to much without bridging the gap through something like the cultivation of proper *hexeis*, states of character or dispositions that are stable, apply in a variety of circumstances, fruitfully guide our experience, judgments, choices, and actions and of course, allow us to flourish.

CONCLUSION

This chapter explored the normative overlap between postphenomenology and character education, that is, shaping moral subjects. In doing so, it identified a gap in the postphenomenological vocabulary that would prohibit it from being fruitfully applied to the analysis of technological mediation in the shaping of students *as* moral subjects; a gap between the reflective and active process of subjectivation and an automatic and passive notion of habit. Fitting the context of education, the notion of *hexis* was proposed as a way to bridge this gap, since it is the same "aspect" of subjectivity that character education aims to cultivate in students. This should allow postphenomenologists to go beyond talking about technological habits or abstract notions of subjectivation, instead enabling questions on technology's role in and impact on the cultivation of *hexeis*, on character. In so doing, it should also expand the possibilities for postphenomenological analysis in the context of education, and possibly beyond.

ABBREVIATIONS

Works by Aristotle

NE: *Nicomachean Ethics*
EE: *Eudemian Ethics*
Cat.: *Categories*
Met.: *Metaphysics*

NOTES

1. The school is but one example provided by Foucault. Such techniques permeated the factory, the military, hospitals, prisons, etc. To Foucault, it is no coincidence that modernity sees these institutions becoming increasingly alike.

2. It is important to recognize here that Aristotle's idea of who can be a citizen is much more restrictive than its modern counterparts.

3. This use of subjectivation as a responsibility of the subject is much more normative and even prescriptive than it is in Foucault's work: "can we say that the care of the self is now a sort of universal ethical law? You know me well enough to assume that I will immediately answer: no" (Foucault, 2005, p. 112).

4. This in spite of Verbeek's admission that virtue ethics is at least more compatible with the idea of technologically mediated subjectivation than utilitarianism or deontology is (2011, p. 63), and some proposals for virtues as relevant for evaluating (the impacts of) technological mediations (Puech 2018; Swierstra and Waelbers 2012; Waelbers 2011).

5. As I understand it, this intuition is similar to the one that led Dorrestijn and Verbeek towards the late Foucault.

6. Note that Aristotle concedes that health or sickness *can* become so engrained in the person that we might call it a *hexis*.

REFERENCES

Aagaard, Jesper. 2021a. "Beyond the Rhetoric of Tech Addiction: Why We Should Be Discussing Tech Habits Instead (and How)." *Phenomenology and the Cognitive Sciences* 20 (3): 559–72.

———. 2021b. "The Passive Subject: A Phenomenological Contribution to STS." *Social Epistemology Review and Reply Collective* 10 (5): 14–19.

Aristotle. 2001. *The Basic Works of Aristotle*, edited by Richard McKeon. New York, NY: Random House.

———. 2009. *The Nichomachean Ethics*. World's Classic. Oxford: Oxford University Press.

———. 2011. *The Eudemian Ethics*. Oxford: Oxford University Press.

Austin, Nicholas. 2017. "Virtue as Habit." In *Aquinas on Virtue: A Causal Reading*, edited bei Nicholas ustin, 23–36. Georgetown University Press.

Bergen, Jan Peter, and Zoë Robaey. 2022. "Designing in Times of Uncertainty: What Virtue Ethics Can Bring to Engineering Ethics in the Twenty-First Century." In *Values for a Post-Pandemic Future*, edited by Matthew J Dennis, Georgy Ishmaev, Steven Umbrello, and Jeroen van den Hoven, 163–83. Cham: Springer International Publishing.

Bergen, Jan Peter, and Peter Paul Verbeek. 2021. "To-Do Is to Be: Foucault, Levinas, and Technologically Mediated Subjectivation." *Philosophy and Technology* 34 (2): 325–48.

Bildik, Cüneyt, and Sertel Altun. 2022. "Undergraduates' Conceptions of Online Education and Technological Mediation." *European Journal of Open Education and E-Learning Studies* 7 (2): 173–89.

Bowen, W Richard. 2009. *Engineering Ethics: Outline of an Aspirational Approach*. London: Springer.

Curren, Randall. 2010. "Aristotle's Educational Politics and the Aristotelian Renaissance in Philosophy of Education." *Oxford Review of Education* 36 (5): 543–59.

Dennis, Matthew, and Tom Harrison. 2020. "Unique Ethical Challenges for the 21st Century: Online Technology and Virtue Education." *Journal of Moral Education* 50 (3): 1–16.

Dorrestijn, Steven. 2012. "The Design of Our Own Lives: Technical Mediation and Subjectivation after Foucault." PhD diss., University of Twente.

———. 2017. "The Care of Our Hybrid Selves: Ethics in Times of Technical Mediation." *Foundations of Science* 22 (2): 311–21.

Foot, Philippa. 1978. *Virtues and Vices and Other Essays in Moral Philosophy*. Berkeley and Los Angeles, CA: University of California Press.

Foucault, Michel. 1973. *The Birth of the Clinic: An Archaeology of Medical Perception*. New York, NY: Pantheon Books.

———. 1977. *Discipline and Punish: The Birth of the Prison*. New York, NY: Vintage Books.

———. 1986. *The History of Sexuality: The Care of the Self*. New York: Pantheon Books.

———. 1990. *The History of Sexuality: The Use of Pleasure*. New York, NY: Vintage Books.

———. 1997. *Ethics : Subjectivity and Truth*. Edited by Paul Rabinow. *The Essential Works of Foucault, 1954-1984*. New York: The New Press.

———. 2005. *The Hermeneutics of the Subject: Lectures at the Collège De France, 1981-1982*. New York, NY: Palgrave-Macmillan.

Han, Hyemin. 2014. "Virtue Ethics, Positive Psychology, and a New Model of Science and Engineering Ethics Education." *Science and Engineering Ethics* 21 (2): 441–60.

Harris, Charles E. 2008. "The Good Engineer: Giving Virtue Its Due in Engineering Ethics." *Science and Engineering Ethics* 14 (2): 153–64.

Hursthouse, Rosalind. 2006. "Virtue Ethics." In *Stanford Encyclopedia of Philosophy*. Meta-physics Reserach Lab Center for the Study of Languages and Information, Stanford University.

Ihde, Don. 1990. *Technology and the Lifeworld*. Bloomington, IN: Indiana University Press.

———. 1998. *Expanding Hermeneutics: Visualism in Science*. Evanston, IL: Northwestern University Press.

Kennedy, Jolie. 2022. "Designing Inclusive Online Learning Environments Through the Lenses of Feminist Technoscience and Postphenomenology: An Educational Design Research Approach." In *Proceedings of EdMedia + Innovate Learning 2022*, edited by Theo Bastiaens, 1155–1158. Online: Association for the Advancement of Computing in Education (AACE).

Kraalingen, Imre van. 2022. "Theorizing Technological Mediation in the Outdoor Classroom." *Postdigital Science and Education*, 5: 754–76.

Kristjánsson, Kristján. 2016. "Aristotelian Character Education: A Response to Commentators." *Journal of Moral Education* 45 (4): 527–34.

Lepe-salazar, Francisco, and Tania Cortes-alvarez. 2023. "FORTAGONO : A Model for the Technological Mediation of the Teaching and Learning Processes." *IEEE Access*: 1.

Lewis, Richard S. 2020. "Technological Gaze: Understanding How Technologies Transform Perception." In *Perception and the Inhuman Gaze*, edited by Anya Daly, Fred Cummins, James Jardine, and Dermot Moran, 128–42. New York, NY: Routledge.

Lockwood, Thornton C. 2013. "Habituation, Habit and Character in Aristotle's Nichomachean Ethics." In *A History of Habit - From Aristotle to Bordieu*, edited by Tom Sparrow and Adam Hutchinson, 19–36. Lanham, MD: Lexington Books.

Lu, Mathew T. 2015. "Hexis within Aristotelian Virtue Ethics." *Proceedings of the ACPA* 88: 197–206.

MacIntyre, Alasdair. 1979. "Seven Traits for the Future." *The Hastings Center Report* 9 (1): 5–7.

———. 1981. *After Virtue – A Study in Moral Theory*. Notre Dame, IN: University of Notre Dame Press.

Malikail, Joseph. 2003. "Moral Character: Hexis, Habitus and 'Habit'." *Minerva - An Internet Journal of Philosophy* 7: 1–22.

Merleau-Ponty, Maurice. 2002. *Phenomenology of Perception*. London: Routledge Classics.

Oele, Marjolein. 2012. "Passive Dispositions: On the Relationship between and in Aristotle." *Ancient Philosophy* 32 (2): 351–68.

Peters, R.S. 1981. *Moral Development and Moral Development*. London, UK: George Allen & Unwin.

Pohlmeyer, Anna, and Pieter Desmet. 2017. "From Good to the Greater Good." In *Routledge Handbook of Sustainable Product Design*, edited by Jonathan Chapman, 469–86. London: Routledge.

Puech, Michael. 2018. "Describing and Valuing Technological Mediation: From Postphenomenological Bridgeheads to Technoethical Outposts." In *Postphenomenological Methodologies – New Ways in Mediating Techno-Human Relationships*, edited by Jesper Aagaard, Jan Kyrre Berg Friis, Jessica Sorenson, Oliver Tafdrup, and Cathrine Hasse, 199–216. London: Lexington Books.

Reijers, Wessel, and Bert Gordijn. 2019. "Moving from Value Sensitive Design to Virtuous Practice Design." *Journal of Information, Communication and Ethics in Society* 17 (2): 196–209.

Rosenberger, Robert. 2012. "Embodied Technology and the Dangers of Using the Phone While Driving." *Phenomenology and the Cognitive Sciences* 11 (1): 79–94. https://doi.org/10.1007/s11097-011-9230-2.

———. 2014. "Multistability and the Agency of Mundane Artifacts: From Speed Bumps to Subway Benches." *Human Studies* 37 (3): 369–92. https://doi.org/10.1007/s10746-014-9317-1.

———. 2021. "The Politics of the Passive Subject." *Social Epistemology Review and Reply Collective* 10 (9): 29–35.

Rosenberger, Robert, and Peter-Paul Verbeek. 2015. "A Field Guide to Postphenomenology." In *Postphenomenological Investigations: Essays on Human-Technology*

Relations, edited by Robert Rosennbergen and Peter-Paul Verbeek, 9–41. Lanham, MD: Lexington Books.

Savater, Fernando. 2001. *De Waarde van Opvoeden*. Utrecht: Bijleveld.

Schmidt, Jon Alan. 2014. "Changing the Paradigm for Engineering Ethics." *Science and Engineering Ethics* 20 (4): 985–1010.

Swierstra, Tsjalling, and Katinka Waelbers. 2012. "Designing a Good Life: A Matrix for the Technological Mediation of Morality." *Science and Engineering Ethics* 18 (1): 157–72.

The Jubilee Centre for Character & Virtues. 2022. "The Jubilee Centre Framework for Character Education in Schools, 3rd Revised Version."

Vallor, Shannon. 2013. "The Future of Military Virtue: Autonomous Systems and the Moral Deskilling of the Military." *International Conference on Cyber Conflict, CYCON*.

———. 2015. "Moral Deskilling and Upskilling in a New Machine Age: Reflections on the Ambiguous Future of Character." *Philosophy and Technology* 28 (1): 107–124.

———. 2016. *Technology and the Virtues: A Philosophical Guide to a Future Worth Wanting*. New York, NY: Oxford University Press.

Verbeek, Peter-Paul. 2005. *What Things Do - Philosophical Reflections on Technology, Agency and Design*. University Park, PA: The Pennsylvania State University Press.

———. 2008. "Obstetric Ultrasound and the Technological Mediation of Morality: A Postphenomenological Analysis." *Human Studies* 31 (1): 11–26.

———. 2011. *Moralizing Technology: Understanding and Designing the Morality of Things*. Chicago, IL: The University of Chicago Press.

Voordijk, Hans, and Farid Vahdatikhaki. 2020. "Virtual Reality Learning Environments and Technological Mediation in Construction Practice." *European Journal of Engineering Education*: 1–15.

Waelbers, Katinka. 2011. *Doing Good with Technologies – Taking Responsibility for the Social Role of Emerging Technologies*. Dordrecht: Springer Science+Business Media.

Walker, David I., Michael P. Roberts, and Kristján Kristjánsson. 2015. "Towards a New Era of Character Education in Theory and in Practice." *Educational Review* 67 (1): 79–96.

Yaron, Gili, Guy Widdershoven, and Jenny Slatman. 2017. "Recovering a 'Disfigured' Face: Cosmesis in the Everyday Use of Facial Prostheses." *Techné: Research in Philosophy and Technology* 21 (1): 1–23.

Chapter 19

Rethinking Ihde's Relations Pedagogically

On Learning and Becoming Who We Are in Relations to Things

Anne Pesch

Relations between human beings, technologies, and the world are a central topic of postphenomenological theory and research. The mediation and transformation of experience and practices are usually at the center of attention. In *Technology and the Lifeworld* (1990) and other works, Don Ihde reflects on various forms of relations that humans enter into with technologies. That these technology relations can presuppose learning processes (e.g., in the *Embodiment* or the *Hermeneutic Relation*) (Ihde 1990, 73), is something that he mentions more or less in passing. Ihde also states, that technology relations not only affect our relation to the world, but also transform *us*: "Technologies transform our experience of the world and our perceptions and interpretations of our world, and we in turn become transformed in this process" (Ihde 2009, 44). These two statements lead precisely into the field of educational sciences.[1] They call for further exploration and differentiation from an educational perspective.

Relations and relationality are crucial in the educational sciences: Human beings learn and develop, are educated and socialized in relation to themselves, others, and the world. Especially, (phenomenological) learning theory, the theory of *Bildung,* and educational philosophy address relationality of this kind.[2] While postphenomenology can learn from a more nuanced look at phenomena like learning and becoming in relations and the concept of responsivity, the educational sciences can benefit from different perspectives on examining the relation to things or technologies, especially regarding mediation and the condition of a comprehensive mediatedness of the relation to the world. Postphenomenology and phenomenological educational sciences seem to be connectable here, especially due to their common grounds in the reception of Husserl, Heidegger, and Merleau-Ponty. There seem to be interesting intersections. At the same time, there is still a great need for negotiation, for example, concerning pathic and responsive relations to the world.[3]

In my paper, I want to bring the perspectives of postphenomenology, phenomenological educational sciences, and learning theory together.[4] In a first step (1), I will take a look at a similar idea that was elaborated both in postphenomenology and pedagogical phenomenology: The description of different ways of relating to things. Here, I will briefly present Ihde's types of relations and his concept of multistability as well as an approach of a pedagogical classic, the Dutch phenomenologist Martinus Langeveld, who discusses "modes of making sense" ("Sinngebungsmodi") and "domains of the intentional object" ("Gegenstandsbereiche"). There are clear similarities that make the thematic overlap visible. In part two (2) and referring to Ihde's two statements that I quoted at the beginning, I will highlight issues where the approaches can benefit from each other with regard to learning in technology relations: After a brief insight into relational learning theory, I will focus on learning in mediation, alterity, and background relations and the role of multistability. In this context, I will address children's learning and will draw a connection to the pedagogical discourse on the "appeal of things." The text ends with a conclusion.

A SIMILAR IDEA: DIFFERENT WAYS OF RELATING TO THINGS AND TECHNOLOGIES

Although there are many differences regarding the focus of postphenomenology and educational phenomenology, there are thematic intersections. An important intersection is the relationship between humans and things or technologies. This is the case since Ihde provides a broad concept of technology and addresses technologies as artifacts. He uses the term to not only to refer to machines, media, and digital technologies but also for tools and artifacts such as thermometers, bicycles and hand axes, or even a stone utilized as technology.[5] Interestingly, postphenomenology and educational phenomenology have produced similar (but not identical) ideas in this area: Both Don Ihde and the Dutch phenomenologist and educationalist Martinus J. Langeveld have dealt with different forms of relations to and via artifacts.

Ihde's Types of Human–Technology Relations and the Concept of Multistability

In postphenomenology, Ihde's distinction of different types of human–technology relations is one of the most influential aspects. As Verbeek elaborated, Ihde distinguishes the different types using schemes based on the concept of intentionality. He addresses technological artifacts (1) as media, (2) as other(s), and (3) as technologies in the background (Verbeek 2005, 123).[6] Thereby Ihde resorts (in a rather free form) to traditions of terms and concepts (such as the "other," etc.) in order to model his types of relating theoretically. By opening up these conceptual perspectives and connections to terminological traditions, he also facilitates theoretical connections in other disciplines, such as educational sciences and their disciplinary exploration of these terms. The different forms of human–technology relations occur at different types of technologies with different frequency or distinctiveness.

(1) In *relations of mediation* (Verbeek 2005), technologies appear as media. *Mediation* takes place via a technological artifact: humans do not perceive or act in unmediated contact with the world, but via a technology, which affects resp. transforms or modifies the relationship between humans and the world. This can affect perception, interpretation, and action. Ihde conceptualizes the effects dialectically in an *amplification–reduction* or *revelation–concealment structure*: on the one hand, perceptual possibilities are revealed, expanded, or strengthened, and on the other hand, aspects of the relation are concealed or possibilities are reduced.[7] *Mediation* is one of the central concepts of postphenomenology and can be found in *embodiment relations* and *hermeneutic relations*. In an *embodiment relation*, a technology is "incorporated" and the body scheme is extended. Human beings and technology merge and it becomes possible to move, perceive, or act via the technological artifact. Examples include riding a bicycle, operating a mouse on a computer, or using an excavator to dig up earth. *Embodiment relations* correspond to the schema: *(Human–Technology) → World.*

In *embodiment relations*, technology withdraws from perception and becomes *transparent*. For example, when riding a bicycle, the focus of attention is on the road traffic and not on pedaling or keeping one's balance.

In a *hermeneutic relation*, a technological artifact offers a representation of the world that is interpreted by humans. For example, we find out about events in the world through television, and we "read" the temperature on the weather app or thermometer. The term *hermeneutics* is used in a broad sense here, as "a special interpretative action within the technological context. That kind of activity calls for special modes of action and perception, modes analogous to the reading process" (Ihde 1990, 80). Ihde conceptualizes this form of relation as: *Human → (Technology–World).*

(2) In *alterity relations*, a technological artifact appears as an Other (thing) or Quasi-Other (person) (cf. Ihde 1990, 98), not as a mediator to the world. Here, the relation is directed to the artifact itself. *Alterity relations* include,

for example, direct interaction with a device such as an ATM (cf. Rosenberger/Verbeek 2015), but also constellations in which we are directed to an "object of fascination" (Ihde 1990, 100) or forms of "relationship" that are accompanied by feelings, which are expressed, for example, in the care and maintenance of a beloved bicycle. Ihde summarizes this form of relation as: *Human → Technology–(– World).*

(3) In so-called *background relations*, technologies operate from the nonthematic background and influence our *being-in-the-world* "without our intervention or even awareness" (Verbeek 2005, 123). Schematically, the relation can be represented as: *Human (– Technology/World)*

In addition to Ihde's different kinds of human–technology relations, the question of how to relate to things or technologies also involves the concept of "multistability." Ihde argues that technologies possess structural ambiguity. Although structural and material properties provide limits, the technologies' embedment and contexts play an important role in what they "are" (Ihde 1990, 69–70). Through sedimentation and habitualization, certain relations become culturally established or "stable." Usually there are one or more dominant "stable" forms of relation and interpretation, as well as less dominant, alternative forms.

Langeveld's "Modes of Making Sense" and "Domains of the Intentional Object"

The Dutch phenomenologist and educationalist Martinus J. Langeveld has also studied different forms of relation between humans and things. His article *Das Ding in der Welt des Kindes* (The Thing in the World of the Child) ([1955] 1968) is an important reference in more recent phenomenological–educational publications on this subject (cf. Stieve 2008; Stieve 2019; Brinkmann 2012; Agostini 2015; Meyer-Drawe 1999). As Langeveld states, things provide a framework for the human–thing relation in their materiality, but at the same time, there are openness and surplus that leave room for different modes of sense-making ("Modi der Sinngebung"). He distinguishes between (1) "open sense-making," which refers to the "world we have in common" (Langeveld [1955] 1968, 143), (2) "nonbinding sense-making," which is structured, but nonbinding toward the world like in a game, (3) "creative sense-making" in art, and (4) "personal sense-making." Each of these modes of sense-making offers more than one way of relating; they provide different "domains of the intentional object" ('verschiedene Gegenstandsbereiche'), which make the "thing" appear differently. Langeveld illustrates this using the example of a slipper.

> We want to undertake a preliminary phenomenological investigation using an everyday example. Let us choose a slipper; what can it be? It can be a slipper, it can also be a hairy something on which one can drool. In the first case it is a simple *object of utility*, in the second *a purely sensual-for-me*. The child can also use the slipper to hammer in a nail. Now the slipper is not itself as an object of use, but it is a *used* object: certain objective properties of the thing—here the hard heel—are selected and applied again in an appropriate and object-oriented ("sachgerichtet") way. [. . .] But the child can also select a property of the slipper that is not used objectively ("sachgerichtet"): the slipper is used, for example, as a small doll's cradle. In this case, the slipper that has been transformed into a cradle remains interpreted in the world of play and is not—like the slipper used as a hammer—included in the open world and used as a tool in an objective ("sachgerichtet") way. However, the slipper can also disappear ("sich verlieren") in the background against which the actuality of life takes place. If it is "registered" there ('verzeichnet'), then it is "just there" and is almost an extra-worldly thing: it does not belong to any actual interpretation ("Aktualdeutung") of this particular child's life. The slipper also comes from a great distance when it is placed and viewed or drawn as a picturesque object. This slipper also comes from far away when we learn that it has been produced in a certain way ("so-und-so"). Moreover, there is the slipper as a memento of the beloved, the little child, the deceased little brother. Or [. . .] it is the father's slipper, and the child recognizes that they belong together. This slipper is a "father-sign," and the child, seeing it, says quite decidedly not "slipper" but "father, father," as if the slipper were a *pars pro toto*. What a slipper can't be! It can even mean far more.—I would like to briefly mention just one more mode of the being-a-thing of an object: the so-called "gift." (Langeveld [1955] 1968, 144–145; my translation)

Ihde and Langeveld

As this quote exemplifies, Langeveld's and Ihde's approaches are similar in distinguishing different forms of relation. They show partial similarities, but the categorizations and perspectives are different. Looking at Ihde's types of human–technology relations, it becomes clear that these types are linked to different "positions" in the intentional relation on the one hand, and to different theoretically and historically established concepts of technology and relation ("other," "medium" . . .) on the other hand. Unlike Ihde, Langeveld does not stress that the slipper is not always given as an "object" as an opposing other. Nevertheless, the examples show great similarities to Ihde's relations: (1) Langeveld's examples address different forms of embodiment (in different modes of sense-making resp. "stability"): The slipper appears as (a) object of utility (put on footwear), (b) (differently) used object (hammer substitute), and a plaything (used doll's cradle). Unlike Ihde, Langeveld does not elaborate on how embodiment takes place. (2) In Langeveld's examples, the slipper also works as a medium in a different kind of hermeneutic relation, when it is "read" as a sign, resp. as a memento and a "father sign."[8] (3) If Ihde's *alterity relation* is understood as a direct relation to an artifact, the relation to the object of learning structurally belongs to that category, as well as the "picturesque object" and the "purely sensual for me," if the child is intentionally directed toward the slipper and its fluffiness.[9] (4) The slipper, which "disappears in the background," refers to the phenomenological concept of *horizon* and Gestalt psychology's *figure-ground* distinction, which are referred to by Ihde as well, as a basis for the concept of *background relation*. Comparing Ihde's and Langeveld's approaches, it becomes clear that Langeveld's text considers the embedding of artifacts in relation to other humans (e.g., as a "father sign," "memento," or "gift"), an important aspect of human relationality that is less present in Ihde's texts. Furthermore, some of Langeveld's examples, such as the memento as a medium of memory, make clear that the role of feelings in artifact relations should be considered beyond *alterity relations* as well.

With regard to *multistability* and modes of sense-making, both Ihde and Langeveld assume an ambiguity of things and a latitude in relations, and stress the role of the materiality of things here. Furthermore, both emphasize the human–thing relationality. Ihde's *multistability* has a stronger focus on the artifact, especially on the different historical–cultural situatednesses and practices in which the artifact is embedded. He focuses mainly (but not exclusively) on established practices in the "common world" in different historical–cultural contexts. On the other hand, Langeveld's modes of sense-making are more focused on the subject's side. They stress the role of sense-making, but vary with regard to different "world"-embeddings with different validity ('Geltung') and differentiate according to the bindingness of the reference to these worlds. They add significant aspects to Ihde's multistability (or stress them more explicitly), like modes of play and art resp. esthetics.[10] Conversely and following Ihde's notion of multistability, the important factor of "stability" and historical–cultural processes of sedimentation in the lifeworld become more apparent.

As has become evident, different forms of human–artifact relations have not only been a subject of postphenomenology but they have also been addressed in phenomenological educational research. In the referred text, Langeveld only interprets one case in detail. Yet, there are notable similarities between Ihde's and Langeveld's forms, although they do not fully overlap. In the next chapter, I am going to connect the previous aspect—different ways of relating between humans and artifacts—with phenomenological learning theory and take a closer look at learning and experience in human–technology relations of various forms. Taking into account learning, it becomes apparent that thing relations should not be thought of as static, since they are undergoing processes of change.

LEARNING AND TECHNOLOGY RELATIONS

The different ways of relating to and through things are present in learning processes. This was also recognized by Don Ihde. Ihde points out (1) that certain technology relations require learning processes (Ihde 1990) and (2) that it is not only the relations to things or technologies that change but *we*, the subject or person, can be transformed in technology relations as well (Ihde 2009, 44). A dialogue between postphenomenology and phenomenological educational science can help us to further explore and better understand learning processes in human–technology relations. Educational science theorizing can benefit postphenomenology with its analyses and theoretical clarifications on learning. Conversely, Ihde's clear emphasis on different forms of intentional relation—especially mediation—can benefit the learning theories of phenomenological educational science.

Relational Learning Theory

In many approaches of phenomenological and hermeneutic educational science, learning is understood as based on experience in relations. The German educational phenomenologist Malte Brinkmann states:

> In contrast to psychological learning theories, relational learning theories focus on the learners' relations and experiences. Learning becomes evident as an experience-based (1), negative (2), bodily (3), social and object-related ("sachbezogen") (4), cultural, generational and educational (5) process. (Brinkmann 2022, 280–281; my translation)[11]

Learning can be formative or transformative: The notion of *transformation* (e.g., Koller 2012, 2017) usually relates to the subject or person and is typically the focus in approaches in the theory of *Bildung*. Processes of *formation* (Brinkmann 2022) involve the "incorporation" of skills and knowledge, schemata and habitus; its notion can be used to describe the everyday, small steps like learning to embody a keyboard as well. The two forms of learning Ihde mentions show that the description of learning processes in technology relations must tackle both scopes.

Pesch (2024) intends to address both these aspects as well, following the German phenomenologist Bernhard Waldenfels (2018) and his interpretation of the connection between *spontaneity, habit, situation, and world* in learning processes. Waldenfels[12] develops this interpretation referring to Husserl and Merleau-Ponty, among others. He describes that learning, on the one hand, refers to habits and processes of *habitualization* and *sedimentation*. The "incorporation of structures" (Waldenfels 2018, 166; my translation) in habitualization has an effect on ability, and in sedimentation, it has an effect on knowledge and the way things are given: "The learned becomes part of the world, it is sedimented in it" ('Das Erlernte geht in die Welt ein, lagert sich in ihr ab'; Waldenfels 2018, 183; my translation). On the other hand, learning involves *spontaneity*, which allows for new creations, latitude ("Spielraum"), and responses to the situation, as well as the possibility of structuring it (Waldenfels 2018, 196). Waldenfels states: "The polarity of spontaneity and habituation embodies two moments of world structure: that we acquire a world through structures, repetitions, and that at the same time we are open to new possibilities that lead beyond these rules and structures" (Waldenfels 2018, 207; my translation). The interplay of sedimentation, habitualization, and spontaneity in Waldenfels' concept of learning also opens up perspectives on the subject and the person in a process of becoming and "becoming different" ("Anderswerden") (Waldenfels 2018, 173).[13]

Learning in Mediation, Alterity, and Background Relations

To describe learning in technology relations, Pesch (2024) combines Waldenfels' remarks on the connection between spontaneity and experience-based learning with Ihde's ideas on mediation.[14] It is pointed out: Learning processes in dealing with technologies involve the technological transformation of world relations that is connected with mediation. These learning processes can be formative and transformative. On the one hand, this means that we get used to the effects of mediation, such as the amplification or reduction: "The modifications in world relations become sedimented resp. habitualized. [. . .] Little by little, the effects become commonplace, encorporated and ready-to-hand ('zuhanden') through repetition" (Pesch 2024, 406; my translation). On the other hand, learning mediation can mean not only formation, but also "reordering" and transformation of relations and possibilities: Being confronted with the effects of mediation may prompt us to change the way we usually relate to the world. It affects the way we can "respond" to situations (Waldenfels 2018), spontaneously using available possibilities. This is also the case on a more general level: Certain images and ideas about the world, technologies, and human beings become sedimented, while others are challenged and changed.

Pesch (2024) illustrates learning in technology relations with the example of learning to ride a bicycle: we learn to embody a bicycle, the movement of the pedals, the maintenance of balance and it can become ready-to-hand ("zuhanden"). At the same time, the effects on the relation to the world are incorporated, such as the changed speed in the movement, and the perception of faster passing things. Even after being able to use the bike in a proper way, new situations can require the learning process to be continued.[15] Also, knowledge and skills are acquired on a level beyond the situation. Mediation can have an impact on the life of the person who is learning to ride a bike. For example, "the increased range of motion can be a step toward independence for children" (Pesch 2024, 406; my translation) when they can visit friends on their own. They participate in the cultural practice of cycling that they share with others. In this respect, too, "a world is acquired." At the same time, new possibilities open up, and the subject or the person can "become different." Being a cyclist can even become part of a teenager's or adult's identity.

Following and expanding Pesch (2024), I want to highlight here, that learning and becoming in technology relations are not limited to mediation. Ihde's different types of relations open the view to the variety of ways in which things influence and participate in learning processes. To give a few brief hints here:

Looking at *alterity relations* with a focus on learning, it becomes clear that this form of Ihde's relations combines heterogeneous aspects, including intentional directedness, quasi-human interaction and emotion/feeling.[16] Emphasizing the aspect of intentionality, a classic form of a learning situation comes into view: the attention is directly focused on the technology/thing and a direct examination takes place. This can be the case, for example, in learning how to handle an object, how to interact with a technology, or in learning through observing it as an "object of fascination" (Ihde 1990, 100). Focusing more on the aspect of learning in interaction with a quasi-other, interactions with dolls or with simulated persons in language learning, for example, can be subsumed. The topic of emotions and feelings is an important aspect in education and learning processes per se (Brinkmann 2019): Feelings are responses that articulate the state of our relation to the world and to ourselves. For example, we might feel curious upon encountering a new technical device, experience disappointment, and anger when technology fails, or feel enthusiastic while riding a bike. Emotions and feelings can also sediment and be habituated, they initiate, articulate, and influence learning processes.

Background relations are influencing learning processes all the time. Technologies and things affect the conditions, are part of the *horizon,* and are therefore involved in making meaning, they affect the *background,* which allows other things to emerge and attract attention by standing out in the first place. They become more explicit, for example, when we learn to respond to being affected by background effects in a certain way, or when changes are occurring.

Learning, Multistability, and the "Appeal of Things"

Thus, things participate in and influence processes of experience-based learning in various ways. However, not only different forms of human-technology relations, but also multistability, "object domains"[17] and modes of making sense play a role here. They open up new perspectives on the thing and allow for different mediations. These can be learned, certain stabilities and contextual references can be established, and, conversely, possibilities for spontaneity can open up.

In the context of Langeveld's example, it is worthwhile to have a closer look at learning in early childhood. Langeveld points out: "the child is in the world for the first time: it does not know things [as adults do]. Things are [. . .] familiar-unknown" (Langeveld [1955] 1968, 148, my translation). For young children, among others, Brinkmann (2022) emphasizes the importance of formative learning, which involves incorporating skills, knowledge, schemata, and habitus. Children learn about the possibilities (and cultural stabilities) of artifacts in contact with other people and through exploration of the different "domains" (Langeveld [1955] 1968) in which things can be given or related to. Especially in the case of conventional uses (stabilities), mimesis plays an important role, and social, cultural, and educational factors are central. Stieve states: "In the process of ontogenesis, the child's relation to its things changes. The relation becomes more and more habitual. However, these learning movements provide insight into adult behaviors as well, emphasizing their openness" (Stieve 2019, 504, my translation).

In phenomenological educational science, this complex is connected with the discourse[18] about the "appeal of things" ("Appell der Dinge") or "prompting character of things" ("Aufforderungscharakter der Dinge") (e.g., Brinkmann 2012; Stieve 2008, 2019). Langeveld discusses this subject in *Das Ding in der Welt des Kindes* (The Thing in the World of the Child) as well:

> We are all familiar with the peculiar appeal of things, which is also so particularly important for children. Some property of a thing appeals to us, and the object speaks to us, as it were, in the gerundive: the object demands of us that we do something with it. [. . .] we know this appeal also from Straus' notion of the pathic relation to the world: the round calls for rolling, the thin for stretching, bending, and whipping, etc. (Langeveld [1955] 1968, 146–147, my translation)

Things are appealing in conventional (socially and culturally imparted and linguistically mediated), pre- and unconventional ways (Stieve 2008, 317).[19] This is related to the horizons or reference contexts of things ("Verweisungshorizonte" and "-kontexte"), which offer multiple possibilities of action or use. Agostini states: "In the form of expectations, needs, and memories, they [the appeals] also refer to both the past and the future and are embedded in a situational field that offers multiple possibilities for action simultaneously" (Agostini 2015, 146, my translation). These appeals, especially the preconventional ones, are closely related to one's corporeality and the materiality or the properties of the thing (similar to postphenomenological multistability).[20] According to Waldenfels, appeals also exist as "foreign relations of things" ("Fremdbezüge der Dinge"): "Things are not only put in order by us, but happen ('widerfahren') to us on the basis of a potentiality that transcends our own experience" (Agostini 2015, 146–147, my translation). Here, things precede expectations and motivate our needs and actions only because "they strike us unprepared in our corporeality as a surplus and demand a response from us" (Agostini, 2015, 146, my translation). In this regard, learning can be understood as responding to the appeals of things and challenges of the lifeworld (Stieve 2019, 503). As is evident here, the discourse on the appeal of things in phenomenological educational science is situated in an approach that emphasizes the significance of pathic and responsive dimensions in human–world relations.

CONCLUSION

Although there are many differences regarding the focus of postphenomenology and educational phenomenology, combining their perspectives with regard to the investigation of learning in technology or thing relations appears to be a rewarding endeavor. In the first part of this text (1), I looked at projects describing different ways of relating to things undertaken by postphenomenologist Don Ihde and phenomenologically oriented pedagogue Martinus Langeveld. Following up Ihde's and Langeveld's thoughts, I focused on learning in thing resp. technology relations in a second step (2) and drew connections to learning theory (i.e., Waldenfels and Brinkmann) and the pedagogical discourse on the "appeal of things." Here, postphenomenology and phenomenological educational science seem to be able to benefit from each other, due to interesting intersections and points of friction.

As became apparent, learning in technology relations is more than adapting to a technology. With the notion of formation and transformation (Brinkmann 2022) and with regard to the aspects of habituation, spontaneity, and relationality (Waldenfels 2018), learning can be modeled as a process that, on the one hand, addresses every day, small steps, and on the other, learning on a more fundamental level. Relations appear not as static, but as changing and evolving, as future relations of self, technology, and world are (trans) formed on the basis of what is learned. Following Ihde, it becomes clear that technologies participate in the learning process in different ways (and have different positions in intentional relations)—as mediators (Pesch 2024), as an opposing other or as an influence from the background. In that sense, we also learn and become who we are in relations to things or technologies, being affected by and responding to background influences, being directly related to things, or being in mediated relations. Moreover, Ihde's remarks on multistability and Langeveld's remarks on modes and "object domains" draw attention to the aspects of openness and variation in these relations. Here it seems to be worthwhile to investigate the role of multistability for children. Children learn about the variability of human-thing relations, as well as the properties and 'givenness' of artifacts, that is by exploring things with various senses and in play. In processes of education and socialization, they become familiar with "stabilities" as culturally and historically situated ways of using artifacts. In these processes, the relations to things become more and more habitual. However, as the discourse about the prompting character of things indicates, unexpected encounters with things, unconventional or preconventional relations, and playful and artistic modes of relating should be given more consideration. This is especially the case for children, but also for adults.

In general, the perspective of phenomenological pedagogy provides a view of the human–world relations that is characterized not only by action and intentionality but also by a pathic and responsive relation to the world. It also offers a stronger focus on relations to other persons. A further dialogue between postphenomenology and phenomenological pedagogy appears very valuable in this context.

NOTES

1. There are diverse academic traditions in the educational sciences in different countries; due to non-English publications, there is not always a direct exchange. For that reason, I would like to introduce approaches as well, which have not been published in English yet, and try to include representatives of different nationalities. However, the German discourse will be more present than others.

2. However, for the most part, relationality is treated implicitly, that is, in research on pedagogical relationships, sociality, conditions, and contexts of genesis and pedagogical action.

3. From the perspective of phenomenological pedagogy, the pathic and responsive represent fundamental dimensions of the relation between human beings and the world (cf. Waldenfels or Levinas). These connections will come up several times in this article. For a detailed discussion, see Redecker's contribution in this volume.

4. For a brief overview of German and English phenomenological educational science, see Brinkmann/Friesen (2018).

5. Ihde states: "There is no 'thing-in-itself'. There are only things in contexts, and contexts are multiple. [. . .] A technological object, whatever else it is, becomes

what it 'is' through its uses. This is not to say that the technical properties of objects are irrelevant, but it is to say that such properties in use become part of the human-technology relativity" (Ihde 1990, 69–70).

6. Here and in the following, I take up Verbeek's (2005) description of Ihde's types of relations.

7. Other postphenomenologists have made further differentiations and extensions here (cf. Rosenberger/Verbeek 2015; Aagaard et al. 2018; Kiran 2015).

8. In Langeveld's example, the thing is also a medium of expression and embodiment (in the sense of "Verkörperung").

9. The example also corresponds to Ihde's notion of bodily-sensual "microperception." A "purely sensual for me" can also be given in a background relation, when it creates a coziness from the background with its hairy, fluffy texture, while the child is directed to other things. In both cases, it is embodied.

10. For instance, in *Embodied Technics*, Ihde also provides a small example of a playful exploration in the acquisition of relations (Ihde 2010, 51).

11. Brinkmann (2022) points out that (1) in the learning process, previous experience is taken up. Learning "by experience" and "as experience" (Meyer-Drawe) can lead to relearning or transformation. The "whole person" (Buck) can be affected and habits, habitus, and "structures of behavior" (Merleau-Ponty) can change. (2) Some learning theories emphasize the role of negative experiences, which are seen as important impulses for (re)learning. Resistances and "Unzuhandenheit" (Heidegger) are therefore seen as productive. (3) However, as Brinkmann points out, an exclusive focus on transformation and reflexivity, as it prevails especially in conventional theories of *Bildung*, falls short. The corporeality of learning processes and formation processes must also be taken into account, in order to include the learning of people with disabilities and young children as well as corporeal, aesthetic and emotional learning processes as a whole. (4) Learning takes place in relation to others and the world and is a (5) culturally dimensioned process that can also be related to generational contexts and educational activities.

12. Bernhard Waldenfels is best known for his works on foreignness and his approach of responsive phenomenology. For Waldenfels, responsivity, "together with intentionality and communicativity, denotes a basic characteristic of all behavior, i.e., all behavior is conceived as a response that responds to foreign offers and demands." He develops his 'responsive phenomenology' with critical reference to medical notions of the organism's response to the environment, to "non-behaviorist theories of behavior, and to 'responsive' features in phenomenology (Husserl's theory of affection, Heidegger's claim-response model, Merleau-Ponty and Levinas in particular), but also in examination of philosophies of dialogue, speech act theory, hermeneutics, discourse theory, Foucault, Derrida, and Lacan" (Unterthurner 2004, 463, my translation).

13. The notion "Anders werden" is also used in Hans-Christoph Koller's educational theory (Koller 2012).

14. The role of experience and learning in human–technology relations has also been tackled by postphenomenologist Robert Rosenberger. Rosenberger wants to describe the force of habit in given relations and uses the notion of sedimentation to

approach this topic (Rosenberger/Verbeek 2015). From the perspective of pedagogical phenomenology, we can expand and shift this view of experience and learning, focusing for example on the "acquisition" of the world, the possibilities of spontaneity, biography, and the becoming of the person.

15. Correspondingly, learning to establish a hermeneutic relation could mean to learn to "read" a technology (Ihde 1990).

16. These do not necessarily co-occur: For instance, emotions can also be present in background relations, or the interaction with voice assistants can be understood as mediation, if the focus is on finding out something about the world or influencing it through voice commands.

17. In Langeveld's example, the thing appears as (1) an object of utility, (2) (differently) used object, (3) a sensual-for-me, (4) a plaything, (5) an object of art, (6) an object of knowledge and discussion, (7) a part of the background, (8) a sign (e.g., that someone is there), and (9) a gift and expression of love. He distinguishes these "modes of making sense": (1) "open sense-making," which refers to the "world we have in common," (2) "non-binding sense-making" in a game, (3) "creative sense-making" in art, and (4) "personal sense-making."

18. I am following Agostini's account of the discourse (Agostini 2015).

19. The pedagogical discourse shows a particular emphasis on the dignity of the preconventional account.

20. Here it becomes clearer than in Ihde's earlier writings that relations to things or technologies can be motivating. It would take further elaboration to draw the connections and differences with Peter-Paul Verbeek's postphenomenological work, including the recourse to actor-network theory.

REFERENCES

Aagaard, Jesper, Jan Kyrre Berg Friis, Jessica Sorenson, Oliver Tafdrup, and Cathrine Hasse. 2018. "An Introduction to Postphenomenological Methodologies." In *Postphenomenological Methodologies: New Ways in Mediating Techno-Human Relationships,* edited by Jesper Aagaard, Jan-Kyrre B. Olsen, Jessica Sorenson, Oliver Tafdrup, and Cathrine Hasse, xi–xxv. Lanham, Maryland: Lexington Books.

Agostini, Evi. 2015. "Zur produktiven Vieldeutigkeit der Dinge in der Erfahrung des Lernens." In *Pädagogische Erfahrung: Theoretische und empirische Perspektiven,* edited by Malte Brinkmann, Richard Kubac, and Severin S. Rödel, 143–158. Wiesbaden: Springer.

Brinkmann, Malte. 2012. *Pädagogische Übung: Praxis Und Theorie Einer Elementaren Lernform.* Paderborn: Ferdinand Schöningh.

———. 2022. "Lernen." In *Schlüsselbegriffe der Allgemeinen Erziehungswissenschaft: Pädagogisches Vokabular in Bewegung*, edited by Milena Feldmann, Markus Rieger-Ladich, Carlotta Voß, and Kai Wortmann. 277–86. Weinheim: Beltz Juventa.

Brinkmann, Malte. 2019. "Bildung of Emotions and Bildung Through Emotions: The Orientative, Evaluative and Bildungs-Effect of Shame." Winterschool "Emotions and Care", Verona, February 5. Accessed September 29, 2023.

Brinkmann, Malte, and Norm Friesen. 2018. "Phenomenology and Education." In *International Handbook of Philosophy of Education*, edited by Paul Smeyers, 591–608. Cham: Springer.

Ihde, Don. 1990. *Technology and the Lifeworld: From Garden to Earth* 1. [print.]. A Midland book. Bloomington: Indiana Univ. Press.

———. 2009. *Postphenomenology and Technoscience: The Peking University Lectures.* SUNY series in the philosophy of the social sciences. Albany: SUNY Press.

———. 2010. *Embodied Technics.* Automatic Press/VIP.

Kiran, Asle H. 2015. "Four Dimensions of Technological Mediation." In *Postphenomenological Investigations: Essays on Human-Technology Relations*, edited by Robert Rosenberger and Peter-Paul Verbeek, 123–140. Lanham, Md. Lexington Books.

Koller, Hans-Christoph. 2012. "Anders werden: Zur Erforschung transformatorischer Bildungsprozesse." In *Qualitative Bildungsforschung znd Bildungstheorie*, edited by Ingrid Miethe and Hans-Rüdiger Müller, 19–34. Opladen, Berlin, Toronto: Barbara Budrich.

———. 2017. "Bildung as a Transformative Process." In *Transformative Learning Meets Bildung: An International Exchange,* edited by Thomas Fuhr, Anna Laros, and Edward W. Taylor, 33–42. Rotterdam: Sense Publishers.

Langeveld, Martinus J. (1956) 1968. *Studien zur Anthropologie des Kindes.* Third edition. Reprint 2012. Forschungen zur Pädagogik und Anthropologie 1. Berlin/Boston: de Gruyter.

Meyer-Drawe, Käte. 1999. "Herausforderung durch die Dinge: Das Andere im Bildungsprozeß." *Zeitschrift für Pädagogik* 45 (3): 329–36. https://doi.org/10.25656/01:5953.

Pesch, Anne. 2024. "Lernen und Anderswerden in Mensch-Technologie-Relationen. Anschlüsse an Don Ihde und Bernhard Waldenfels." *Vierteljahrsschrift für Wissenschaftliche Pädagogik* 100 (3): 393–411. https://doi.org/10.30965/25890581-10003008.

Rosenberger, Robert, and Peter-Paul Verbeek. 2015. "A Field Guide to Postphenomenology." In *Postphenomenological Investigations: Essays on Human-Technology Relations,* edited by Robert Rosenberger and Peter-Paul Verbeek, 9–41. Lanham, Md. Lexington Books.

Stieve, Claus. 2008. *Von den Dingen lernen: Die Gegenstände unserer Kindheit.* Boston: Brill.

———. 2019. "Diesseits und Jenseits des Konstruierens (2010): Phänomenologisch-gestalttheoretische Ansätze zur leiblichen Präsenz der Dinge." In *Phänomenologische Erziehungswissenschaft von ihren Anfängen bis heute: Eine Anthologie*, edited by Malte Brinkmann, 485–506. Wiesbaden: Springer VS. First published 2010.

Unterthurner, Gerhard. 2004. "Responsivität." In *Wörterbuch der phänomenologischen Begriffe*, edited by Helmuth Vetter, Klaus Ebner, and Ulrike Kadi, 463. Hamburg: Felix Meiner.

Verbeek, Peter-Paul. 2005. *What Things Do. Philosophical Reflections on Technology, Agency, and Design*. Second edition. University Park: Pennsylvania State University Press.

Waldenfels, Bernhard. 2018. *Das leibliche Selbst: Vorlesungen zur Phänomenologie des Leibes*. Seventh edition, edited by Regula Giuliani. Frankfurt am Main: Suhrkamp. First published 2000. DeepL Write and DeepL Translator were utilized to refine the language quality in this article.

Index

About the Contributors

Jesper Aagaard is an associate professor at the Department of Psychology and Behavioural Sciences at Aarhus University. His research focuses on "technopsychology," how the everyday use of (particularly digital) technologies affects the human mind. Furthermore, because tech companies have a vested interest in highlighting the benefits of such use, his research deliberately focuses on its downsides (e.g., distraction, phubbing, Zoom fatigue).

Håkon Jakobsen Aaltvedt has master's degrees in music technology and pedagogy from the Norwegian University of Science and Technology. He has experience working both as a freelance audio engineer/musician and as a head teacher in music design at Kristiania University College, Trondheim. He is currently doing a PhD in Educational Sciences at the Norwegian University of Science and Technology, studying how 1:1 technologies are shaping pedagogical practices in Norwegian primary education.

Jan Peter Bergen specialized in postphenomenology as a post-doctoral researcher at the University of Twente after completing a dissertation in applied ethics of technology with a grounding in phenomenology. After some time dedicated to philosophy education at Delft University of Technology, he is now an assistant professor in the section of Philosophy at the University of Twente. His work, next to teaching philosophy and ethics of technology to engineering students, centers around postphenomenology, ethics, and (moral) subjectivity.

Markus Bohlmann is a postdoctoral researcher at the Department of Philosophy in Muenster, he specializes in philosophy education and the philosophy of technology. In 2014, he received his PhD in philosophy of education in

Muenster. He conducted several empirical studies on students′ conceptions of philosophy and has been a high school philosophy teacher for five years. With his book *Bildung—Philosophie— Digitalisierung* (Metzler 2022), he introduced postphenomenology to a wider audience in Germany. In the same year, he was visiting the Georgia Institute of Technology in Atlanta for a research stay at the invitation of Robert Rosenberger and Michael Hoffmann. Bohlmann leads the Focus Group on Didactics of the Digitality Research Working Group of the German Society of Philosophy (DGPhil) and has been elected Executive Director of the German Society for Philosophy and Ethics Didactics (GPED) since 2020.

Patrizia Breil took on a postdoctoral position at the Center for Digitalization in Teacher Education at the University of Tübingen following her doctoral thesis on bodily alienation in phenomenological and pedagogical theory. She is now a research assistant at the collaborative research center "virtual life-worlds" at the Ruhr University Bochum. She works on (post)phenomenology of the body, philosophy of education, and media philosophy.

André Epp has a visiting professorship for Educational Science at the Brandenburg University of Technology Cottbus-Senftenberg. Further, he received his doctorate at the University of Hildesheim on subjective theories of teachers and his habilitation at the University of Education on the biographical professionalization of teachers and (youth) social workers in comparative comparison. His research interests are teacher education research, biographical research, methodologies and scientific theoretical foundations of qualitative research, educational theory and research, comparative pedagogical professional research, transitions in the educational system, and non-formal and informal learning.

Kristy Forrest is currently completing a PhD at the University of Melbourne on the potential applications of the philosophy of technology to education, with a particular focus on teacher thinking, following her Masters in Education in Pedagogy and Assessment. She also works as a classroom teacher in English and Philosophy and middle leader as head of Professional Practice in St Catherine's School in Melbourne and is interested in bridging the research and practice gap in education.

Lesley Gourlay is a professor of education in the department of Culture, Communication, and Media at University College, London. Her scholarship focuses on technologies and knowledge practices of students and academics, with a particular emphasis on textual practices and the digital. Her recent theoretical work has focused on sociomaterial and posthuman perspectives on engagement in the university, exploring themes of space, inscription, nonhuman agency, and digital media. She is a recipient of a Leverhulme Major Research Fellowship (2021–2024), and is writing a new monograph for Bloomsbury on the topic The Datafied University: Documentation and Performativity in Digitised Education."

Stacey O. Irwin is a professor of media arts production at Millersville University of Pennsylvania, USA. Postphenomenology focused work includes *Digital Media: Human-Technology Connection*, the co-edited *Postphenomenology and Media: Essays on Human-Media-World Relations*, chapters in several Lexington Postphenomenology series books, and many years of conference presentations about human–technology–world relations. Stacey creates, teaches, and studies media and is a co-curator of the website postphenomenology.org. and a keeper of the Postphenomenology Post Facebook page.

Annie Kurz is an interdisciplinary artist and designer, a lecturer, and a doctoral researcher at the University of Art and Design Offenbach am Main, Frankfurt (Hessen State University of Art and Design). During her years living and working in New York City, her art and research interests in digital technologies gravitated toward utilizing methodologies developed by Don Ihde and postphenomenology. Through a visit to Stony Brook University in 2014, she discovered the philosophical direction, which she identified as best equipped to understand situated knowledge concerning technologies and thus compatible with the perspectives of a designer. Postphenomenology also turned out to be helpful to describe phenomenological discoveries of fellow students and herself participating in the international art and research project SkypeLab (skypelab.org). She subsequently became a research assistant at SkypeLab, co-coordinated the project, and co-curated exhibitions. Her recent work, forthcoming book, and doctoral thesis are preoccupied with the so-called phenomenon of "digital detoxing" and apps related to the practice. She considers her personal background growing up in Romania, where technological developments at that time were far behind Western standards, as one of the formative reasons for her artistic and philosophical interests trying to decode the "magic" of technologies.

Ilya Levin received his PhD in Computer Engineering in 1987 from the Institute of Computer Technologies, Latvian Academy of Science. From 1987, he was the head of the Computer Science Department in the Leningrad Institute of New Technologies (in the former USSR). From 1993 to 1997, Ilya Levin was the head of the Computer Systems Department in Holon Institute of Technology, Israel. In 1997, he was a research fellow in the Computer Science

Department of the University of Massachusetts. For four years, between 2003 and 2006, Ilya Levin was an associate professor of the School of Engineering at the Bar Ilan University. In 2014, 2018, and 2022, he was a visiting professor at Ca' Foscari University of Venice, Italy, and Aix-Marseille Université, France. Presently, Prof. Ilya Levin is a full professor in the School of Education of Tel Aviv University. His recent research interests include Computer Design, Cultural Studies of Information Society, Philosophy of Digital Technology, and Artificial Intelligence in Education. Prof. Levin is the author of around 200 research papers, both in Computer Engineering and Humanities.

Dan Mamlok is a lecturer (assistant professor) in the School of Education at Tel Aviv University. His primary areas of scholarship include philosophy of education, sociocultural studies of education, technology in education, and aesthetic education. His research explores the sociocultural aspects of integrating technology and education and specifically deals with questions pertaining to democracy, education, and citizenship.

Anne Pesch, at the Brandenburg University of Technology, explores the connections between postphenomenology and phenomenological pedagogy, with a particular focus on learning processes in human-technology relations. In addition to her research in postphenomenology, Pesch is also active in the fields of philosophy of education, philosophy of the body, aesthetics, and anthropology. She works as a research associate in the project "NET-KUBI"—Network Spaces of Cultural Education—at the TU Dortmund University.

Anke Redecker is a private lecturer at the University of Bonn. She worked as an interim professor for transcultural ethics at the University for Applied Sciences, Darmstadt, former interim professor for educational justice at Pädagogische Hochschule Karlsruhe and research assistant at the University of Bonn. She received her habilitation in educational science and her doctorate in philosophy. Her research focuses are education theories and (transcultural) ethics, diversity education, pedagogical and philosophical anthropology, moral education and anti-discrimination, and dialogic and investigating learning (in the digital age).

Eliott Rooke wrote his thesis on embodiment relations and perceived body boundaries after first being drawn to postphenomenology as a doctoral student. Rooke currently works as a lecturer in Human Geography at the University of Exeter, United Kingdom. His work centers on the intersections between bodies, technologies, and skill with a particular interest in how emergent events and technologies disrupt skilled performance.

Robert Rosenberger is the president of the Society for Philosophy and Technology and a professor in the School of Public Policy at the Georgia Institute of Technology. His edited and co-edited books include *The Critical Ihde*, *Postphenomenology and Imaging*, *Postphenomenological Investigations*, and *Philosophy of Science: 5 Questions*. His monographs include *Callous Objects: Designs Against the Homeless*, and *Distracted: A Philosophy of Cars and Phones*.

Tomoki Sakata recently completed his PhD research on the culture philosophy of Ernst Cassirer and is interested in epistemology, comparative cultural studies with a focus on Japan, and the philosophy of technics and AI which underlies his current project on the ethics of the smart city. Over the last year, he dealt with Watsuji's community ethics and Japanese social conformity regarding the use of technology. His future research features the Goethean concept of nature.

Andreas Stock is a doctoral student at the Institute for Educational Science with a focus on extracurricular fields at the Karlsruhe University of Education. He received his Master's degree at the University of Education on the implications of an esthetic perception according to Merleau-Ponty for educational science. His research interests are phenomenological and ontological research, philosophical theory of education, cultural education, critical theory of social transformations, and processes of subjectivation.

Joris Vlieghe is an associate professor of philosophy and theory of education at KU Leuven. With Naomi Hodgson and Piotr Zamojski, he published *Manifesto for a Post-critical Pedagogy* (Punctum Books 2018), and with Piotr Zamojski, he published *Towards an Ontology of Teaching. Thing-centered Pedagogy, Affirmation and Love for the world* (Springer 2019). In line with his interest in the figure of the teacher and the educational meaning of studying, his current research explores how study practices can offer a response to the issue of how to live well together with the world's human and non-human inhabitants in times of ecological catastrophe. He is also interested in the impact of digital technologies on education, and on the future of the school, especially when a culture of the screen is (rapidly) substituted for a culture of the book.

Galit Wellner is an associate professor at Holon Institute of Technology (HIT) and an adjunct professor at Tel Aviv University. Galit studies digital technologies and their inter-relations with humans. She is an active member of the Postphenomenology Community. She published peer-reviewed articles and book chapters, and edited special issues of Techne and some collections.

Her book *A Postphenomenological Inquiry of Cellphones: Genealogies, Meanings and Becoming* was published in 2015 by Lexington Books. She translated to Hebrew Don Ihde's book *Postphenomenology and Technoscience* (Resling 2016). She co-edited *Postphenomenology and Media: Essays on Human–Media–World Relations* (Lexington Books, 2017) and *The Philosophy of Imagination: Technology, Art, Ethics* (Bloomsbury Academic, forthcoming). Her research on AI led her to consult on AI regulation for the Israeli Ministry of Innovation, Science and Technology.

Martin Wilmer is currently completing a teaching degree in philosophy and social sciences after studying philosophy and political science in Aachen and Muenster. He works on philosophical and political education as well as classical German philosophy (esp. J.G. Fichte).

Sou Hee Yang is an assistant professor at the Faculty of Social Sciences, Waseda University in Tokyo, specializing in comparative legal research with a focus on criminal law. She holds a JD degree from Columbia University and is licensed to practice law in New York. Yang's research takes an interdisciplinary approach, incorporating insights from various social science fields to inform her work.

www.ingramcontent.com/pod-product-compliance
Lightning Source LLC
Chambersburg PA
CBHW071945040325
22979CB00004B/30
* 9 7 8 1 6 6 6 9 3 9 1 4 9 *